SMALL ARMIES, BIG CITIES

SMALL ARMIES, BIG CITIES

Rethinking Urban Warfare

edited by
Louise A. Tumchewics

LYNNE RIENNER PUBLISHERS

BOULDER
LONDON

Published in the United States of America in 2022 by
Lynne Rienner Publishers, Inc.
1800 30th Street, Suite 314, Boulder, Colorado 80301
www.rienner.com

and in the United Kingdom by
Lynne Rienner Publishers, Inc.
Gray's Inn House, 127 Clerkenwell Road, London EC1 5DB
www.eurospanbookstore.com/rienner

Library of Congress Cataloging-in-Publication Data
Names: Tumchewics, Louise A., 1986– editor.
Title: Small armies, big cities : rethinking urban warfare / edited by Louise A. Tumchewics.
Other titles: Rethinking urban warfare
Description: Boulder, Colorado : Lynne Rienner Publishers, Inc., [2022] | Includes bibliographical references and index. | Summary: "Explores how today's professional armies can overcome the challenges of—and even find advantages in—conducting urban operations"— Provided by publisher.
Identifiers: LCCN 2022010046 (print) | LCCN 2022010047 (ebook) | ISBN 9781955055307 (hardback) | ISBN 9781955055666 (ebook)
Subjects: LCSH: Urban warfare.
Classification: LCC U167.5.S7 S63 2022 (print) | LCC U167.5.S7 (ebook) | DDC 355.4/26—dc23/eng/20220301
LC record available at https://lccn.loc.gov/2022010046
LC ebook record available at https://lccn.loc.gov/2022010047

British Cataloguing in Publication Data
A Cataloguing in Publication record for this book
is available from the British Library.

Printed and bound in the United States of America

The paper used in this publication meets the requirements of the American National Standard for Permanence of Paper for Printed Library Materials Z39.48-1992.

5 4 3 2 1

Contents

1

Fighting Wars in Cities

Louise A. Tumchewics

THE WORLD'S URBAN AREAS, WITH THEIR MAZES OF DENSELY POPULated residential spaces, streets and alleyways, unseen layers of sewers and subways, sturdy concrete walls, soaring skyscrapers, and sprawling shantytowns, present challenges for military operations. The multiple dimensions of urban spaces—horizontal, vertical, subterranean, and, increasingly, virtual—are imbued with historic, cultural, religious, social, economic, and political significance, presenting a dynamic complexity that the physical austerity of arctic, desert, or jungle warfare does not provide.

The perils and pitfalls of urban operations have been recognized for millennia. Military theorists from Sun Tzu onwards have cautioned against operating in built-up areas, advising it only as a last resort. This sage advice has typically been followed by militaries, who have preferred throughout history to fight in open terrain. This principle continues in contemporary doctrine, where fighting in cities is presented as a last resort rather than a first option, and armies continue to organize and equip for fighting in unconstrained areas.

Over the past three decades, however, we have witnessed a number of urban battles—Sarajevo, Grozny, Baghdad, Mosul, Fallujah, Marawi, Sadr City, Raqqa, Idlib, Aleppo, and most recently, Mariupol, among many others—that have reemphasized the inherent difficulties of operations in built-up areas. These battles have demonstrated the ferocity and destruction that have characterized urban operations throughout the ages. In the past ten years, mobile phones and social media platforms have brought a real-time global audience to these urban battles, allowing the world to witness the intense suffering of civilians whose neighborhoods and even homes become battlefields.

Professional Western armies such as those within the North Atlantic Treaty Organization (NATO) observe international humanitarian law (IHL), also known as the laws of armed conflict. Respect for laws and norms surrounding the protection of civilians and cultural sites and for the restrictions on the methods and means of warfare inform armies' rules of engagement.[1] IHL does not prohibit urban operations. However, reconciling urban warfare with the observance of IHL is particularly difficult. Urban landscapes, with dense infrastructure and civilian populations, complicate efforts. Yet difficulty does not eliminate responsibility of respecting IHL. As has been demonstrated most recently in Syria and Ukraine, operations executed without due consideration for IHL, and the indiscriminate use of explosive weapons, turn the urban battle into a humanitarian catastrophe.[2]

Armies may find themselves confronting the myriad difficulties of urban operations more often because urban battles may become more common as a greater percentage of the world's populations moves to cities and towns. Trends in global demographics indicate that the process of urbanization, which started with the Industrial Revolution, is rapidly accelerating. The population living in cities, high-density places of at least 50,000 inhabitants, has more than doubled over the last 40 years, going from 1.5 billion inhabitants in 1975 to 3.5 billion in 2015. It is projected to reach 5 billion and almost 55 percent of the world population by 2050.[3] Future conflicts, whether counterinsurgencies, peer confrontations, or near-peer confrontations, are likely to incorporate an urban element.

Military establishments and academics are aware of this demographic development and the potential implications for contemporary armies. There has been a relatively recent expansion in literature on the multiplicity of challenges posed by urban operations, by scholars and practitioners alike.

Alice Hills's landmark interdisciplinary study of urban warfare, *Future War in Cities: Rethinking a Liberal Dilemma*, highlighted the diversity of military operations in urban areas and the need for Western military forces to develop an integrated, multidisciplinary conceptual framework for understanding the strategic challenges of operations in urban areas.[4] Hills's subsequent writings make a case for considering the strategic dimension of urban operations, rather than focussing narrowly on the tactical.[5]

Michael Evans's writings also encourage Western military analysts to engage with urban studies experts to develop a more sophisticated understanding of urban dynamics within the realm of strategic studies. This interdisciplinarity will enable development of doctrine for urban operations, incorporating appropriate rules of engagement that conform to international humanitarian law. He also refutes the contemporary military focus on megacities, arguing that medium-sized cities with strategic importance and their own myriad complexities may present challenges as significant as their megacity counterparts.[6]

Other authors, such as Louis Di Marco, Alec Wahlman, and Colonel John Antal, and studies conducted by the RAND corporation and US Marine Corps, have drawn upon historical case studies to explore the heterogeneity of urban operations and describe the tactical challenges of close-quarters combat, not unlike the contents of this book.[7] Anthony King looks into the changing—and unchanging—tactics and characteristics of urban battles of the twenty-first century to examine the evolution of urban operations.[8] Charles Knight discusses the prevalent use of high explosives in modern urban combat and changing attitudes toward force protection and civilian casualties.[9]

Stephen Graham, a geographer, critiques the securitization of urban space and militaries' technology-based approaches to resolving urban security issues.[10] David Kilcullen's writing discusses how cities—particularly the megacities of the developing world—provide havens for guerrilla and insurgent forces.[11] He considers the city a system, and he proposes models for an approach to urban control operations.

The complexities of urban warfare have also stimulated robust debate and discussion in professional military education platforms such as the Modern War Institute at West Point, the independent War on the Rocks and *Small Wars Journal*, and the Australian Army's Cove, to name but a few. The growing body of literature on various aspects—strategic, operational, and tactical—of urban operations is a fertile ground for increasing our conceptual appreciation and understanding of this physical and human environment.

The steady expansion of urban areas around the world contrasts with the continued reduction in the size of NATO member states. Our reference points for modern urban operations are often the seminal battles of World War II—Stalingrad, Berlin, Aachen, Manila, Hue, or Grozny—battles involving vast armies or army groups. But what about smaller armies? How have they fared in urban battles? How did they find advantage in the urban environment? And what lessons from these historical experiences have continuing relevance for today's professional armies? These are the questions we seek to answer in this book.

Defining Terms

Small is a relative term, particularly when it comes to armed forces. For the purpose of this book, we define *small* as a division (10,000) or less. This may not be the entire deployable army but can also be a unit from a larger force, such as a battle group or brigade combat team that has limited recourse to the resources of its parent force. *Small* can also refer to an army's range of capabilities and budget, typically limited when mass is also restricted. *Army* in this book refers to an organization armed and trained for war on land, though not specifically the armed force of a nation-state or political party, thus allowing us to explore the experiences of non-state actor groups. Battles included in the book were selected for their geographic and chronological variety and relevance to the research questions at hand. Although there is a long chronology of urban operations, the time span for this book is limited to the past seventy-five years for greater familiarity, access to primary sources, similarity to current military structures and capabilities, and similar observance of laws of armed conflict (LOAC) and IHL, as set out in the Geneva Conventions and Additional Protocols.

The definition of *urban* varies widely according to country, with some considering an agglomeration of 200 people to be an urban area, whereas others consider 50,000 to be the threshold. In early 2020, the United Nations (UN) endorsed the Degree of Urbanisation methodology as a means of comparing urban areas around the world.[12] Using this method of classification, some of the examples in this book would be considered cities (populations of 50,000 or more), while others would fall in the category of towns or semi-dense areas (populations of 5,000).[13]

Urban operations can include any of the many missions that fall within the full range of military operations: humanitarian aid operations, peace support operations, policing, counterinsurgency, or counterterror operations, against unconventional forces or violent extremist organizations, or a major combat operation against a peer or near-peer adversary. The case studies in this book are limited to insurgency and counterinsurgency operations and major combat operations. This reflects the character of urban conflicts over the past 75 years, the time period from which our cases were selected.

Structure of the Book

This book is not intended as a complete history of urban operations, nor as a guidebook to future urban operations. Rather, it aims to inform and stimulate thinking on how small armies have fought in cities, reflecting on the enduring nature and changing character of urban conflict through a number of case studies. Focusing first on the particulars of urban warfare then on the operational arts, the chapters unfold as follows.

In Chapter 2, Wendy Pullan introduces the city as a collection of social, cultural, and personal spaces, infused with layers of meaning and identification beyond grid references. She discusses the characteristics inherent in urban spaces that make them so inhospitable to military forces, and the consequences of conflict on the fabric of the city and the social terrain.

Although cautioned against in military theory and doctrine, there is a long history of urban operations. In Chapter 3, Paul Latawski provides an overview of urban operations from medieval sieges to the present day. He explores the major historical trends in urban warfare, the development of theory and doctrine for urban warfare, and the impact of technology on urban warfare.

Alex Neads discusses the totemic value of cities in Chapter 4, and how the symbology of authority invested in certain key elements of a city outstrips the practical benefit of fighting in, or holding, urban terrain.

Focusing on the battle for Donetsk Airport in 2014, he illustrates how the strategic value of a city, or a piece of ground, can initially draw in a force, and how the reason for fighting becomes expanded to encompass other political aspirations and meanings, leaving an armed force with little option but to continue a battle of attrition.

Urban counterinsurgency presents a particular challenge to contemporary armies, and one that is anticipated to persist. Looking at the case study of the Battle of Marawi in Chapter 5, Louise Tumchewics discusses the advantages that the defender can find in urban environments and, conversely, the limitations the urban environment presents to irregular forces.

The themes of insurgency and counterinsurgency continue in Chapter 6, in which Patrick Finnegan addresses the role of space and identity in urban operations, focusing on the case study of the small Northern Irish town of Crossmaglen. Though relatively small, Crossmaglen was a complex, closed society that presented an enduring challenge to security forces. Drawing on interviews with former Provisional Irish Republican Army members and security force personnel, he explores the social intricacies of the town and the role of "place" in shaping identity and giving legitimacy to a cause.

Numerous historical examples indicate that urban operations can inflict high numbers of civilian casualties. In Chapter 7, Matthew Wiseman highlights the importance of managing the civil-military relationship in cities to advance outcomes and reduce casualties. He looks at the Battle of Groningen and how the relationship between civil authorities and military leadership contributed to consideration of targeting, weapons used, the maintenance of civil order, postbattle reconstruction, and transition to peacetime civilian leadership.

Civilian populations and adversaries are the primary intended audiences of influence operations. Steve Tatham discusses the importance of influence operations to achieve the desired political outcome of an urban campaign in Chapter 8. He explains how influence operations incorporate more than the transmission of information. They can involve physical actions, political influence, and legal changes. Moreover, influence operations extend well beyond the kinetic phases of an operation. They encompass outcomes, population movements/resettlements, regenerations, political transitions, and the creation of historical narratives. Tatham mines the case studies cited earlier in the book as well as contemporary examples from Syria and Iraq to demonstrate where influence operations have succeeded or failed, and what capabilities armies require to conduct successful influence operations.

Then, in Chapter 9, Maj. John Spencer (retired) posits that nonkinetic intrusive operations, such as the use of T-walls in Sadr City, interrupt the flow of hard and soft systems within a city to achieve outcomes without engaging in house-to-house battles. He outlines the capabilities required for nonkinetic intrusive operations based on the 2008 experience in Iraq and details how these operations could be better integrated with influence operations (described in the preceding chapter) to achieve desired outcomes while sustaining fewer casualties.

Technology is often promised to replace mass, increase precision, and reduce or even eliminate the requirement for frontline personnel and attendant risks, an attractive prospect when there is political aversion to the use of force. Lt. Col. Paul Lushenko and John Hardy explore the utility and limitations of drone strikes in urban operations in Chapter 10.

In Chapter 11, Tyrone Groh discusses the use of proxy or surrogate forces to compensate for armies' lack of mass, an appealing solution for armies seeking to increase mass while minimizing risk. He considers the interaction among armies, their proxies, and the social terrain and addresses how surrogates can be used to maximize advantage.

Recounting the Battle of Jaffa, Louise Tumchewics questions in Chapter 12 whether there are alternatives to mass in maneuver operations in urban spaces. She discusses command and control compression in the urban environment, and how training can amplify the abilities of a relatively small force.

Finally, in Chapter 13, Lt. Gen. Andrew Graham incorporates the lessons and theory from the previous chapters and his reflections on his own experience as a corps commander in Iraq to offer suggestions for urban campaigning in cities of any size in the twenty-first century.

Notes

1. Antouly, "Urban Warfare," 110.
2. "Conduct of Military Operations in Urban Areas," International Humanitarian Law Research Initiative, Harvard University, May 2004, 1.
3. OECD, "Cities in the World: A New Perspective on Urbanisation," https://www.oecd-ilibrary.org/sites/d0efcbda-en/index.html?itemId=/content/publication/d0efcbda-en.
4. Hills, *Future War in Cities.*
5. Hills, "Deconstructing Cities: Military Operations in the Urban Era"; Hills, "Fear and Loathing in Falluja."
6. Evans, "Future War in Cities"; Evans, "Case Against Megacities"; Evans, *City Without Joy.*
7. DiMarco, *Concrete Hell.*
8. King, *Urban Warfare in the 21st Century.*
9. Knight, "Urban Warfare Capability: A Background to the Challenges and a Call for Professional Debate."
10. Graham, *Cities Under Siege*; Graham, *Cities, War, and Terrorism*; Graham, *Vertical.*
11. Kilcullen, *Out of the Mountains.*
12. Lewis Dijkstra, Ellen Hamilton, Somik Lall, and Sameh Wahba, "How Do We Define Cities, Towns, and Rural Areas?" *Sustainable Cities* (blog), World Bank, March 10, 2020, https://blogs.worldbank.org/sustainablecities/how-do-we-define-cities-towns-and-rural-areas.
13. Ibid.

2

Urban Arenas of War

Wendy Pullan

FORMAL, STATE-BASED MILITARY ORGANIZATIONS DO NOT LIKE TO WAGE war in cities. The topography is dense, diverse, and complex, infrastructures are oriented to the needs of civilians, and traditional armed forces are faced with the uncertainties of urban life and guerrilla responses. State-of-the-art military technology is often less effective in city settings. Moreover, the overwhelming reality of urban warfare is civilian lives, often not well protected, easily shattered and regularly misunderstood in terms of their geographies and specific identities.[1] For the small, professional armies of liberal democracies, the prospect of fighting in cities is fraught with moral and operational problems.

Nonetheless, wars have changed, so that urban settings have now become a regular feature of much armed conflict, and in some cities everyday life has to some extent adapted for different types of combat. With respect to many recent conflicts, it is clear that cities are increasingly the epicenter for war. This is often persistent and long-term, where war and violent conflict affect the city in ways that go beyond conventional military attacks to strike at fundamental urban configurations and distort expectations of urban life and culture. To a large extent, what is being attacked and often destroyed are the processes, practices, and fabric that make a city understood as a city. Yet where and how this happens is often not recognized, and as urban wars are likely to be globally ubiquitous for some time, it is more than necessary to better understand them. Using examples, this chapter will address certain key aspects of cities that may be altered, distorted, or destroyed by war and violent conflict to the extent that core values of those cities may be fundamentally altered. With these observations, the underlying questions—from the standpoint of the city and urban life—

will be how and why such conflict can present not just long-term but existential problems for them.

Such a discussion needs to be seen against a background of change for both war and cities in the latter part of the twentieth and beginning of the twenty-first centuries. Mary Kaldor and others have considered radical shifts in the character (and some would argue, nature) of war to what may be called "new wars," where there is a blurring between war, seemingly unsolvable conflicts, organized crime, and large-scale violations of human rights.[2] In these situations, it is increasingly difficult to distinguish between states of war, armed combat, and violent conflict. Kaldor argues that rather than ideological or geopolitical reasons, these are now fought primarily in terms of identity politics and directed against civilians in order to control, displace, or reestablish populations.[3] Occupation by foreign forces, often long-term and nearly always contested, is common. In new wars, combatants may be states acting for themselves, in transnational situations, or as proxies of global powers, or they may be nonstate individuals and groups, such as paramilitaries, migrants, minority factions, regional ethnicities, or the representatives of superpowers. It would be fair to say that one's view of such irregular armies depends upon political orientation: one person's insurgents, terrorists, or thugs may be another's freedom fighters. Investigations carried out by the Centre for Urban Conflicts Research[4] show that such conflicts may be long-term or indeterminate and may be characterized by cycles of violence interspersed by periods of relative peacefulness. The structure of hostilities is not necessarily a linear one where war is declared, fought, and ended by an agreement of some sort, leading to a period of reconstruction. Thus, the idea of "postconflict" becomes questionable, and in today's wars, peace and conflict may have a fair degree of simultaneity.

At the same time, armed conflicts and even full-scale wars are increasingly played out in urban settings. As much as the changing nature of war is still debated,[5] the metropolitan location of many recent hostilities creates conditions that we can begin to characterize. The world's population is becoming overwhelmingly city-bound; in 2010, more than 50 percent were urban dwellers and the number is increasing.[6] It is not surprising that cities are targeted and bear the brunt of deliberate attack, or that conflicts based upon ethno-national and racial disputes, economic inequalities, and religious differences may be generated from within dense urban settings. Violence may emerge from both exterior threats and internal strife, making urban conflict wide-ranging and complex.[7] Military organizations play different roles as active combatants, occupiers, peacekeepers or representatives of a particular faction or power. Stephen Graham emphasizes the technologies of surveillance, control, and simulation in the increased military interventions.[8] Certainly these are significant drivers in urban situations, but they are very widespread through all aspects of life today and not restricted to cities. Rather, I would suggest that there are factors specifically pertinent to and implicated in the structure and experience of cities, and I examine some of them in this chapter.

Cities are not only the setting for war; they may also take on a more active role in waging war, giving them a degree of agency that may initiate, intensify, or prolong conflict in order to reinforce political positions or cement land grabs and power plays. Cities are regularly used to create "facts on the ground." Architecture, planning policies, urban practices, mobility, and infrastructures may be employed strategically, and city spaces and objects are regularly engaged in ways beyond their normal functions to feature in demonstrations and battles. The conquest and occupation of various urban squares, from Tahrir Square in Cairo (2011) and Independence Square in Kiev (2014) to the streets, malls, and flyovers of Hong Kong (2019–2020), are recent examples where large public spaces became tactical populist centers as well as finely tuned conquered territory.[9] In some cases, parts of cities have become instrumental for long-term urban wars; for example, streets, infrastructures, and particular civilian buildings such as hotels not only became part of the strategic topography of Beirut during the civil war but also represented shifting sovereignties throughout the hostilities.[10] War in cities is often inconsistent, with different areas variously affected, exacerbating widespread fragmentation; when fighting is raging in one part, residents may be living relatively normally in other areas, although with curtailed mobility and other limitations. Urban demographics provoke fault lines that play a role in the nature of combat, determining patterns of destruction, ethnic cleansing, and plans for reconstruction.

These two major global shifts—in wars and in cities—have an impact upon each other. Certainly, war changes urban structures and processes long after battles have finished, but the city will have altered preconceptions about the nature of battle, molding it to greater or lesser extent by particularities of urban population, culture, and space. As much as they might prefer otherwise, many national military forces have reluctantly come to accept the new urban wars, focusing increasingly upon how to fight in cities rather than whether to do it. Their efforts may be confounded by other perpetrators of violent urban conflict involving paramilitaries, terrorists, militias, mafias, and as John Mueller puts it, "opportunistic predation waged by packs—often remarkably small ones—of criminals, bandits, and thugs engaging in armed conflict."[11] The extent to which organized national armed forces engage with these groups depends on the individual case and itself is not a focus of this chapter. But I do wish to point out that violent and armed conflict regularly reflects and even confirms existing urban dissonance that has already brought uncertainty to city populations. Pervasive and long-term conflict often intermingles with other urban conditions to create fissures in everyday life and fabric that can rupture and change the nature of cities. This may not be the direct result of battle hostilities but, instead, part of a complex series of knock-on effects. My chapter interrogates some of them.

As this book shows, there is an increasing body of empirical research and scholarly assessment that addresses the problem of fighting wars in cities from the point of view of the combatants. At the same time, civilians and civilian casualties are, rightly, of crucial concern. However, in the context of waging war in cities, it is also important to consider the city as a whole as the arena of violence, destruction, and long-term upheaval. War interrupts the most basic trajectories and relationships of cities and civic life. In this, a seemingly simple question presents itself: What effects do wars and violent conflicts have on cities as a whole, in the long-term and, ultimately, in their nature as cities? This chapter will not address the problems of how to wage urban war, nor even, at least not directly, how to minimize the cost to civilians; rather my concern here is to consider the nature of the city when it is put in such peril. This is a large subject, but there are two key themes—urban order and urban violence—that reappear regularly and can be used in reciprocity, almost as two sides of the same coin. I shall concentrate upon Middle East examples that reflect not only my own research over many years but also the locus of many urban wars today.

Urban Order

Michel Foucault and many following him have focused on spatial ordering as fundamental to power and knowledge, resulting in the domination of various systems of control. Although it is easy to see how a causal interpretation of order may be particularly relevant in cases of urban conflict and war, it is also worth noting that the translation from certain general characteristics of spatial configuration as power to the complex settings of a city may be more nuanced. Certain specific examples of spatial power such as those so well demonstrated by Foucault in the modern formulation of prisons can be applied in some part to some cities, as we see in the periodic bombardment and long-term blockade of Gaza. The unequal power structures are distinct in various manifestations, from the underground tunnels dug to transport banned goods, to the ruined buildings unreconstructed due to an embargo of concrete, to the perimeter fences, walls, and checkpoints that define Gaza's overall space and mobility.[12] The blockade has become the overriding feature of existence for all of its citizens and for many illustrates Giorgio Agamben's idea of "bare life," that which is removed from the higher aims and possibilities of the state yet remains in the eye of that sovereignty.[13] Foucault reminds us that resistance is always a possibility,[14] and this too can be found in Gaza in events such as regular mass street demonstrations. The intensive use—fishing, recreation, military training, etc.—of the stretch of Mediterranean beach gives some relief beyond the prison of the ruined streets, although this bit of coastline is also prone to attack and suffers from water pollution. Methods of survival that can lead to creative resilience and resistance are evident,[15] and some forms of civic life and order exist even in this severely damaged and disgracefully neglected strip of land. The interrelated conditions and possibilities of incarceration and resilience are more complicated than simple power relations and can emerge even in the most unpromising of places.[16] My intention here is not to excuse any aspect of Gazan suffering but to point out that what can be called *urban order* is deep-set and persistent in cities even under intolerable circumstances.

Urban order can be understood as the relationships embodied in the physical and spatial fabric and registered over time to support and articulate broad forms of civic life. Peter Carl calls it *hierarchy* but warns that what is meant by that "is not an arrangement of power, but rather of participatory intensity, where the settings of conflict, negotiation, accommodation, and collaboration are reflected in the topography according to their importance to the city's meaning as a whole."[17] Despite great diversity among cities, urban order is dependent upon the key relationships that make a city somehow recognizable as a city, and much of that will be made by participation in it. Recognition will be tied to a particular place or time, but certain manifestations exist, at least in partial form, from one city to the next over wide geographies and chronologies; examples of continuity can be seen in the relationships between the shops and a high street, the way people enjoy cafes on an urban square, the clustering of activities around a bus stop or underground station, the arrangement of stalls in the area of a market. Not all is visual; for instance, there is a powerful urban order in the Islamic world in the siting of a mosque with respect to the sound of the call to prayer. There is a dynamic between the familiar and the extraordinary, and expectation will play a role.

These links may be most evident in horizontal dimensions of the city—this is usually how people meet and interact—but stratifications over time make cities layered and complex in vertical dimensions that may or may not be visible. Likewise, political, economic, and social stratifications also contribute to an order that is revealed to greater or lesser extent and often idiosyncratically. People live within and cut across these levels, and the simultaneity of space and experience is a key aspect for the order of cities. In this simultaneity, order is complex, fluid, and challenging to instrumental patterns or systems. Temporality is critical, often embedded in well-established customs and conventions of everyday urban life.[18] Warren Magnusson makes the important observation that people live on many scales simultaneously and suggests that not only is urban order generated from the bottom up, but in no simple way, there is also some sort of urban sovereignty.[19] Urban institutions, many of which are archaic and reinvented over time, can incorporate aspects of sovereignty as integral parts of everyday life.

Urban order is not idealized or static and certainly not neutral. Elements of disorder exist as part rather than in opposition, and cities are built on the absorption of difference rather than an overarching unity[20] or purified identity.[21] It can be claimed that disorder is one of the significant constituent parts of urbanism.[22] Clearly, disorder and fluidity are important qualities for the diversity of cities. In simple terms, cities are "messy" places where order is partial and imperfect; as much as it may exist, a unified version remains incomplete and aspirational. There is an ongoing dynamic of what is particular and what is more universal that Carl sees in terms of ethical value, where the city, primarily through its institutions, has the capacity to sustain and situate the individual within the wider domain.[23] This will depend upon the efficacy of the institutions, and the ethical dimensions of a city may be more aspirational than experienced except in fragmentary ways. The sense of what is partial or incomplete helps to identify the city as a dynamic field of relationships rather than an absolute entity so that the dream of an ideal city, like the circular or star-shaped plans that were common in Renaissance urban thinking, is just that, a dream. Urban experience can to some extent be conceptualized, but not as an intellectual synthesis alone; rather, city life is experienced on the ground as a part or a fragment of the whole, "a qualitative and communicative reality that is only to a limited extent accessible to reflective understanding."[24]

Praxis in all of its complexities and variations is critical to experience urban order, despite the intellectual desire to simplify and codify. The reciprocity between practical experience and conceptual knowledge is critical but often ignored; James C. Scott blames this on the development of abstract thinking and schemes formulated by the Enlightenment project, arguing that "formal schemes of knowledge are untenable without some elements of the practical knowledge that they tend to dismiss."[25] He points to the use of the cadastral map to be one of the most potent manifestations of this, with the power to transform as well as summarize.[26]

This problem, as it developed with and became symbolized by modern mapping, raises several issues with respect to common approaches of formal military organizations to cities. Perhaps most obviously, modern armed forces are creations of the state, established in large part to act as enforcers of the state's significant policies and rooted in its sovereignty; in normal conditions, they have little to do with civic life, where urban diversity is more often set to resist homogeneity. Clearly cities are not small states. Scott notes that although the state has no monopoly on utilitarian simplification it does aspire to control on the legitimate use of state force.[27] Today much of the friction between city and state has been translated into the problems of identity politics and in many of the new wars, formal armies sit at an uncomfortable junction between large-scale ethno-national identities produced and propagated by modern states and long-term neighborly relations gone sour in cities. More specifically, military knowledge of cities derives from the planning of states rather than the experience of cities. It is based upon certain technologies that have been developed in order to enhance the instrumentality of state control: the grid system of the cadastral map, aerial photos and drone imaging, and satellite navigational systems that treat every part of every city equally and the same to create a balanced and complete picture. These systems privilege the conceptual, usually through the visual image, over the experiential, and they present an understanding of the city that is in many ways foreign to those who live there and know it on the ground.

There is much about cities that is potentially treacherous for the modern state and its organizations, such as the army. As older structures of human life and governance without Westphalian sovereignty, cities may be in some ways incompatible with states. Magnusson maintains that cities are essentially uncontainable, putting them at constant odds with statism and its controls.[28] Although the more abstract state is often regarded by urban citizens as remote and sometimes threatening, the city is a concrete place embedded in a more familiar praxis experienced on a quotidian basis. Arguably, with fewer claims to unity, the city is more inclusive, whereas the nation-state is "part of a vast field of 'Othering'—distinguishing 'us' from 'the others.'"[29] Such urban diversity can be exhilarating but in some circumstances intimidating. Premodern cities are often perceived as particularly chaotic and incomprehensible for strangers, although residents use them comfortably. Urban order exists, not necessarily where the stranger, or invader, would look for or hope to find it, but rather in points of community significance and well-established familial, religious, or economic structures that are embodied by physical structures rather than generated by them. Even modern, so-called rational cities develop highly local peculiarities and hidden characteristics over time that may delight but often baffle outsiders.[30] Although significant continuities link one city to the next, at a deeper level, each city is idiosyncratic. Both these points—continuity and idiosyncrasy—have ramifications for any sort of desire for a common approach to cities.

As new wars move into urban settings, the relationship between state and city is reflected in particular tensions. With good reason, there is considerable concern about state-sponsored violence inside cities that may extend to increased militarization of specifically civic organizations, such as the police. At the least, there are gray areas between military operations and policing, and intensified militarization can create alien forces unidentifiable to residents. In more extreme cases, militarization of the police can overrun democratic procedure, legitimize the use of heavy weaponry inside the city, and threaten civic rights and human well-being.[31] On the flip side of the coin, other bodies may seize authority in the vacuum created by failed or neglectful states, forming a state within a state, sometimes with considerable success among the local population. New wars are characterized by a variety of fighting organizations, from formal armies to the more indigenous groups. Paramilitaries, insurgents, mafias, gangs, and even demonstrators may come from within the urban milieu and may have longstanding connections, a strong sense of how to operate on urban ground, and strong attachments to that ground that make their operations worthwhile or accepted.

The success of Hezbollah in the Shiite south of Lebanon is one example where this combined political party (sometimes known as "the Party of God") and paramilitary group has effectively created and become master of a twin city to Beirut, known as Dahiya.[32] After the bombing of southern Lebanon, including Hezbollah's al-Manar television station, in the 2006 war with Israel, Hezbollah rebuilt civilian structures, using architecture, reconstruction, and what, in Dahiya, was the creation of a city to assert its own hegemony as well as offering the beleaguered population a route back to normal life rather than living in displacement or squalor. As the dominant authority, Hezbollah took over not just recovery but policing and security, government, culture, health and welfare, and jobs, effectively creating a state, and as Lara Deeb and Mona Harb argue, through their institutions helped to set Shiite norms for morality.[33] Before that, Dahiya had developed as one of the impoverished suburbs of Beirut, a destination for mostly Shia who fled the countryside. The rest of Beirut tended to wash their hands of it, and more prosperous residents long regarded the area as a "'Shi'ite ghetto,' a rebellious space defying state [Lebanese] authority."[34] Today Dahiya can be considered a territory of civilian enterprise made to function through religio-military means. Religiosity is an integral feature and figures significantly for both its leadership and population as the defining feature of this location in Beirut, yet it shifts according to pragmatic needs, in a combination that has probably led to its relative success.[35] An urban culture has developed in Dahiya that fulfills the needs and desires of many of the Shiite population unavailable to them in the rest of Beirut.[36] Dahiya was born out of battle and nurtured by local practices that recognized specific needs; arguably, a new and different urban order has been formed that is successful in its own terms and has little use for a weak Lebanese state.

Urban Violence

In urban wars, attacks on infrastructures such as water, gas, transport, electricity, sewage, and housing will have immediate and debilitating effect. Modern cities cannot survive without these systems. However, there are other forms of destruction that are more insidious, hitting the city at an ontological rather than a systematic level, tearing apart not only its physical functions but also its basic configurations and relationships and our expectations and perceptions of them. This rupturing of the urban order will usually be long-term with knock-on effects and with responses to both destruction and recovery often contested. It may be the result of good intentions gone wrong, or there may be more malicious objectives.

The term *urbicide*, the killing of cities, refers to purposeful and premeditated urban destruction. Martin Coward describes urbicide as not only deliberate destruction but also erasure of the urban fabric and way of life.[37] The emphasis is on the overall extinguishing of an urban culture. Rather than collateral damage, urbicide has a clear aim where a city is attacked in ways that question its continued reason for being. Although there are attempts to disguise it within the exigencies of waging war or the demands of security, urbicide is increasingly being identified as a strategy for the political aims of urban wars. The term was first used in war to describe the devastation of Bosnian cities in the 1990s.[38] Since then it has been used in a number of settings, including retroactively in reference to the reduction to rubble of central Beirut in the civil war, for much of the obliteration of Syrian cities in 2011–2016, and in the persistent attacks on Palestinian towns and cities over many years.[39] In some cases, only certain structures are attacked, but these are carefully calibrated to cause the most thorough and lasting damage to their respective urban populations, often being symbolic in meaning and critical for the existing urban order. Carl's understanding of urban order as an ethical order seems particularly relevant here.

Urbicide does not drive all urban wars. In some cases, humanitarian concerns are observed as much as possible. But even where there is concern for urban life and environment and restraint is used, the urban order can be severely damaged, even if inadvertently. When violent destruction is not only extreme but fundamentally alters the urban order, it may be seen as a tipping point. Caroline Moser has carried out important research on the urban tipping point, determining it as the moment of violence.[40] I would suggest that we can also find a point of no return later on in conflict as a result rather than a cause of violence; this is when urban order is affected, sometimes irrevocably. Rather than a relatively isolated point of attack, there can be a series of events that go right through destruction to politically motivated urban renewal, connecting what Graham refers to as "acts of building and physical restructuring, on the one hand, and acts of all-out organised war and place annihilation, on the other."[41] For perpetrators, urbicide will be planned, perhaps as a process, but for the general population and other observers it may be identified only retrospectively. As war destruction becomes increasingly common and severe in cities across the globe, it is worth learning to recognize what a tipping point may look like and where evidence for it may be found.

One aspect of urban order that can be badly savaged in war is the reciprocity between public and private. In cities, public and private domains exist as part of a spectrum rather than oppositionally or absolutely: neither is complete without some portion of the other. The civilian home occupies the more private end of the urban spectrum. The shelling and bombing of residential neighborhoods carries particularly vicious connotations, to the extent that such destruction has been designated by the specific term of *domicide*.[42] The Fourth Geneva Convention stipulates that destruction of civilian property is forbidden "except where such destruction is rendered absolutely necessary by military operations."[43] Unfortunately this can be interpreted broadly, abused for political ends, and in some cases may be purposeful urbicide—hence domicide—especially when neighborhoods or buildings associated with specific ethnic or religious groups are targeted. Such actions destroy the home as a place of safety, refuge, and dignity; but beyond that, they thrust private dwellings into an unacceptable position of national or international prominence, where they become statistics, pawns in the conflict, something to be negotiated. Not only are surviving residents forced out, but the houses themselves are often turned inside out. Syrian cities show street after street of such devastation. Facades have been ripped off to expose interior spaces where what were lovingly choreographed personal spaces become displayed like the back of a macabre doll's house. One displaced Syrian relates, "the front wall of my mother's building was bombed open to show everything she owned. She would have preferred that the whole house be knocked down."[44] That which was most private becomes most public as these interiors appear on news reports across the world and live for posterity on internet sites. Eventually personal belongings are removed or stolen so that only gaping dark holes remain. Whole streets of such buildings transform the city into something that is deserted, threatening, screaming in its silence, and the facades, usually the public face of the street, become more absent than present. Such neighborhoods are ruptured and no longer integral to a community, often abandoned and anonymous for years to come.

How the wrecked neighborhoods of Syrian cities will be rebuilt and what kind of life might return to those streets are difficult to predict. The damage is extensive throughout the country, with Aleppo and East Ghouta each having lost approximately 35,000 buildings.[45] The knock-on effects are substantial, and in no way is this a clear-cut case of reconstruction.

Nearly 12 million residents have become displaced, either internally (about 6.7 million) or as refugees to other countries (about 6.6 million).[46] Yet, the problem exceeds scale. To a large extent, dispersion was haphazard or secretive, making the tracing of property owners extremely difficult or impossible. Most owners are not present for consultation in the rebuilding of their neighborhoods, and when they are available, usually not consulted. Too often rebuilding is regarded as an essentially economic enterprise, massive and expensive, but one that can be lucrative and treated as a spoil of war. Large international consortia are assembled to rebuild, both to finance reconstruction and to reward loyalties during war.[47] Such rebuilding is usually formulated around large-scale international development, with emphasis on foreign audiences and tourism, profitability, security, control, and ease of construction with little concern for community, heritage, or human rights. Increasingly, this sort of process is being regarded as a form of urbicide with intentional destruction of particular urban cultures.[48] The Marota project in Damascus is one such project.[49] The buildings lack association with any particular place and are constructed as complexes with limited access for easy surveillance and control. These scenarios are designed for developers not urban citizens, replacing modest neighborhoods with wealthier residents. Comparisons have been made with the Solidere project that rebuilt downtown Beirut after Lebanon's civil war.[50] But even with the considerable effort to make Solidere a place- and heritage-bound reconstruction,[51] its realization resulted in neglect of the city's diverse social structures. The result is noninclusive and overly sanitized and contributes to further division in an already fragmented city.[52]

No city stands still, and fluidity between constancy and change is constitutive of urban order. At times, war destruction may provoke needed modernization. Some scholarship has pointed to the possibilities of extensive war destruction doing away with sectors of cities based upon colonial planning or other aspects of ethnic, racial, or religious segregation.[53] Nonetheless, although radical change can have desirable outcomes, caveats are necessary, at least as much as they underline a controversial and complex subject that cannot be deliberated in this chapter. First, it is impossible to start a city again at zero. Extensive and severe destruction can make a situation worse, and usually it is the poor and disadvantaged who suffer. Second, although the end of colonialism has meant reform in some cities, it is not uncommon to find colonial segregation going on to become class ghettoization. Third, whatever benefits emerge from modernization, much will be lost in sweeping reconstruction. For example, the ambitious plans for London, devised while World War II was still raging, offered safe, hygienic, spacious housing in the suburbs to those who were dispossessed by the destruction of so much of the city's East End; but many complained of the loss of their communities that had been made vibrant by their very density. The process of building and rebuilding cities is rooted in a metabolism that is layered and complex, and although instrumental solutions address certain problems, much of what makes the city is rooted in intangible factors. And finally, there is evidence of certain city sectors being specifically targeted in war bombing; such neighborhoods are often regarded as "informal," inhabited by minority or impoverished populations who have built without official planning, and in removing them by war there may be the intention to rebuild for residents that will be closer to the political and economic vision of the authorities. In these cases, destruction is determined to be so extensive and complete as to render the city "clinically dead."[54] Such conditions are welcomed by some so that the only solution is complete demolition and rebuilding. This is a form of combined urbicide and population cleansing based on particular identities or class structures.[55]

Not all minorities are driven out; division and fragmentation are a feature of many urban wars that are fought by different ethnic, political, or religious factions on their own terms or as proxies for or allies of larger powers. Despite widespread use of sophisticated electronic means for surveillance, tracking, and documentation of people and activities, physical means for restricting mobility, separating populations, and segregating parts of the city into enclaves remains key to much violent urban conflict. Buffer zones, barriers, checkpoints, walls, and fences are introduced in and around cities, as strategic operations to gain or secure territory or sometimes as attempts to separate local populations from fighting and killing each other. Yet, in many cases these physical divisions exacerbate existing ethnic or religious tensions by imposing hard barriers along preexisting, soft fault lines, or they reflect decisions made in war that may be strategic for the combatants but are irrelevant or damaging for civilian life. The complex and problematic situation of Baghdad since 2003 illustrates this well.

Like many Middle Eastern cities, Baghdad is made up of quarters where a relatively workable modus vivendi evolved among different sectors of the population. But as insurgencies developed after the US invasion, the city fragmented into partisan enclaves identified by particular authorities or ethnicities, some clearly barricaded and relatively secure, such as the American-dominated Green Zone, and others with populations that shifted in order that they take refuge with their own ethnic or religious groups as conditions worsened.[56] As hostilities grew, the enclaves were reinforced by concrete walls and checkpoints erected by the US occupying forces, both as a direct part of military strategy to maintain or regain control of the city as well as to separate embattled populations. For some years Baghdad was in many ways abandoned by the Iraqi state, and "ethno-sectarian violence became a means of communication."[57] As a city, fragmentation into warring sectors was its most dominant characteristic.

Baghdad's Sadr City, the base for Shiite cleric Muqtada al-Sadr and his Mahdi Army Jaish al-Mahdi, as well as home to other sectarian groups, suffered high levels of violence between 2004 and 2008. The US Army reasoned that entering the area by conventional attack would be problematic, dangerous, provocative, and counterproductive and instead, in 2008, chose to construct walls along al-Quds Street, one of the main markets, and around the southern quarters to restrict insurgency mobility and to blockade the area without entering it.[58] The primary purpose of the action was explained by the United States as the need to stop rocket attacks on the Green Zone and reassert Iraqi government control of Sadr City. It was deemed a success, employing new types of urban warfare, especially concrete siege walls, to avoid high levels of civilian deaths or heavy destruction.[59]

The urban impact, however, has been more complex and enduring. Based on a poorly conceived and instituted modernist plan, Sadr City was already isolated.[60] It had become miserably overcrowded, with poor public facilities, and the area was badly stigmatized for being "the poorest of the poor reviled by most Iraqis as uneducated, as rabble."[61] Sadr City was neglected and angry, and help for the Shia residents was forthcoming only from the Sadrists. The siege walls further isolated an already problematic area, turning its residents into combatants; for some time vilified as "other," they now became considered as terrorists who must be kept behind their walls.[62] Sadr City is a gated slum, keeping the residents in rather than danger out. Although conditions for Shiite Baghdadis improved with the coming of new Shiite leadership, well over a decade later the stigmatization of Sadr City remains and the violence is long-term. Sadr City sets its own agendas and is still ruled by drug gangs, paramilitaries, currency traders, and arms dealers; in the central market, one can buy a kilo of tomatoes or a fully operational tank.[63]

The legacy of the US siege of Sadr City is a good example of military expediency and urban viability working at cross purposes. The wall-building operations have been assessed for their impacts on future urban warfare and deemed a new, nonkinetic (nonviolent) strategy for urban war, potentially minimizing casualties (see Chapter 9).[64] Such urban barriers have found some support in civilian circles as well, as an emergency action to break cycles of sectarian violence; David Kilcullen describes them as a "tourniquet," creating "gated communities" in Baghdad.[65] But in places like Sadr City, the line between gated community and gated slum is very thin indeed, and even Kilcullen points out that the technique may have side effects and "can only be temporary." Herein lies a particular problem, for such urban divisions are rarely temporary, as we know from other fissured cities (Nicosia, Belfast, Jerusalem, etc.) where division has been seen as a solution to violence. It may be relatively easy to erect the barriers but very difficult to get rid of them when they inevitably become political issues and residents become dependent upon the perception of security offered by suspending the conflict instead of addressing its roots. Violence can blow up again at any time.

Baghdad has become a city of inward-looking and fearful sectarian enclaves, divisions, and no-go zones. Although there has been an attempt to remove some of the walls, mobility remains dependent upon checkpoints, negotiating a frustrating and sometimes dangerous terrain, with long detours, traffic jams, and confusion.[66] Iraqis complain of being like "caged animals":[67] the T-walls and watch towers—concrete prefabricated modules of varying heights that due to their inverted T-shape can be dropped into place by crane without the need to dig foundations—are a relatively cheap and rapid construction system adopted from the Israeli separation barrier in Palestine. The infrastructure is a transferable technology, but its origin exacerbates the "caged" analogy for Iraqis.[68] Today, walls and checkpoints remain ubiquitous as defining features of Baghdad's urban landscape. The overall capacity for security is debatable; as recently as January 2021, one response to a suicide bomber was, "How could the terrorists infiltrate Baghdad? The city is surrounded by checkpoints, militias and security forces."[69]

The role of walls in cities over time is a large subject that is well beyond the scope of this chapter. Nonetheless, it is worth remembering that historically cities have usually been walled and such boundaries played a significant role, providing not only defense but also definition and status, most significantly representing the city as a place for its citizens. Although most city walls have disappeared, their structuring capacity for the city remains, as is evident in the Paris Périphérique, the dual relationship between the City of London and Palace of Westminster, and the expanding sectors of Istanbul, each walled zone reflecting its historical growth outward. Today municipal boundaries rather than walls form important legal definitions for urban spaces. Borders and boundaries—roads, fences, signposts, buffer zones, even changes in the curbstones—remain an important means of structuring space—legally, socially, economically, culturally, ritually, politically, and psychologically—and without them we would be unable to function. These infrastructures, large or small, continue to orient cities to create topographical realities. City walls and other sorts of boundaries that function well usually had or have a degree of porosity to be part of the wider urban context and as such, part of the wider urban order. A wall or border needs to function as such, but concurrently it must be integral to the city. Most significantly, such permeable walls are multivalent; they have a number of compounded uses and meanings in the city that emerge and are validated over time. In doing so they become recognized as part of the whole city.

Some city walls and barriers do not reflect urban situations developed organically over time and instead are imposed. Often they are rooted in contestation and are there to divide and segregate rather than draw together. Many such imposed barriers can be seen as acts of despair when all other negotiation has failed. Imposed inner city walls and borders are too often used as a "quick fix," whether as a military tactic, a (re)assertion of political authority, or to separate warring groups. Unfortunately such solutions often persist over time. The city of Nicosia has one of the oldest of such divisions in place when, after some years of ethnic hostilities culminating in the Turkish invasion in 1974, it was split by a buffer zone to separate Turkish Cypriots in the north and Greek Cypriots in the south. This hard division is still there today, maintained for nearly half a century by UN forces. No civilians are allowed to enter the buffer area, and today the imposed division is known by the locals as "the zone." In the historic center, the zone follows the alignment of Hermes Street, formerly the commercial district where bilingual Turkish and Greek Cypriots did business.[70] It was where people came together and like any urban meeting ground was a fault line for both exchange and conflict, rich in urban diversity but with the potential for dissonance and violence. Today, Hermes Street is long gone, replaced by the empty buffer zone that is an international border. Such urban barriers are relatively easy to put into place and readily accepted at times of great crisis, but almost impossible to remove.

Too often these urban divisions become political markers, and not a few are international borders. A frontier through the middle of a city creates a very different situation than at the periphery of a state. Most obviously this is because the city center is dense, full of people, places, and activities, whereas a state border tends to fall in minimally inhabited land. Most large concentrations of population tend to shrink away from borders.[71] Moreover, an imposed border through the middle of dense urban life and fabric is located exactly where it should not be: as much as cities do have borders, they exist at the urban periphery, not the center. As we see in Nicosia, such moves can have long-term consequences. As I have suggested elsewhere, these ill-conceived divisions create a "frontier urbanism" where civilians face each other against hostile borders inside the center of cities.[72] They may be supported by armed forces, paramilitaries, or occupying armies, but because the division is in the center of a city, it will inevitably involve civilians becoming combatants. The frontiers themselves—such as walls, fences, buffer zones, barriers, or checkpoints—are infrastructures designed specifically to enhance such confrontations. The transfer of such an imposed frontier from the periphery to the center is an inversion of urban order, perhaps with irrevocable consequences, and as such is an act of profound violence on the city.

Conclusion

Recent examples show that the relationship between cities and war is deep-set and complex; urban violence and urban order are intermingled and in some ways reciprocal. As cities are increasingly the arena for armed conflict, the problems of civilian injury and death and the destruction of the physical fabric are the most obvious and immediate causes for concern. But in many cities that suffer heavy levels of armed conflict and war, violence is long-term and the devastation caused by armed battle is only the beginning of more enduring ruptures to urban order. Rather than distinct and isolated events, the various stages of conflict, including early evidence of discord, intense or protracted combat, and the long aftermath of recovery or stagnation or both are part of a continuum, and in many cases these different facets act cyclically or simultaneously as they influence and intermingle with the everyday life of the city. There is no such thing as a speedy military action without long urban ramifications. Armies and other fighting forces are increasingly embedded in the cities, where they are in varying degrees responsible for urban survival; at the same time, they may also be witness to urban resilience and, in some cases, resistance. But they will inevitably be destructive. When they depart, the city left behind may be transformed in various ways, but certainly it will not be the same city as when they arrived.

The long-term effects of armed combat and protracted violence in cities is hard to predict and will need more dedicated research, but because these conflicts affect people on their home ground, they inevitably have great impact. One already apparent feature of urban wars is that wealthier and better educated residents are able to remove themselves, and the phenomenon of middle-class displacement is increasingly common. This leaves the urban poor at home to fight or simply to survive. Urban centers such as Gaza, Dahiya, and Sadr City have experienced different outcomes of war on their soil, and it is unlikely that further conflicts can be ruled out. Urban poverty is regularly exacerbated by violent incursion, not relieved by it, and the hard realities of war often shatter existing fragile situations. Such cities become more volatile with war not necessarily because they are poor but because the social, cultural, and political settings that constitute their everyday reality are part of larger disputes. Most of the tactics used in these hostilities—the bombing of residential areas, imposed hard divisions in urban centers, the extensive destruction of public streets—attack and disable key aspects of the urban order to create different realities, a new urbanism of the frontier, where violent antipathy is the most common language. It is a growing reality of any urban battleground and one with which armed forces will become increasingly familiar.

Notes

1. Brathwaite and Konaev, "War in the City."

2. Kaldor, *New and Old Wars*. For a variety of opinions, see also Duffield, *Global Governance and the New Wars*; Münkler, *The New Wars*; Berdal, "The New Wars' Thesis Revisited."

3. Kaldor, "In Defence of New Wars."

4. Centre for Urban Conflicts Research, home page, https://www.urbanconflicts.arct.cam.ac.uk. See, for example, research carried out from 2003 to 2013 on a number of cities that have experienced war or violent conflict in Europe and the Middle East: Conflict in Cities (CinC) Briefing Papers 1–11, https://www.urbanconflicts.arct.cam.ac.uk/publications/briefing-papers-english.

5. Strachan and Scheipers, *The Changing Character of War*.

6. By 2050 it is expected to be 68 percent. UN Department of Economic and Social Affairs, "68% of the World Population Projected to Live in Urban Areas by 2050, Says UN," May 16, 2018, https://www.un.org/development/desa/en/news/population/2018-revision-of-world-urbanization-prospects.html.

7. The literature on the subject is substantial. For the targeting of cities, see Graham, *Cities Under Siege*. For conflicts that emerge from inside cities, see Pullan and Baillie, *Locating Urban Conflicts*.

8. Graham, *Cities Under Siege*, 62ff., passim.

9. Hatherly, "Architects of Revolt"; Pullan, "Violent Infrastructures, Places of Conflict."

10. Fregonese, *War and the City*, chap. 6.

11. Mueller, "War Has Almost Ceased to Exist," 308.

12. The Gaza Strip is an area of 365 square meters on the Mediterranean between Israel and Egypt. Gaza City is at its center, but the whole strip is one of the most densely populated places in the world, and much of it is covered in urban sprawl. Gaza is ruled by the Palestinian group Hamas but is actually controlled by its more powerful neighbors; in three wars since 2008, it has been heavily bombed by Israel. Fifty-six percent of the population lives below the poverty line. UNCTAD, "Economic Costs of the Israeli Occupation for the Palestinian People: The Gaza Strip Under Closure and Restrictions," August 13, 2020, https://unctad.org/system/files/official-document/a75d310_en_1.pdf.

13. Agamben, *Homo Sacer*.

14. Foucault, "Space Knowledge and Power."

15. Conflicts in Cities (CinC), *Coping with Conflict*.

16. See, for example, Katz, "Between *Bare Life* and *Everyday Life*."

17. Carl, "*Convivimus Ergo Sumus*," 25.

18. Pullan, "Justice as the Urban Everyday."

19. Magnusson, *Politics of Urbanism*, 118–119.

20. Isin, *Being Political*, chap. 1.

21. Sennett, *Uses of Disorder*.

22. Ibid.

23. Carl, "*Convivimus Ergo Sumus*."

24. Vesely, *Architecture in the Age of Divided Representation*, 83.

25. Scott, *Seeing Like a State*, 7.

26. Ibid., 87.

27. Ibid., 87–88.

28. Magnusson, *Politics of Urbanism*, 24–25.

29. Therborn, *Cities of Power*, 23.

30. The question of individuality in smart cities is beyond the scope of this essay, but pertinent to this point.

31. See, for example, Hills, "Deconstructing Cities"; Salter, "Toys for the Boys?"; Kappeler and Kraska, "Normalising Police Militarisation"; Mendel, "Urban Violence and Militarization."

32. Beirut's southern suburbs, or Dahiya, has been Hezbollah's ruling center since the Lebanese civil war, when they won a battle in 1989 to become dominant in the area. Dahiya's population grew exponentially in 2006 when Shia from southern Lebanon fled there to escape Israeli bombing. It remains an essentially unrecognized city with a population of about 500,000; 85 percent are Shia (excluding the Palestinian refugee camps next to Dahiya).

33. Deeb and Harb, *Leisurely Islam*.

34. Ibid., 48.

35. Harb (2011).

36. Deeb and Harb, *Leisurely Islam*.

37. Coward, *Urbicide*.

38. Bogdanović, "The City and Death"; see also Herscher, "Warchitectural Theory."

39. There are many examples cited in a special issue on urbicide in *Theory & Event* 10, no. 2 (2007). See also Fregonese, "The Urbicide of Beirut?"; Fregonese, *War and the City*; Graham, "Bulldozers and Bombs"; Graham, "Lessons in Urbicide"; Abujidi, *Urbicide in Palestine*; Graham and Gregory, "Urbicide."

40. Moser and Horn, *Understanding the Tipping Point*, 3.

41. Graham, "Cities as Strategic Sites," 33.

42. Porteous and Smith, *Domicide*, chap. 3.

43. Geneva Convention IV, 1949, Article 53, "Geneva Convention Relative to the Protection of Civilian Persons in Time of War of 12 August 1949," https://www.un.org/en/genocideprevention/documents/atrocity-crimes/Doc.33_GC-IV-EN.pdf. Further commentary emphasizes the problem but retains the clause; see International Committee of the Red Cross (ICRC), "Practice Relating to Rule 50."

44. Resident of Aleppo, interviewed in Beirut, April 17, 2018.

45. Homs, Raqqa, and Hama have more than 10,000 damaged or severely damaged structures each; for details, see REACH-UNITAR, *Syrian Cities Damage Atlas*.

46. UNHCR, "Syrian Refugee Crisis Explained."

47. Daher, "Paradox of Syria's Reconstruction."

48. Pullan, "Conditions of Urbicide"; DiNapoli, "Urbicide and Property Under Assad."

49. Syrbanism, "Marota City."

50. Hamadeh and Bassil, "Demolishing Human Rights."

51. Khalaf, *Heart of Beirut*.

52. There are many critiques of Solidere; see, for example, Nagel, "Reconstructing Space, Re-creating Memory."

53. Kipfer and Goonewardena, "Colonization and the New Imperialism."

54. Jansiz, "Let's Ditch Homs."

55. Graham, "Cities as Strategic Sites," 33; DiNapoli, "Urbicide and Property Under Assad"; Pullan, "Conditions of Urbicide"; Pullan and Azzouz, *Destruction and Reconstruction*.

56. This process is explained by Derek Gregory in Gregory, "Biopolitics of Baghdad." He makes the important point that although the groups were identified through ethnicity and religion they were driven by political concerns, specifically the right to rule Iraq (p. 5).

57. Gregory, "Biopolitics of Baghdad," 6.

58. See accounts from different points of view in Gregory, "Biopolitics of Baghdad"; Spencer, "Stealing the Enemy's Urban Advantage."

59. John Spencer ("Stealing the Enemy's Urban Advantage") has stated: "The role and effectiveness of concrete in reducing violence across Baghdad cannot be overstated" (p. 3).

60. Sadr City is based on a 1958 grid plan by Constantinos Doxiadis, which was devised to alleviate severe housing shortages and was populated largely by influxes of the rural poor. Since its inception as al-Thawra or Revolution City, Sadr City has been the home to various agitating political factions. The area is mostly made up of cinder block houses, based on what Doxiadis presented as a rational, Western plan. Although one can question whether such planning was appropriate for a Middle Eastern city, Sadr City was overpopulated from the outset.

61. Enders, "Behind the Wall," 122.

62. Gregory, "Biopolitics of Baghdad"; Dagher, "Baghdad Is Safer"; Izady, "Urban Unplanning."

63. Interview, Iraqi architect, via Zoom, November 23, 2020; see also Al Jazeera, "Growing Slums and Misery"; Rubin, "Our Patience Is Over."

64. Spencer, "Stealing the Enemy's Urban Advantage"; Johnson, Markel, and Shannon, *2008 Battle of Sadr City*; Bowers, "Future Megacity Operations."

65. David Kilcullen, "The Urban Tourniquet—'Gated Communities' in Baghdad," *Small Wars Journal* (blog), April 27, 2007, https://smallwarsjournal.com/blog/the-urban-tourniquet-gated-communities-in-baghdad; cf. the argument against the tourniquet by Gregory, "Biopolitics of Baghdad."

66. OCHA Services, Relief Web, *Urban Baghdad*; Izady, "Urban Unplanning."

67. Quoted in Gregory, "Biopolitics of Baghdad," 13.

68. See Dagher, "Baghdad Is Safer," for further reference to Palestine.

69. Ahad and Chulov, "Two Suicide Bombers."

70. Bakshi, *Topographies of Memories*, part II.

71. Anderson and O'Dowd, "Borders, Border Regions and Territoriality."

72. Pullan, "Frontier Urbanism," 15–35.

3

The Evolution of Urban Warfare

Paul Latawski

SINCE THE END OF THE COLD WAR, WESTERN MILITARY ESTABLISHments have been increasingly focused on the problem of warfare in the urban environment. Mirroring this interest has been a burgeoning literature in recent years on the challenges of urban warfare.[1] This surge in interest in urban warfare has given rise to a number of assumptions about the changing character of armed conflict in the urban environment. The first of these assumptions is that with the global population increasingly urbanized the urban environment will become the central arena of future warfare—particularly the megacity.[2] It is an assumption that sees armies engaging in the urban environment as almost an inevitability. The second assumption is that the urban adversaries will be irregular rather than conventional armed forces. The work of David Kilcullen nicely encapsulates this school of thought.[3] The third assumption is more deeply rooted and conflicts with the futurology of the inescapability of fighting irregular adversaries in cities. It is the view that armies have a strong aversion to getting mired in the urban environment, which absorbs large amounts of manpower and material.[4] On the surface, this last assumption does resonate with states whose armies have diminished in the wake of post–Cold War drawdowns. As armies have been getting smaller, cities have been getting bigger. Predicting future war, however, is a very inexact business. Warfare in the urban environment may or may not be inevitable, adversaries may or may not be irregular, and armies may or may not have a choice as to where they fight. For armed forces preparing for future conflict, the fixation on the urban environment has produced an important debate that cannot be ignored even if it is one that is prone to "reductionism" or "determinism."[5] This tendency toward sweeping generalizations and ideas of inevitability is symptomatic of a debate that is not well grounded in an empirically derived understanding of urban warfare in the broad sweep of history.

With a view that history can inform the present and help us in assessing future directions, the aim of this chapter is to provide a historical context to the current debate on urban warfare. In doing so it does not suggest that history offers ready-made templated solutions to current challenges nor does it assign the role to history of providing evidence to support currently fashionable thinking. Rather the intention here is to identify continuities in urban warfare and those discontinuities that reflect its changing character. The adoption of this overtly Clausewitzian framework of analysis is intended to bring historical understanding about what is new and what is not new regarding the challenges of urban warfare. In looking at the historical evolution of urban warfare, the chapter will draw on a range of historical examples to consider trends and issues in its conduct. Serving as a kind of illustrative "backbone" and running as a thread in later sections will be an examination of the British Army's historical experience and thinking regarding urban warfare.

Siege: The Oldest and Most Enduring Form of Urban Warfare

The prevalence of the siege in history challenges a narrow definition of urban warfare that would see it only as fighting in the city, street by street and house by house. The siege, or fighting in the city, is best understood as two different methodologies for conducting urban warfare. Each shares the common goal of determining which military force will take or retain possession of an urban settlement. Contemporary military glossaries no longer contain *siege* as an entry. That definitional well-spring of the English language, the *Oxford English Dictionary*, defines a *siege* as "the action, on the part of an army, of investing a town, castle, etc., in order to cut off all outside communication and in the end to reduce or take it."[6] The closest contemporary military terminology related to the siege is *blockade*. *NATOTerm: The Official NATO Terminology Database* states that the function of a blockade is "to prevent a hostile entity's freedom of movement and contact with external support."[7] As late as the nineteenth century, military writers regularly defined the meaning of *siege* in instructional manuals in military education establishments. In 1824, General Charles Stanislas de Malortie de Martemont, a "Professor of Fortification and Artillery" in the Royal Military Academy, Woolwich, stated that "the object of a siege is to take possession by force of a fortified town, the conquest of which political motives, or military views, render important for the besieger."[8] Seventy years later, James Mercur, a professor of civil and military engineering at the United States Military Academy, West Point, defined a siege in the following manner: "by a regular siege is meant a systematic and more or less deliberate attack upon a fortified place, in which the besieger aims to invest the place and capture its fortifications . . . ending only with . . . the surrender of the garrison."[9]

Ancient Foundations

The conduct of sieges from a tactical and technical point of view was well established and relatively unchanged from ancient times well into the early modern period. With urban settlement associated with wealth and central political authority, the need to protect cities led to their fortification. For the defender of a city, curtain walls with the towers interspersed to provide enfilading fire of projectile weapons built on the perimeter of the city became the norm. Particularly weak or vulnerable points such as gates had more elaborate and in-depth structures for protection. A major vulnerability of the defenders in a city was the ease with which essential resources could be denied to the city through blockade. Irrespective of how strong the perimeter defensive fortifications of the city were, without resupply of food eventually the garrison (and the noncombatant urban population) would succumb to starvation. The siege of Calais (1346–1347) is a good example of a besieging force starving the city into submission.[10] Time was therefore an important factor in conducting siege warfare for both sides.

The tactics of the defender or besieged force, however, were not entirely passive. With adequate warning, preparations for the defense to endure a lengthy siege included stripping the surrounding countryside of resources, particularly food, to increase the sustainment problems of the force besieging the city. Once under siege, launching sorties to attack the besieging forces to disrupt their efforts to assault the city was another means of conducting an active defense. Maintenance of morale and vigilance against internal threats were other issues the defenders of a city had to consider. A fourth century BC work on "how to survive under siege" by Aineias the Tactician outlined the importance of reliable soldiers manning critical points such as gates and the need to be vigilant for "plotters" and to maintain the unity of the urban population behind the defense.[11]

For an attacker investing a city, the objective of the siege was to capture it. The least costly course of action open to the attacker would be to take a city by surprise, ruse, or deception. The use of the Trojan horse in the Greek siege of Troy circa 1250 BC is perhaps the most famous, if historically unproven, example of the use of deception.[12] Failing a quick capture of the city, the central aim was to cut off the city from resupply, making a siege a waiting game until looming starvation of the garrison and inhabitants forced surrender. The besieging force, however, was also exposed to difficulties in sustainment, the impact of weather and climate for prolonged periods that could lead to malnourishment, disease, and weakened morale due to inactivity and boredom. The threat of an outside force relieving the siege could also not be discounted. The Great Siege of Malta in 1565 saw many of these factors contribute to the Ottoman Turks ending their efforts to take the center of power of the Knights of Malta.[13]

Resisting or conducting a siege was a major engineering effort. The construction of a city's fortifications required high levels of engineering expertise and resources that had to be applied on a large scale. For besiegers of a city, the engineering challenges were similarly large if different from the defenders of a city. Surmounting walls required escalation utilizing ladders, earthen ramps, or siege towers. Undermining a wall to induce collapse was achieved through mining and was another important way of creating a breach in defenses. The art and science of siege operations was well established already in ancient times and produced many authors articulating both theory and practice. The legions of the Imperial Roman Army included specialist engineers with the knowledge to construct siege machines and pioneers whose tools, skills, and labor supported siege operations.[14] Thus, in the long span of history, military engineering support was an important thread running through siege warfare. Over time, practices and siege machinery were refined, but the basic capabilities remained the same until the introduction of gunpowder.

Siege warfare was conducted to achieve strategic effect. The Roman siege and capture of Carthage (149–146 BC) put an end to a major adversary of Rome in the Punic Wars that lasted for over a century. It gave rise to the expression 'Carthaginian peace' meaning the total subjugation of an adversary and their end as an independent political and cultural entity. The capture of Constantinople (1453) brought an end to the Byzantine Empire and saw its replacement by the Ottoman Empire. Similarly, the Spanish capture of Granada (1491) marked the end of the Reconquista and expulsion of the Moors from Spain.[15] With all three of these examples, the result had a strategic impact in seeing the definitive end of an empire or kingdom.

Siege warfare also developed its own set of brutal rules and customs. If a city did not capitulate before the siege started, then it could expect no quarter. When a city was taken by assault or belated surrender, its inhabitants, combatant and noncombatant alike, would be subject to the most extreme treatment. It was the exception for a more generous treatment of the inhabitants of a city captured by siege. The norm for the defeated side was to experience massacre, pillage, rape, enslavement, or deportation of the population and the sack of the city.[16] Neither besieged nor besieging military forces had much regard for the noncombatant population. In medieval times, the garrison of the city would drive out noncombatants because they were nonproductive mouths to feed. The besiegers, however, often did not let them pass through their lines and kept the noncombatants trapped between the two warring sides in order to pressure and punish the city for not surrendering.[17] These harsh practices were considered part of the customs of war well into the nineteenth century. Promulgated at the height of the Civil War by the US Army, the Lieber Code can be considered a precursor to the contemporary law of armed conflict. Nevertheless, Article 18 of the Lieber Code stated that "when a commander of a besieged place expels the non-combatants, in order to lessen the number of those who consume his stock of provisions, it is lawful, though an extreme measure, to drive them back, so as to hasten on the surrender."[18] In the history of siege warfare, the customary laws of war "justified violence rather than restrained it" when it came to determining the fate of combatant and noncombatant urban populations.[19]

Advent of Artillery: Early Modern Siege Warfare

The introduction of gunpowder and artillery in early modern warfare changed the conduct of sieges, even if the new technologies operated within existing tactical approaches. The end of the fifteenth century marked an important turning point in the history of siege warfare.[20] The descent of Charles VIII of France on Italy in 1494–1495 with an army equipped with the then state-of-the-art artillery demonstrated how fortifications could be quickly breached and taken in comparatively short periods of time with firepower.[21] The introduction of the mortar as an area weapon meant in addition to facing malnutrition or starvation during a siege, the population of an urban settlement could now be subjected to the perils of bombardment. Armies developed specialist artillery units whose role was to operate heavy artillery pieces designed to destroy or breach fortifications. Urban fortifications that had long employed the curtain wall and towers at the core of their perimeter defense were found to be too weak to withstand artillery fire, making them obsolete. To resist the impacts of artillery projectiles, more angular bastions that could mount artillery became necessary. Moreover, cities were no longer the only venues of defensive systems. Separate fortresses built in belts on frontiers or astride likely avenues of attack became more common.[22] This is not to say that fortified cities were no longer besieged, but not every city could justify the large investment in new, artillery-resistant fortification. The changes ushered in by the advent of effective artillery gave new life to the science of military engineering. By the seventeenth century, a body of fortification experts with treatises on how to make a locality more resistant to artillery during a siege and how to breach fortresses successfully emerged across Europe.[23]

Nineteenth Century Glimpse into the Future: Fighting in Cities

The nineteenth century marked an important time of change in the conduct of urban warfare. The British Army's experience from the beginning of the nineteenth century reflected the ways in which urban warfare was entering a period of transition. The onset of change can be illustrated by British efforts to gain control of the Spanish Viceroyalty of Rio de la Plata (present day Argentina and Uruguay) in 1806 and 1807. During this campaign, the British Army became embroiled in an urban battle that had many of the attributes of those seen later in the twentieth century. In July 1807, a British force led by Lieutenant General Sir John Whitelocke launched an attack on the city of Buenos Aires. The city itself had a regular grid pattern with uniformly sized city blocks. The construction of houses typically included one large door, a few small exterior windows, open courtyards, and flat roofs bordered by meter-high parapets that made them natural fortresses.[24] With the Plaza Major at the heart of the defense, barricaded streets and emplaced artillery covered approaches to the central square. The number of men under arms available to defend Buenos Aires exceeded 9,000, with militias providing a significant part of overall strength.[25]

The British plan to attack Buenos Aires was complex and called for thirteen columns to be launched from the city's western suburbs to drive eastward until they reached the Rio de la Plata. The British columns were then to occupy buildings as defensive positions. Whitelocke's plan dissipated his force and effectively devolved command and control down to each column. Believing that the superior training of his troops gave a decisive edge, his plan emphasized speed over methodical preparation. Indeed, there was no artillery support, and the troops were ordered to advance with fixed bayonets and unloaded muskets to avoid firefights lest the attack become stalled.[26] For the assault, Whitelocke had 6,000 troops at his disposal, the bulk of which he committed to the attack with only a small reserve.[27]

Two hours before sunrise on July 5, 1807, the British assault began. In a day's fighting, British columns advancing toward the Plaza Major suffered heavy losses, with many columns being cut off and forced to surrender. In Whitelocke's dispatch published in the *London Gazette* on September 13, 1807, he highlighted the way in which each house had been turned into an effective fortress where ill-trained militia stymied the attack of his well-trained troops.[28] British command and control at any level proved difficult.[29] Despite making some inroads into the city, Whitelocke saw the situation as irretrievable. In the one-day action, the British force suffered losses of 316 killed, 674 wounded, and 208 missing for an overall total of 1,198.[30] Of the 6,000 British troops committed to the battle, roughly 20 percent ended up as casualties. In comparison, Spanish forces lost 200 dead and 400 wounded.[31] As a consequence of the battle, British forces evacuated Buenos Aires and the entire River Plate region.

British experience in fighting in cities in the nineteenth century was not limited to Buenos Aires. The British effort to regain control of Delhi, in what the British called the "Indian Mutiny" or what in India today is referred to as the "First War of Independence," was a hybrid urban battle that mixed siegecraft with fighting in the city. Labeled by historians as the "siege of Delhi" (June 8–September 21, 1857), the British force was only strong enough to hold a ridge running southwest to northeast that dominated the northern perimeter of Delhi. This strong position, however, would serve as a springboard for the British force to breach the defenses of the city and take it by assault.[32] It was just under three months before the British force was able to undertake offensive operations with the arrival of the heavy artillery of the siege train on September 4. In the meantime, the British force grimly held against repeated attacks on the ridge position while heat and disease kept "one-third of the whole force virtually *hors de combat.*"[33] The environmental degradation of a besieging force living in the field for a prolonged period remained a feature of siege warfare.

British offensive operations against Delhi began on September 8 with the preparatory bombardment of selected points in the northern Delhi defenses. The aim was to create a major breach near the Kashmir gate for an infantry assault. By the evening of September 13, the Delhi Field Force commander, Brigadier Archdale Wilson, ordered the assault to go ahead in the early hours of September 14. Three columns of the British force assaulted the area of the Kashmir gate, a fourth column attacked the suburb of Kishenganj, and a fifth column formed the reserve. Despite the breaching effort by siege artillery, the troops still had to contend with a ditch and to climb ladders to reach the top of the walls.[34] After a day's fighting, the Delhi Field Force only managed to gain possession of a narrow lodgment just within the city walls.[35] Securing a foothold in Delhi came with a heavy price as the official return of casualties recorded 282 killed, 878 wounded, and 19 missing for a total of 1,170 casualties.[36] With the assaulting columns having an overall strength of approximately 5,000 men, the casualties suffered on September 14 amounted to nearly a quarter of the British force.[37] With his assaulting force exhausted and soldiers breaking into stores of beer and spirits, Wilson had to act swiftly to restore discipline, ordering stocks of alcoholic drink in the lodgment to be destroyed.[38] In preparation for an advance through the city, gunners and engineers worked to establish artillery on the captured bastions to support an infantry advance into the interior of the city.

The Delhi Field Force had taken 7 days from the beginning of the preparatory bombardment to the establishment of a lodgment in the city. It would now take another six days to clear and control Delhi. The two most critical days were September 17 and 18, when the Delhi Field Force fought a methodical urban battle. With attacks by infantry moving down streets and in the open only leading to heavy casualties, the British force adopted the tactical methodology of "sapping gradually from house to house."[39] As one British officer later commented, the advance "though slow, was made secure."[40] Attacks were also supported by artillery wherever guns could be employed. By September 20, the fortified king's palace fell, marking the moment when Delhi returned to British control.[41] Between September 15 and 20, the Delhi Field Force lost only 52 killed, 123 wounded, and 1 missing for a total of 176 casualties.[42] This was a remarkably low cost compared to the losses suffered in the assault of September 14.

The remorseless British advance succeeded in breaking the will of the Indian defenders, who along with the noncombatant population began to escape the city while parts of its perimeter remained open. As the last of the major objectives fell to the Delhi Field Force on September 20, control of the troops once again disintegrated, leading to a short sack of the city that resulted in large-scale physical destruction and loss of Indian lives.[43] Following restoration of control, retribution began by the British authorities against the rebels in the belief that strong punitive action was necessary to restore British authority.[44] While aspects of the battle for Delhi pointed to future urban combat in the city, actions after British success were very much consistent with the historical pattern of brutal treatment of the defeated forces and population of a city. In the suppression of the Indian Mutiny, Delhi had a totemic importance that made the city a strategic objective for both sides. As one British officer who served at Delhi recounted just over a quarter of a century later: "upon its issue practically rested the fate of the British Empire in India."[45]

The British Army, however, was not the only army to be introduced to the challenges of fighting in cities. During the peninsular war, the successful second French siege of Saragossa (December 1808–February 1809) saw the Spanish garrison continue to fight house by house after a successful French breaching assault.[46] The Mexican-American War (1846–1848) had examples of urban combat that echoed the British experience in 1807. The Mexican general Pedro de Ampudia in Monterey built an urban defense that was virtually identical to that in Buenos Aires nearly 40 years earlier. Despite the formidable defenses and resoluteness of Mexican soldiers, the US Army, by sapping through buildings as the British did in Delhi, eventually prevailed in a 3-day battle (September 21–24, 1846).[47] The US force also used artillery "in concert with the infantry."[48]

At the Barricades: European Urban Insurgency

In Europe during the nineteenth century, armed urban insurrection and civil strife drew armies into cities to suppress disorder that threatened governments across the continent. Rapid social and economic change generated political aspirations among wider segments of society in Europe, creating revolutionary demands for political enfranchisement and representative government that threatened the status quo. Three revolutionary episodes in Europe saw major urban uprisings in 1830, 1848, and 1871, with Paris experiencing all three.[49] The Revolution of 1848 witnessed the most widespread unrest, which led to military interventions in Berlin, Paris, and Vienna as well as other cities in Central Europe.

Better trained and equipped government forces confronted and prevailed over the urban insurgents, but the insurgents' quest to level the disparities in fighting ability between the insurgent and soldier was to lead to the rise of the ubiquitous street barricade. With ease of construction, the barricade provided an obstacle that was an impediment to movement and gave cover from fire, making it the improvised urban fortification of choice for insurgents. It served, as one author has argued, as an iconic symbol of revolution and a tool for social and political mobilization for the urban insurgency.[50] With Paris serving as the European epicenter of urban insurgency in 1848, French urban unrest led to the "spread of barricade consciousness" across Europe, ensuring its widespread adoption and cementing its status as a symbol of revolution.[51]

Nineteenth-century urban insurgency gave rise to some theoretical thinking on fighting in urban areas from both the perspective of radical movements and constituted governments. The French socialist and republican activist Louis Auguste Blanqui, who experienced the events of 1830 and 1848 in Paris, produced the pamphlet *Instructions pour une prise des armes* on urban insurgency, seeing it as a viable means to power, but one that required organization along military lines.[52] Another French radical figure, General Gustave Paul Cluséret, whose military career in the French Army, in the Union Army during the US Civil War, and on the side of the communards in 1871 brought practical military experience to the problem of urban insurgency. The second volume of Cluséret's *Mémoires,* published in 1887, reads like a doctrine manual for the urban insurgent, providing tactical methods adapted to simple organizational structures. Cluséret's weapon of choice was the bomb because it could be easily made in secrecy.[53] The question of suppression of urban insurgency, not surprisingly, also received some attention by military thinkers in nineteenth century France. Marshal Thomas Robert Bugeaud, best known for leading France's military effort to gain control of Algeria, completed a study of tactical aspects of urban combat against insurgents, titled *La Guerre des rues et des maisons,* just before his death in 1849.[54] Shaping his tactical thinking were his experiences as a junior officer in the second siege of Saragossa.[55] Outside France, the German Imperial Army completed a study of urban insurgency in 1907 based on the events of 1848 and 1871 that identified the urban proletariat as a potential source of urban insurgency.[56] Despite the increasing prevalence of urban insurgencies, conventional military thinking in the nineteenth century still treated fighting in cities as an aberration rather than an emerging norm.[57]

The Siege, Firepower, and Fortification

The siege remained the dominant form of urban warfare in nineteenth-century military thinking. For the first half of the nineteenth century, the conduct of sieges followed well-established practices. Siege artillery opened a breach in perimeter fortifications, followed by an often costly infantry assault, and concluded with the victorious assaulting troops sacking the city with attendant atrocities. The British Army's sieges in the Peninsular War, in most cases, followed this pattern, with the siege of Ciudad Rodrigo (January 1812) and third siege of Badajoz (March–April 1812).[58] What did change in the nineteenth century was that artillery had become the dominant instrument in siege warfare. Although mining was still occasionally utilized, the ascendancy of artillery was the result of its perceived efficiency as reflected prominently in theoretical works in the period. General Sir Charles William Pasley, a Royal Engineer officer, produced one of the best guidebooks on artillery in siege operations, *Rules for Conducting the Practical Operations of a Siege*, which became the standard British Army reference work on the subject.[59] Artillery effectiveness, however, did not necessarily mean a swifter end of a siege. In the British and French siege at Sevastopol (1854–1855), the city's defenses were based on earthen field fortifications that proved resistant to artillery fire. It took six bombardments of British and French artillery before a successful breach and assault ended a siege that had lasted nearly a year.[60]

With the rapid technological change associated with the industrial revolution, the power of artillery grew enormously in the second half of the nineteenth century. The range and destructiveness of artillery was the result of the introduction of rifled barrels, better barrel construction, breach loading and improved shell design, fuses, explosives, and propellants.[61] Moreover, the improvement in indirect fire proved highly destructive to fortifications.[62] In the 1880s, the Royal Artillery conducted experiments to assess the effectiveness of its siege artillery and gained greater understanding of the effect of indirect fire on fortifications. The Royal Artillery concluded that "the attack and defence of a modern fortress was primarily a contest between the two artilleries."[63] What these changes meant was that the fortifications were becoming less resistant to the growing power of artillery. As a consequence, compelling strategic reasons drove the construction of fortifications irrespective of cost and risk of obsolescence. The Austro-Hungarian fortress town of Przemyśl from the late 1870s onward was intended to block Russian military access to Dukla Pass to the south and hence through the Carpathians to the Danubian basin.[64] Anti-access and area denial (A2AD) was clearly not a military concept invented in the twenty-first century, but A2AD encapsulated the important role driving investment in fortress Przemyśl.

The impact of the industrial revolution had a more profound and widespread effect on urban warfare than the rapidly changing technology it spawned. Industrialization and rising agricultural productivity drove a process of dramatic urban growth. In developed countries the percentage of urban population grew from 10.7 percent in 1800 to 35.7 percent by the eve of World War I in 1914.[65] These changes meant that the footprint of cities grew, they had larger populations, and the commerce and industry they possessed made them economically more important. The industrialized city became a provider of the military sinews of war and thus a potential target of military action. In the US Civil War, the capture of Atlanta in 1864 by General William T. Sherman indicated how the industrialized city now became a military objective. Atlanta was the southern rebellion's "second most important industrial center," and Sherman's intent to destroy its industrial capacity was akin to the thinking that underpinned the Allied bomber offensive against Germany during World War II.[66] Thus Sherman's action in destroying Atlanta's industry and transportation infrastructure presaged the deliberate targeting and destruction of industrial urban centers in twentieth-century total war.[67]

The increasing economic value of the city resulting from the industrial revolution highlighted the need for finding new means for the city's defense. In the second half of the nineteenth century, the problem of defending London generated a major British debate with the perceived threat of France invading England. As Colonel Millington Henry Synge wrote in his *The Defence of England: A Military Sketch* published in 1872, "the interest of the defence of England, centres peculiarly on London," describing it as "the most important, the first commercial city of the realm, and the wealthiest city of the world."[68] As early as 1860, the government led by the Viscount Palmerston dismissed the idea that it was possible to fortify London.[69] With conventional military thinking believing London the central objective of an invasion, the debate on finding a solution to the defense of the capital city rumbled on to the end of the nineteenth century.[70] There was also the recognition that the regular army was unlikely to be sufficiently large to provide enough manpower and that the defense of London would rest on the locally raised volunteers.[71] The logic of relying on volunteers mobilized in London led two contributors in the debate to suggest a radical approach to London's defense. In an address at the Royal United Service Institution in 1860, Colonel Shafto Adair stated that fighting in the houses and streets of London was the "ultimate defence."[72] In 1886, a Royal Engineer officer, Major H. Elsdale, echoed more categorically Adair's thinking: "The real and ultimate defence of London, in my view, lies in its vastness, and in the almost unlimited capabilities it presents for street fighting on a gigantic scale, coupled with the extreme difficulty of surrounding and investing it . . . what number of invaders will be required to take London, supposing the vast city to be defended house by house."[73]

The debate on the problem of defending London meshed both old and new ideas. It examined the problem by looking backward to siege and fortification but also anticipated new directions by seeing the potential in fighting an adversary in the city. This British debate encapsulated the ambiguities and transitional character of the conduct of nineteenth century urban warfare.

Emerging Humanitarian Norms: Laws and Customs of War

Despite the persistence of brutal treatment of the inhabitants of cities in siege warfare, the nineteenth century witnessed important progress in the evolution of customs and laws of war toward more humane treatment of urban populations. The nineteenth century was important for laying the foundation of the contemporary laws of armed conflict (LOAC). For example, the Geneva Convention of 1864 was an important first step in the promulgation of international conventions to limit the effects of war by giving categories of individuals such as the wounded, medical personnel, and prisoners of war legal protections.[74] More directly relevant to urban warfare in its implications was the Hague Convention of 1899, which stated that "the right of belligerents to adopt means of injuring the enemy is not unlimited."[75] In line with the idea that there were limits on belligerents in the conduct of war, the Hague Convention of 1899 imposed constraints when attacking towns or cities in its Articles 25 to 28. Among the limitations were prohibitions on "attack or bombardment" of undefended urban settlements (Art. 25), giving warning before "commencing a bombardment" (Art. 26), and according protection to "edifices" of religious, cultural, or scientific purposes along with hospitals and other medical facilities (Art. 27). Article 28 explicitly prohibited "the pillage of a town or place, even when taken by assault."[76] This last article addressed what was one of the most egregious features of many centuries of siege warfare. These changes represented important steps in establishing international norms as they pertained to the urban aspect of warfare but did not yet create a legal regime with "powers to indict and punish."[77]

Twentieth Century: Urban Operations During the World Wars

If the nineteenth century represented a period of major transition in urban warfare, then the twentieth century firmly established an important change in its conduct. Although the siege did not disappear completely, urban warfare had shifted to principally fighting in towns and cities. The few instances of something approaching a major urban battle were a quintet of sieges conducted in the first year of World War I: Liège (August 1914), Namur (August 1914), Maubeuge (August–September 1914), Antwerp (September–October 1914), and Przemyśl (September 1914–March 1915). The first four examples in Belgium and France demonstrated how heavy siege artillery could make short work of the siege.[78] Przemyśl, however, was an exception. The garrison of the fortress town capitulated after a failed breakout attempt and when the lack of food supplies threatened starvation for the garrison and the town's people.[79] During the course of the war, on the Western Front, towns and villages near the fairly static lines of contact got caught up in the fighting and were usually obliterated by artillery fire; on the Eastern Front, the war of movement usually swept by urban areas rather than fought over them. On the surface, the conduct of urban warfare during World War I merely represented a continuation of nineteenth-century military practice and thought, but this would dramatically change by the middle of the twentieth century.

Thinking About Fighting in Cities: The Irregular Threat

In the first four decades of the twentieth century, the focus of military thought on fighting in cities was on urban rebels or insurgents. The importance of this body of thought is that it produced some of the earliest examples of proto doctrine for urban operations. The first example was the product of officers in the British Indian Army. Two prize-winning essays penned on the eve of the war for the United Service Institution of India (USII) encapsulated this military thinking on fighting in cities against internal rebels or insurgents. Published in the *Journal of the USII* in 1915, the set topic for the essay competition was "the tactics of street fighting as applied to Eastern countries."[80] The "Gold" prize-winning essay by Lieutenant Colonel W. F. Bainbridge argued that "street fighting" between two conventional powers was very unlikely but that it was a realistic prospect in the case of "a hostile population or turbulent section thereof."[81] Thus Bainbridge's essay focused on "grand tactics" and "tactics" necessary to overcome the irregular urban adversary in an internal security context. What is remarkable about Bainbridge's essay is that it represents perhaps one of the earliest British military attempts to understand the nature of fighting in the urban environment, and it articulates a kind of tactical doctrine for meeting its challenges. Indeed, Bainbridge argued that there was a gap in the published doctrine regarding "street fighting." He noted that the British Field Service Regulations Part I provided chapters on "mountain and bush fighting" but offered no specific doctrine on fighting in cities.[82] The "Silver" prize winner, Major C. L. Norman, closely drew on the experience in Delhi in 1857 in his essay and attempted to turn that nineteenth century experience into practical tactical principles for internal security operations in cities.[83] Both of these essays indicated that the limited attention given to fighting in cities was linked to internal security of the Empire rather than preparation to meet the challenge of a European peer adversary. British interest in operations in the city in the interwar period remained centered on internal security. Major General Sir Charles W. Gwynn's famous *Imperial Policing* used a number of examples in the urban environment, including Amritsar, Khartoum, Shanghai, and Peshawar.[84]

World War I may not have experienced much in the way of fighting in urban areas, but in its aftermath, civil unrest and urban fighting broke out, driven by left-wing revolutionary insurgents seeking to exploit postwar political vacuums. Three notable examples included the "Spartacist" revolt in Berlin (January 1919), the Hamburg uprising (October 1923), and a coup attempt in Tallinn (December 1924). In all three cases, the urban insurrections were quickly crushed. These events, however, triggered wider military interest in the problem of fighting insurgents in an urban environment. In the United States, official concern over the possibility of a radical revolution led to the drafting by the US War Department of War Plan White in 1919–1920.[85] The genesis of the plan was in the frenzied climate of the postwar Red Scare that fueled official fears of radical revolution in the United States.[86] The central driver of the White Plan in 1919–1920 was to counter the threat of a socialist revolution led by left wing groups exploiting a radicalized pool of industrial workers. Such a threat, according to US military intelligence estimates, existed in major urban areas of the United States.[87] By 1923, this plan was revised and renamed Emergency Plan White, and versions of it existed throughout the interwar years with an implicit urban focus.[88] Similarly, in 1919, the British Army was introduced to the planning challenges of meeting revolutionary industrial action on a national scale that would largely take place in the city.[89]

In Europe, the postwar urban unrest inspired a book that qualifies as one of the earliest examples of military doctrine for fighting in urban areas. Colonel Stefan Rowecki, one of the interwar Polish Army's prominent military thinkers produced his most important and best-known book, *Walki Uliczne* (*Street Fights*), in 1928.[90] In *Walki Uliczne*, Rowecki set out to consider the challenges of how to suppress a major urban revolt as a military problem. The kind of urban unrest examined by Rowecki involved the armed resistance of insurgents and not simply mass protest. Rowecki clearly had been influenced by the post–World War I communist urban insurgencies in Germany, Bulgaria, and Estonia. *Walki Uliczne*, in its 286 pages, provided a comprehensive doctrinal articulation of the issues and military activities associated with urban warfare against an insurgent force.[91]

Although little-known outside of Poland, Rowecki's *Walki Uliczne* was a prescient book of tactical doctrine for fighting in the city. It also demonstrated how fighting in cities against insurgents prior to World War II was viewed as the most likely feature of urban warfare in the future.

New Norm: Fighting in Cities During the Second World War

World War II is very much the major source of contemporary military experience and understanding of fighting in cities. One estimate suggested that 40 percent of all combat took place in the urban environment during World War II.[92] With urban settlements of all sizes and shapes experiencing urban combat, fighting in the city came of age as the dominant form of urban warfare. The table at the end of this section provides twelve examples of major urban battles of World War II in Asia and Europe (see Table 3.1). The selection of urban battles chosen is to highlight the broad issues that fall out of urban warfare experience of World War II. What does this selection tell us in general about urban warfare in World War II? The first observation is that the urban adversaries were overwhelmingly conventional armed forces. There are very few examples of urban battles in World War II that involve irregular forces. Warsaw (August–October 1944) and Athens (December 1944–January 1945) are the notable exceptions. Second, although World War II was a global conflict, more urban fighting took place in the European theater than in the Asia-Pacific theater, which experienced fewer urban battles.

One of the striking and misleading features of the list of urban battles in Table 3.1 is the preponderance of exceptionally large urban battles. Rather symbolically on the list, Shanghai in the early stages of the Sino-Japanese war in 1937 heralded the onset of the very large urban battle and Berlin marked it at the close of the war in Europe. All of the major combatants during the war fought at least one major urban battle with the exception of France and the United Kingdom. Certainly, there has been a tendency among historians to devote their attention to the exceptionally large urban battles, such as Stalingrad and Berlin. The reality, however, was that enormous diversity in scale characterized urban operations during World War II. Although important to the understanding of urban warfare, these major battles in large urban conglomerations may in fact be products of the conditions of total war with its massive mobilization of people and resources and ideological intensity that fueled extremes in belligerents' behavior in the conduct of war. Lessons therefore may have less enduring importance to future urban operations. There has been far less study of the hundreds of smaller-scale urban actions in towns and villages. This area is now receiving more attention. A recent examination of 51st Highland Division operations in February and March 1945 in Germany suggests that although there are common elements when fighting in the urban environment, tactical solutions adopted can be determined by the scale of the urban settlement.[93]

Because fighting in urban areas became ubiquitous during World War II, the tactical perspective of urban combat came to dominate analysis. This tactical focus on urban combat is understandable given the learning curve that armies went through in gaining tactical proficiency in the urban environment. Soviet tactical adaptation in the Battle of Stalingrad (August 1942–February 1945) was an impressive example gained through costly effort.[94] Perhaps the most common tactical lesson was the large numbers of troops required when fighting in a city. Armies during the war besieged (isolated) urban areas, conducted both offensive and defensive operations, and performed what would today fall under the rubric of stability operations. Tactical realities encountered in the war included high levels of destruction, the impediment of rubble, obstacles, snipers, improvised explosive devices, and the vulnerability of armored fighting vehicles. Overcoming an adversary highlighted the need for direct fire and optimization of weapons mix such as the use of submachine guns in confined spaces. No doubt this list could be made much longer. These tactical challenges suggested the need for well-crafted doctrine, but urban doctrine was in fact still in its infancy.

The British Army arguably had more experience and opportunity in the nineteenth century than most major powers to think about fighting in the urban environment as it embarked upon what was to be its most encompassing global armed conflict. Nevertheless, it lacked, until 3 years into the war, a doctrine pamphlet devoted to the problem of urban warfare. *Fighting in Built-Up Areas*, Military Training Pamphlet No. 55 (MTP 55), published by the War Office in April 1943, was the British Army's first formal urban doctrine. This pocket-sized pamphlet of forty-four pages had a tactical focus covering attack and defense in the urban settlement, minor tactics and methods, and training. The "general considerations" section in MTP 55 laid out the assumptions that underpinned army thinking on urban warfare while acknowledging that "much fighting in this war has revolved about, or taken place in, towns."[95] One of MTP 55's key assumptions was that "built-up areas favour the defence."[96] Regarding the characteristics of the urban environment, MTP 55 stressed that "no other battlefield includes ground both so open and close" and whose landscape could be radically changed through destruction and the generation of rubble.[97] Moreover, it summarized the special challenges and their impact on urban fighting:

> In densely built-up areas it is possible to climb thirty, fifty, perhaps a hundred feet in as many seconds; it is possible to by-pass an enemy by going directly over or under him. Built-up areas thus possess a third dimension not present in open warfare which, combined with the abundance of cover, involves a constant drain on manpower, creates great difficulties of cohesion and control, and calls for a maximum of ingenuity from everyone engaged.[98]

The deductions that the doctrine drew from the physical characteristics were a list of issues that shaped the tactical battle: [99]

- difficulty of control
- value of short-range weapons
- restricted visibility
- conspicuousness of movement
- difficulty of locating fire
- restricted maneuver
- value of the first shot
- darkness
- fire (the element)
- looting
- importance of height

Most of these are self-explanatory, but three need further elaboration: darkness, fire (the element), and looting. Believing that urban combat favored the defender, MTP 55 also took the view that combat at night shifted the advantage: "darkness is the ally of the attacker rather than of the defence."[100] The use of "incendiarism" (fire) was seen as a useful means of "dislodging the enemy from a building" but was recognized as a "double-edged weapon, which can do more harm than good."[101] The issue of looting was seen as something that undermined discipline and focus on the job at hand. As MTP 55 warned: "the ordinary battlefield offers few amenities calculated to distract a soldier; a built-up area, unless completely devastated, offers many."[102] Although not addressing a potential problem on the scale of a sack of a city by licentious soldiery, the fact that doctrine had anything to say on looting was something of a throwback to the experiences of earlier centuries.

The presence of civilians was another general consideration MTP 55 addressed. The pamphlet recognized that "the problems of controlling and administering the civilian inhabitants will nearly always arise, and may be complicated by a flow of refugees into built-up areas and by fifth columnists."[103] On this last point, the presence of hostile civilians was assumed whether on the territory of an occupied ally or enemy territory. A major theme in the doctrine was force protection while operating in the urban area. There was also identified a "civil affairs" responsibility and a requirement to address civilian issues in the battlespace and that "liaison between civil and military authorities must be arranged and practiced." MTP 55 identified the need for civilian resources to "be utilized to the fullest extent." Indeed, the employment of civilians and civilian resources to support military operations was acceptable practice "in defence or occupation of a town."[104]

The remainder of MTP 55 was devoted to the employment of weapons and tactics for urban combat. It did not see urban combat as the role for specialists but argued that "fighting in built-up areas must be considered an integral part of every soldier's training."[105] Although the basic framework for employing weapons and tactics was applicable to fighting in built-up areas of whatever scale, MTP 55 only refers to "towns and villages," implicitly excluding cities from consideration. Although the doctrine was brief in length by today's standards, it was a practical guide for tactical action.

Engineering support for urban operations remained important. MTP 55 identified the role of engineers as being "to remove booby traps, carry out major defensive preparations in captured buildings, demolish buildings that are liable to collapse, and to improve communications."[106] More widespread combat in urban areas did not lead to specialist capabilities designed specifically for the urban environment. Nevertheless, some new engineering capabilities proved useful in urban warfare. In the British Army, the requirement to remove obstacles in beach assaults led to specialist armored engineering vehicles. The Assault Vehicle Royal Engineers (AVRE), as it was initially called, had capabilities such as a low-velocity mortar for destroying obstacles that had utility in the urban setting.[107] The German Army had engineering units such as the 500 Engineer Battalion that operated a small wire- or radio-guided tracked mine for clearing obstacles or conducting demolitions. The Leichter Ladungsträger Goliath (Goliath Light Charge Carrier), or simply Goliath, was employed during the Warsaw uprising (August–October 1944).[108] Like the AVRE, Goliath was not built specifically for urban combat.

The use of armor in the urban environment proved from the earliest days of the war in Europe to be problematic. The attempt of the German 4th Panzer Division to take Warsaw by coup de main on September 9, 1939, provided a vivid illustration of the vulnerability of armor in a built-up area. The German armored fighting vehicle losses in the failed assault numbered sixty-three destroyed or disabled, the highest single loss of German armor in any engagement during the attack on Poland.[109] The proliferation of handheld antitank weapons as the war progressed only made armor more vulnerable when operating in confined urban areas. Armor, however, brought valuable direct fire capability to urban combat and protected mobility in open urban spaces. Armor did prove viable in urban combat as part of a well-choreographed combined arms tactical group. The successful US Army battle to take the German city of Aachen (October 1944) was a good example of the utility of armor in the urban environment when employed in tactical combined arms groups.[110]

In fighting the urban battle during World War II, artillery undoubtedly became the central instrument of support to infantry fighting in a built-up area. In the use of massed indirect fire, there were contrasting views on the value of the fire given the high levels of destruction. The picture of wartime use of indirect fire in urban operations suggests the employment of a wider range of artillery tactics dependent on the situation.[111] British and US forces believed excessive destruction impeded mobility and delivered disappointing results. In contrast, the Red Army generally accepted "maximum destruction." There were, however, common elements in the use of artillery in urban fighting among the allies. The employment of artillery in the direct fire role, often in close support of infantry was a ubiquitous feature of urban fighting in World War II.[112] It often saw medium and even heavy artillery in this role. For example, the attack on Fort Dufferin included medium artillery firing at comparatively close ranges in the direct fire role in the British 14th Army's battle for Mandalay in March 1945.[113]

"Expending shells rather than lives" was very much the driving force for massive application of firepower against the lethality of urban defense during World War II. The battle to take Manila (February–March 1945) demonstrated the importance of artillery in urban combat. Initially operating with tight rules of engagement for use of artillery, mounting casualties led to a change in approach. Employing a mixture of indirect and direct fires, the US XIV Corps developed combined arms assault techniques that relied on lavish use of artillery fire.[114] The last Japanese stronghold in Manila to be cleared was the old walled city of Intramuros. In support of two battalions of infantry were 120 artillery pieces ranging in caliber from 75 to 240 millimeter. In the two-hour preparatory concentrations fired on February 23, 1945, eleven battalions of artillery fired 7,896 rounds (230 tons of munitions) at targets in the Intramuros. Each subsequent day until Intramuros was finally cleared on March 3 saw high ammunition expenditure rates by artillery in the direct-fire role.[115] The tenacity of the Japanese urban defender might have been exceptional in Manila but not the role of firepower in terms of World War II urban offensive operations.

Beyond the tactical battle, the conduct of urban warfare had a strategic dimension during World War II. With Britain (and the United States) embarking on "strategic bombing" of its opponents, a new dimension was added to urban warfare. The British strategic bombing campaign for the most part targeted urban areas, with the objective of "the progressive destruction and dislocation of the German military, industrial and economic system."[116] Strategic bombing was made possible by the advent of the long-range bomber that allowed massed air strikes to be conducted against industrial targets. The results of bombing of German urban industrial areas may have been less damaging to industrial production than originally thought by the advocates of strategic bombing, but it did reflect a form of military operations that sought strategic effect by targeting urban areas.[117]

The operational level also found a place in the conduct of urban warfare. Stalingrad provides an excellent example of using an urban battle as part of an operational design. By fighting in Stalingrad, the Soviet high command used the battle as a tactical means to an operational end. By fixing the

German 6th Army in the urban battle for Stalingrad, the Red Army was able to launch Operation Uranus, a counteroffensive that struck on the more thinly held sectors of the front outside the city. The outcome was the envelopment and destruction of German forces in the city.[118] In the last year of the war, Germany practiced a "fortress strategy," particularly on the Eastern Front, by attempting to hold urban centers to act as "breakwaters" to weaken the momentum of Soviet offensive operations and tie down troops. One study of this defensive German fortress strategy argues that in the case of Festung Königisberg it did cause adaptation of the Soviet operations and slow the speed of the advance into Germany.[119]

During World War II, the siege continued to play a significant role in urban warfare. In most, but not all cases, they were adjuncts of urban operations whose ultimate decision rested on fighting for a city block by block and house by house. In some instances, the siege itself proved ultimately decisive in the outcome of the urban battle. The capitulation of Warsaw in 1939 after three weeks of being besieged by the German Army was the result of the garrison running out of ammunition and the exhaustion of food supplies for the city's population.[120] The example of the battle for Budapest (1944–1945) illustrated the more common pattern of siege being combined with urban combat. The isolation of the city and Soviet operations to fight through it determined the outcome of the battle that lasted over 100 days. The cost of the prolonged siege and urban combat, however, was enormous for both sides.[121]

The loss of human life, both military and civilian, in urban combat during World War II was immense and not altogether surprising given the scale at which the urban environment featured in wartime operations. Normative constraints of international humanitarian law seemed absent in the conduct of belligerents driven by the hard logic of total war. The body of LOAC that concerned urban warfare had not developed beyond the Hague Convention of 1899 and its successor in 1907. Indeed, Articles 25 to 28 (described earlier in this chapter) were identical in the 1899 and 1907 Hague Conventions. What World War II demonstrated was the degree to which LOAC was underdeveloped and inadequate for the unprecedented loss of life and physical destruction that occurred in urban areas. The impact of World War II was notable less for the application of LOAC than for acting as a driver in its further development.

Cold War Urban Operations: Contradictory Impulses

The competition between the United States and Soviet Union and their respective camps was a dominant factor shaping the international security environment during the Cold War. The post-1945 world also witnessed the

Table 3.1 Urban Battles of World War II

Battle	Date	Nature of Combatants	Tactical Activities
Shanghai	August–November 1937	Conventional forces	Offensive and defensive
Warsaw	September 1939	Conventional forces	Siege—offensive and defensive
Stalingrad	August 1942–February 1943	Conventional forces	Siege—offensive and defensive
Ortona	December 1943	Conventional forces	Offensive and defensive
Warsaw	August–October 1944	Insurgent force vs. conventional forces	Offensive and defensive
Aachen	September–October 1944	Conventional forces	Offensive and defensive
Budapest	December 1944–February 1945	Conventional forces	Siege—offensive and defensive
Manila	February–March 1945	Conventional forces	Offensive and defensive
Königsberg	January–April 1945	Conventional forces	Siege—offensive and defensive
Groningen	April 1945	Conventional forces	Offensive and defensive
Breslau	February–May 1945	Conventional forces	Siege—offensive and defensive
Berlin	April–May 1945	Conventional forces	Offensive and defensive

rise of national liberation movements that fueled insurgencies and accelerated the process of decolonization and the emergence of new states. These broad characteristics pushed urban warfare in new directions. Europe may have been the center stage of the Cold War confrontation between the two camps, but it was in the global arena where armed conflict became more prevalent. Thus, Asia and parts of the Middle East and North Africa were where urban warfare now took place. Of the eight examples discussed in this section, conventional force-on-force urban combat, in a manner recognizable in World War II terms, took place in three of them: Seoul, Long Son (Vietnam), and Khorramshahr (Iran). The siege virtually disappeared during the Cold War period, with Beirut being among the few examples. The core idea of the siege, that of isolating an urban area to achieve military effect, was still practiced in a truncated form, as in the case of Algiers. The rest of the examples highlighted the growing prevalence of the irregular insurgent adversary in urban warfare. Military activity of conventional forces in the urban environment was not only contending with counterinsurgency but sometimes adversaries who mixed their conventional and insurgent forces, as in Hue during the Vietnam War. Urban operations varied in scale, with no Cold War example reaching the enormous size of the largest of World War II battles.

Urban warfare in the Cold War period continued to have a tactical myopia whether the challenge was preparation in Europe to meet the Soviet threat or the problem of the urban insurgent in the global arena. Despite the continuing tactical focus, urban warfare continued to have a role at the operational level of war. The US amphibious landing at Inchon (Operation Chromite) in September 1950 during the Korean War and subsequent battle for Seoul was an example of operational level maneuver where the seizure of Seoul severed the North Korean Army's communications, leading to the destruction of large elements of the force and its expulsion from South Korea.[122] Similarly, the large-scale cordon and search operation conducted in Nairobi by the British Army in April–May 1954 (Operation Anvil) had an operational level effect of causing large-scale disruption to the Mau Mau insurgent organization and its resources in the Kenyan Emergency.[123] Urban warfare in the Cold War also witnessed strong joint integration with air and naval forces and coalition partners in urban areas in the maritime littoral environment. Such examples include Athens, December 1944–January 1945; Surabaya, October–November 1945; Seoul, September 1950; Suez, November 1956; and Hue, January–February 1968.

The British operational experience broadly reflected the trends in urban warfare during the Cold War. Table 3.2 lists eight examples of British urban operations from the Cold War period. Athens and Surabaya chronologically took place during World War II but are included in the Cold War period because postwar political rather than wartime military issues drove these two urban battles. Moreover, Athens and Surabaya had the distinction of being the largest urban operations ever conducted by the British Army. Apart from Suez in 1956, insurgent rather than conventional adversaries dominated the British urban operations experience. Given the uniformity of the British Army's urban adversaries, the Cold War would suggest that the urban insurgent would become the driving force for the development of British Army urban doctrine. This was not, however, to be the case. The Soviet threat in Europe would instead be the arena most influential in shaping British Cold War urban doctrine.

With little discernible change, MTP 55, *Fighting in Built-Up Areas*, published by the War Office in April 1943, remained the basis of postwar British Army urban doctrine for a considerable period of time during the Cold War. MTP 55 was reprinted with amendments in April 1945 and may not have been formally replaced until the early 1980s.[124] When the venerable MTP 55 was eventually replaced, it had taken on a distinctly Cold War Northwest Europe focus. Army Field Manual, Volume V, Part 5, *Fighting in Built-Up Areas*, 1983, echoed its predecessor in its tactical emphasis, devoting two chapters to offensive and defensive operations. The doctrine also considered defensive operations within a city. However, the 1983 doctrine showed its Cold War purpose in a chapter on "Fighting from Villages." This

Table 3.2 Urban Battles of the Cold War Period

Battle	Date	Nature of Combatants	Tactical Activities
Shanghai	August–November 1937	Conventional forces	Offensive and defensive
Warsaw	September 1939	Conventional forces	Siege—offensive and defensive
Seoul	September 1950	Conventional forces	Offensive and defensive
Seoul	March 1951	Conventional forces	Offensive and defensive
Budapest	November 1956	Conventional forces and insurgent forces	Offensive and defensive
Algiers	September 1956–September 1957	Conventional forces and insurgent forces	Cordon and search; nodal defence
Hue	January–March 1968	Conventional forces and insurgent forces	Offensive and defensive
Long Son	March 1979	Conventional forces	Offensive and defensive
Khorramshahr	September–November 1980	Conventional forces	Offensive and defensive
Beirut	June–August 1982	Conventional forces and insurgent forces	Siege—offensive and defensive

chapter looked at how villages in northern Germany could be used as strong points to impede a Soviet armored advance through surrounding open ground. The chapter also examined Soviet urban tactics as an aid to preparing a defensive position in a German village.[125] In 1986, a revised version of *Fighting in Built-Up Areas* appeared that had a more overtly defensive focus in the context of fighting in German villages. It also had an expanded section on Soviet urban tactics.[126] British urban doctrine thus served the purpose of preparing the British Army for urban operations against a peer adversary in future conventional war. Urban operations against irregular forces were not formally covered in doctrine even though they were a strong feature in Cold War British Army operations. Such doctrine that existed for urban operations against irregular forces was produced in theater. A good example of this was the pamphlet produced by General Headquarters Far East Land Forces devoted to "Internal Security in a City."[127]

Given the shift toward irregular adversaries in Cold War urban warfare, the role of massed firepower of artillery was dramatically reduced except in the few examples of urban battles against conventional adversaries. The US Army and Marine Corps had the most significant Cold War experience in firepower-intensive battles for Seoul in 1950 and Hue in 1968. The clearing of the Korean capital Seoul in September 1950 saw extensive use of firepower delivered by artillery, airpower, and armored vehicles.[128] In the battle for Hue, US artillery fired some 52,000 rounds, with US Navy gunfire support adding another 7,670 rounds from destroyers and cruisers and the US Air Force expending 600 tons of munitions in support of operations to retake the city.[129] Where urban battles reflected conditions found during World War II, firepower was used on a large scale.

The insurgent adversary may have become more prevalent in the Cold War, but largely in the rural environment. Latin America, however, was the exception in seeing a shift in insurgency from the countryside to the city. The reason for the rise in urban insurgency in Latin America was the failure during the 1960s of rural insurgencies through the lack of external and internal support along with effective counterinsurgency campaigns.[130] This shift had a major conceptual spur from Carlos Marighella, a Brazilian Marxist-Leninist politician and founder of the National Liberation Action (Ação Libertadora Nacional), who wrote *Minimanual of the Urban Guerrilla* (*Minimanual do Guerrilheiro Urbano*). This was written like a tactical doctrine manual for the urban insurgent. Marighella did not see the urban environment as the only arena but one that was a springboard to spread insurgency into the countryside.[131] Despite the shift of insurgency in Latin America to the urban environment, in the end it proved no more successful than earlier rural efforts.[132]

The further development of the law of armed conflict that followed the experience of World War II had implications for the conduct of urban warfare. The Geneva Conventions of August 12, 1949, represented the first major treaty-based additions to the body of LOAC produced in the late nineteenth and early twentieth centuries. The approach taken in the Geneva Conventions and additional protocols of the 1970s was to extend the protection to categories of people such as the wounded, sick, prisoners-of-war, and civilians. Unlike the earlier body of international humanitarian law, where specific actions—such as the bombardment of an undefended urban locality—were addressed, the approach of the 1949 Geneva Convention and Additional Protocols was to strengthen general principles for protected persons to be used in a variety of circumstances.[133] Principles such as distinction, military necessity, and proportionality in military action applied equally to protected persons whether warfare occurred in a desert or in an urban area. These principles were also balanced against the principle of military necessity. As Article 28 of the "Geneva Convention Relative to the Protection of Civilian Persons in Time of War" stated: "The presence of a protected person may not be used to render certain points or areas immune from military operations."[134] Thus, the application of LOAC principles requires interpretation and judgment in the light of operational circumstances. Although the Additional Protocols to the Geneva Conventions provided more clarity in application of principles, in the specific context of urban warfare, developments in international humanitarian law during the Cold War did not provide much new guidance tailored solely to the urban environment.[135]

Conclusion

In Gregory J. Ashworth's view, the city has not had a "central role in military science."[136] This viewpoint, however, may be overstating the absence of military consideration of urban warfare. Looking to history to provide a broader knowledge and understanding reveals that there are a number of reoccurring themes in the evolution of urban warfare. Military challenges of operating at the tactical, operational, and strategic levels have continued to be imposing not the least in terms of command and control in the urban environment. Adversaries typically have been identified by a few simple categories, such as conventional and irregular, but in whatever category they have been placed they have demonstrated the ability to innovate, adapt, and pose new threats. The dynamism of adversaries has been matched by changing technology. It has shaped and reshaped the conduct of war in the urban environment, proving to be an enduring source of discontinuity in urban warfare practice. The experience of "civilian" noncombatants in urban warfare reflects a continuity of suffering despite the development of conventions, norms, and laws of armed conflict seeking to mitigate its worse effects on noncombatants. Theorists and doctrine writers have endeavored to provide conceptual and practical frameworks for the conduct of urban warfare, recognizing both the advantages brought by urban terrain and the fiendishly difficult military problem it represents.

The most striking feature in the historical evolution of urban warfare is the prevalence of the siege. As one study of siege warfare in the ancient period claimed, it is "older than civilization itself."[137] The siege can be said to have been practiced from the earliest periods of human civilization to the present, and for many centuries it was a central military activity in war. The degree of continuity in siegecraft from the ancient to the early modern period was exceptionally large. Throughout this period, the isolation and blockade of an urban settlement was conceptually at the core of siege warfare. Investing a city by an attacker to degrade the defense of a city deliberately targeted the ability of a city's population to sustain itself. This aspect gave siege warfare a totality that had an impact on combatant and noncombatant in equal measure. The pervasiveness of the siege in the recent conflicts in Syria and Iraq suggests that its utility in urban warfare is not over, nor is its more brutal aspects on urban populations.

Political, social, and economic change, particularly in Europe and North America, began to reshape the urban environment's place in warfare. In particular, the industrial revolution led to a major demographic shift from rural to urban. The transformation of the urban landscape in the nineteenth century was arguably a more far-reaching event than the rise of the megacity that has attracted so much attention at the dawn of the twenty-first century. Cities became more populous, physically larger, and of enormous economic importance.[138] The rapid social change wrought by the industrial revolution created unrest in urban areas from politically disenfranchised and exploited parts of society.[139] Thus, the internal security problem of armies having to quell urban rebellion or mass disorder grew in military importance. The nineteenth century also began to witness on a limited scale a shift to armies being confronted by adversaries who chose to fight in the city. With this development, improvised fortifications based on existing structures and makeshift obstacles demonstrated their utility; the city itself became the fortification.[140] Although the siege continued to permeate military thinking, new technologies meant that the fortification of a city on its perimeter had diminishing military value. With artillery becoming the weapon of decision in sieges, the perimeter fortification of cities became too costly and ultimately ineffective by the end of the century. These developments fueled efforts to mitigate the worst effect of siege warfare in the development of laws and customs of war.

The twentieth century had its own major currents that gave rise during the two world wars to the unprecedented mobilization of societies and resources in strongly ideologically driven armed conflict. The advent of total war brought such high stakes, such as national survival, that constraints in warfare became utilitarian, with total war blurring the distinction between combatant and noncombatant. Industrial warfare made the city a natural target and battleground. By World War II, combat regularly took place in urban settlements of all shapes and sizes. Consequently, this period had a formative effect on the tactics and doctrine used by armies to this day.[141] After 1945, the irregular insurgent adversary gained more prominence in urban warfare, adding to military challenges in discriminating between combatants and noncombatants. In the Cold War, however, urban combat between conventional forces persisted and dominated development of doctrine. Twentieth century trends in LOAC also led to a new emphasis on developing humanitarian norms in the laws and customs of war that had an impact upon the conduct of urban warfare.

After the end of the Cold War, the belief that the urban environment will "dominate warfare in the 21st century" has occupied an important place in Western debates on future war.[142] The interest in the megacity as a future battlespace is but one facet of persistent military interest in future warfare in the urban environment.[143] The obsession with megacities, however, has produced its critics who have argued that the megacity may not be the future center of gravity for urban combat.[144] There is also the practical matter of finding the military resources for meeting such a large-scale urban military contingency as posed by the megacity. At the end of the Cold War, Western armies experienced dramatic reductions to their establishments, making the prospect of large-scale urban combat in the megacity a challenging prospect for Western military planners. This conundrum of big cities and small armies is not likely to go away in the future. While the urban environment is being seen as the new battleground of the twenty-first century, something very old is also trending in the conduct of urban warfare. The reemergence of the siege in urban warfare has been widespread and has been paralleled by the maintenance of the comparatively new norm of fighting in the city. Similarly, the historic horrors visited upon resident civilian populations in urban warfare persist even as the body of LOAC provides normative constraints on urban warfare's conduct.[145] While irregular adversaries have dominated conflict in the post–Cold War urban environment, the possibility of conventional force-on-force operations in the urban environment remains real enough. If anything, the post–Cold War experience demonstrates that urban warfare is going to continue to be shaped by a mixture of elements that both reflect historical continuity and include new features driven by its dynamic and ever-changing character.

Notes

1. Much of the current literature on urban warfare is captured in an excellent recent bibliography: *Urban Operations: A Bibliography*, Sandhurst Bibliographies No. 1 (Central Library, Royal Military Academy Sandhurst, September 2018). Other dated but still useful bibliographies include the following: Lester L. Miller Jr., *City Warfare: A Bibliography of Periodical Articles*, Special Bibliography Number 85 (Fort Sill, OK: US Army Field Artillery School Library, June 1, 1982); Dianne S. Tapley, *Military Operations on Urban Terrain (MOUT) Bibliography*, Donovan Technical Library (Fort Benning, GA: US Army Infantry School, September 1981); and D. Robert Worley, Alec Wahlman, and Dennis J. Gleeson Jr., *Military Operations in Urban Terrain: A Survey of Journal Articles* (Institute for Defence Analyses, October 2000).
2. See, for example, Felix and Wong, "Case for Megacities."
3. See Kilcullen, *Out of the Mountains* (2013).
4. Evans, *City Without Joy*, 3–5.
5. Evans, "Lethal Genes," 547–548.
6. *Oxford English Dictionary*, Volume XV, 2nd edition (Oxford, UK: Clarendon Press, 1989), 437.
7. NATOTerm, Official NATO Terminology Database, accessed March 21, 2020, https://nso.nato.int/natoterm/Web.mvc.
8. De Malortie, *Treatise on the Attack and Defence*, 1.
9. Mercur, *Attack of Fortified Places*, 12.
10. Bradbury, *Medieval Siege*, 157–159.
11. Whitehead, *Aineias the Tactician*, 50–58.
12. Davis, *Besieged*, 3–9.
13. Bradbury, *Medieval Siege*, 234–240.
14. Kern, *Ancient Siege Warfare*, 310–311.
15. Davis, *Besieged*, 28–29, 84, and 87.
16. Bradbury, *Medieval Siege*, 296–334; and Kern, *Ancient Siege Warfare*, 323–351.
17. Bradbury, *Medieval Siege*, 309.
18. Article 18 in "Instructions for the Government of Armies of the United States in the Field (Lieber Code), April 24, 1863," https://ihl-databases.icrc.org/applic/ihl/ihl.nsf/ART/110-20018?OpenDocument.

19. Kern, *Ancient Siege Warfare*, 324.
20. Mallett, "Siegecraft in Late Fifteenth-Century Italy," 255.
21. Black, *Fortifications and Siegecraft*, 57–58; and Duffy, *Fire and Stone*, 9.
22. Duffy, *Fire and Stone*, 13.
23. Black, *Fortifications and Siegecraft*, 99–100.
24. Duncan, *History of the Royal Regiment*, 171–172.
25. Hughes, *British Invasion of the River Plate*, 157.
26. Ibid., 178–179.
27. Ibid., 205.
28. Official account of Lieutenant General Sir John Whitelocke, *London Gazette*, September 13, 1807, 1207.
29. "Recollections of the British Army," 503.
30. Whitelocke, *London Gazette*, 1209.
31. Hughes, *British Invasion of the River Plate*, 205, 219.
32. Medley, "The Siege of Dehli," 8.
33. Ibid.
34. Llewellyn, *The Siege of Delhi*, 111; and Field Marshall Lord Roberts of Kandahar, *Forty-One Years in India*, 224–226.
35. Field Marshall Lord Roberts of Kandahar, *Forty-One Years in India*, 240–241.
36. "Numerical Return of Killed, Wounded, and Missing," 414.
37. Field Marshall Lord Roberts of Kandahar, *Forty-One Years in India*, 226.
38. Ibid., 243.
39. Vibart, *Sepoy Mutiny*, 143.
40. Medley, "Siege of Delhi," 16.
41. Ibid.
42. "Numerical Return of Killed, Wounded, and Missing," 418.
43. Griffiths, *Narrative of the Siege of Dehli*, 174.
44. Vibart, *Sepoy Mutiny*, 150.
45. Medley, "Siege of Dehli," 3.
46. Esdale, *Peninsular War*, 159–163.
47. Grant, *Personal Memoirs*, 73.
48. Beall, "United States Army," 169.
49. See House, *Controlling Paris*.
50. Traugott, *Insurgent Barricade*, 178–224.
51. Ibid., 140–142, 176.
52. Blanqui, "Blueprint for Insurrection," 154, 158.
53. Cluséret, "Street Fighting," 57–60.
54. Bugeaud, *La Guerre des rues et des maisons*.
55. Ebel, "Du maintaintien de l'ordre à la guerre des rues."
56. Kitchen, *Military History of Germany*, 164.
57. House, *Controlling Paris*, 1.
58. Myatt, *British Sieges*, 63–109.
59. Pasley, *Rules for Conducting the Practical Operations of a Siege*.
60. Jocelyn, *History of the Royal Artillery*, 165–173, 419.
61. Dastrup, *Field Artillery: History*, 37–53.
62. Bailey, *Field Artillery and Firepower*, 211.
63. Callwell and Headlam, *History of the Royal Artillery*, 277–279, 281.
64. Tunstall, *Written in Blood*, 5–11; and Watson, *Fortress*, 4–8.
65. Bairoch and Goertz, "Factors of Urbanization," 286.
66. Williamson and Hsieh, *Savage War*, 443, 464.
67. Report of Major General W. T. Sherman, in Ainsworth and Kirkley, *War of the Rebellion*, 584.
68. Synge, *Defence of England*, 197–198.

69. Statement by the Viscount Palmerston, House of Commons, July 23, 1860, Hansard, Volume 160, https://hansard.parliament.uk/Commons/1860-07-23/debates/71743528-078d-45d2-a06d-365718c0b2bb/Committee?highlight=fortifications#contribution-52f458d6-b73a-4a8d-a17b-3ab532cdc47c.

70. Adair, "Defence of London," 292–293; and Elsdale, "Defence of London and of England Part I," 601–602.

71. Hamley, *National Defence: Articles and Speeches*, 138.

72. Adair, "Defence of London," 303.

73. Elsdale, "Defence of London and of England Part II," 656.

74. Quataert, "War-Making and Restraint by Law," 143.

75. Laws and Customs of War on Land, The Hague Convention, July 29, 1899, Library of Congress, https://www.loc.gov/law/help/us-treaties/bevans/m-ust000001-0247.pdf.

76. Ibid.

77. Quataert, "War-Making and Restraint by Law," 161.

78. Keegan, *First World War*, 96–97, 116, 138–139.

79. Watson, *Fortress*, 204.

80. Bainbridge, "Gold Medal Prize Essay 1913–1914," 1–21; and Norman, "2nd Gold Medal Prize Essay 1913–1914," 161–178.

81. Bainbridge, "Gold Medal Prize Essay," 6.

82. Ibid., 19.

83. Norman, "2nd Gold Medal Prize Essay," 161–162.

84. Gwynn, *Imperial Policing*.

85. Laurie and Cole, *Role of Federal Military Forces*, 328–334.

86. Murray, *Red Scare*.

87. Laurie and Cole, *Role of Federal Military Forces*, 330.

88. Ibid., 334–338, 341.

89. Jeffrey, "British Army and Internal Security," 377.

90. Rowecki, *Walki Uliczne*.

91. Ibid.

92. Slater, "Field Artillery and the Urban Battle," 9.

93. Mann, "The Rest of the Day Was Perfectly Bloody," 118–119.

94. Stone, "Stalingrad and the Evolution of Soviet Urban Warfare," 205–206.

95. Military Training Pamphlet No. 55, *Fighting in Built-Up Areas*, The War Office (UK), April 1943, 1.

96. Ibid.

97. Ibid., 2.

98. Ibid., 2–3.

99. Ibid., 3–5.

100. Ibid., 4.

101. Ibid., 4–5.

102. Ibid., 5.

103. Ibid.

104. Ibid., 6.

105. Ibid., 1.

106. Ibid., 11.

107. Chamberlain and Ellis, "Churchill and Sherman Specials," 105–109. For an example of its employment, see Mann, "The Rest of the Day Was Perfectly Bloody," 122.

108. Komorowski, *Bitwa o Warszawę*, 85–86.

109. Głowacki, *Obrona Warszawy i Modlina na tle kampanii wrześniowej*, 84.

110. DiMarco, *Concrete Hell*, 62–64; and Doubler, *Closing with the Enemy*, 97–98.

111. Bailey, *Field Artillery and Firepower*, 89–90.

112. Boucher, "Artillery Attacks in Stone Villages," 709; and Menshikov, "Artillery in Street Fighting," 471.

113. Bean, "Come Ye Back to Mandalay," 134, 137.

114. XIV Corps (US), "Japanese Defense of Cities," 20–23.

115. Randall, "Battle for Manila," 451–455; and Smith, *United States Army in World War II*, 291–306.

116. Bottomley, "Letter to Air Chief Marshal Sir Arthur Harris," 160.

117. Overy, *Bombing War*, 399–409.

118. Walsh, "The Battle of Stalingrad," 85.

119. Willems, "Defiant Breakwaters or Desperate Blunders?" 370–371.

120. Rómmel, *Za honor i ojczyznę*, 354.

121. Ungváry, *Battle for Budapest*, 311–315.

122. Schnabel, *United States Army in the Korean War*, 177.

123. Bennett and French, *Kenya Papers of General Sir George Erskine*, 121.

124. Military Training Pamphlet No. 55, *Fighting in Built-Up Areas* (Reprinted with Amendments [No. 1], 1945), The War Office (UK), April 1945.

125. See Army Field Manual (UK), Volume V, Part 5, *Fighting in Built-Up Areas*, Army Code No. 71346 (Pt5), 1983.

126. See Army Field Manual (UK), Volume V, Part 5, *Fighting in Built-Up Areas*, Army Code No. 71346 (Pt5), 1986.

127. See "Internal Security in a City," General Headquarters (GHQ) Far East Land Forces (FARELF), July 1959.

128. Wahlman, *Storming the City*, 155–168.

129. Ewing, *Vietnam Studies*, 142.

130. Moss, *Urban Guerrillas*, 148–158.

131. Carlos Marighella, "Minimanual of the Urban Guerrilla," in Jay Mallin (ed.), *Terror and Urban Guerrillas: A Study of Tactics and Documents*, (Coral Gables: University of Miami Press, 1982), 70–115.

132. Moss, *Urban Guerrillas*, 208; and Taw and Hoffman, *Urbanization of Insurgency*, 17.

133. See *The Geneva Conventions of August 12, 1949* (Geneva: International Committee of the Red Cross, 2002) and *Protocols Additional to the Geneva Conventions of 12 August 1949* (Geneva: International Committee of the Red Cross, 1977).

134. Article 28, "Geneva Convention Relative to the Protection of Civilian Persons in Time of War," in *The Geneva Conventions of August 12, 1949*, 165.

135. For example, see Protocol I, Chapter II, Article 51, in Geneva Conventions, *Protocols Additional to the Geneva Conventions of 12 August 1949*, 35–37.

136. Ashworth, *War and the City*, 200.

137. Kern, *Ancient Siege Warfare*, 9.

138. Bairoch and Goertz, "Factors of Urbanization," 285.

139. Fuller, *Conduct of War*, 79–80.

140. This shift is stressed in DiMarco, *Concrete Hell*, 25.

141. Ibid., 26.

142. Ibid., 214. See also Hills, *Future War in Cities*, xv; and Kilcullen, *Out of the Mountains* (2013), 262.

143. See Felix and Wong, "Case for Megacities"; Harris et al., *Megacities and the United States Army*; Konaev, *Future of Urban Warfare*; Matthews, "Megacity Warfare"; and Williams and Selle, *Military Contingencies*.

144. Evans, "Case Against Megacities," 43.

145. Power, "Siege Warfare in Syria," 9.

4

The Totemic Value of Cities

Alex Neads

ON JANUARY 20, 2015, AT ABOUT HALF PAST THREE IN THE AFTERNOON, a major explosion ripped apart the New Terminal building at Sergei Prokofiev International Airport. Situated at the northern edge of the city of Donetsk in Ukraine's eastern Donbass region, the airport had become the focus of heavy fighting between Ukrainian troops and separatist militiamen from the Donetsk People's Republic (DPR)—a breakaway Russian-backed statelet. By late January, the Ukrainian military had been all but driven from the airport, with the last defenders holed up in the New Terminal's shattered mezzanine galleries. Once the steel-and-glass centerpiece of a modern airport, the fighting had long since carried away the top two of the New Terminal's original seven stories. The day before, separatist troops had used repurposed antiship mines to blow out the last of the New Terminal's interior walls, forcing the surviving defenders down into the basement. This latest blast collapsed the terminal's uppermost floor, crushing the Ukrainian soldiers trapped in the structure below. The remaining defenders surrendered the following morning, ending an 8-month-long battle for control of the airport. During the course of this fighting, Donetsk Airport emerged as a potent symbol of national resistance for both sides. In Ukraine, the airport's defenders were lauded as superhuman "cyborgs," while its capture became an acid test for the viability of the separatist cause. This totemic importance, together with the unusual intensity of fighting, has lent the battle a reputation as the "little Stalingrad" of the Donbass war.[1] Indeed, the fighting there continued long after the airport's functional utility had been destroyed, persisting unabated even in defiance of the Minsk I cease-fire signed by both parties.

Urban fighting has an established reputation for political symbolism, evoking images of bloody quagmires such as Stalingrad and Grozny, or strategic missteps such as Mogadishu and Fallujah. Yet, the causes and consequences of urban symbolism in war are far from universally accepted. On the one hand, urban symbolism has been regarded as a regrettable if sometimes inexorable driver of strategic *engrenage*—a spiral, an inescapable series of events, a slippery slope. As Barry Posen laconically observed, the political imperatives behind urban struggles like Stalingrad "were probably stupid, but the decision to engage in battle was made at the highest political levels."[2] On the other, the impact of urban symbolism on the conduct of operations has been dismissed altogether as an "urban myth" borne of "Stalingraditis"—an overinflated concern for the intrinsically political qualities of urban spaces.[3] Intriguingly, while the fighting at Donetsk Airport appears to have been laced with symbolic meaning, it was far from the only hard-fought urban battle of the Donbass war, nor arguably was it the most militarily significant. During the summer of 2014, for example, Ukrainian troops made a similar stand against separatist rebels at Luhansk Airport, holding out for almost 5 months until rescued in a dramatic Ukrainian raid. Simultaneously, Ukrainian soldiers fought a bitter rear-guard action at the town of Ilovaisk; their defeat marked the high-water line of government efforts that year. Yet, in stark contrast to the fighting at Donetsk Airport, neither battle displayed any profound symbolic importance, at the time or since.[4]

This chapter seeks to understand the development of symbolic meaning at Donetsk Airport and the extent to which this accounts for the intensity of the fighting there. If the future of warfare lies in the megacity, as some predict, then explaining how and why ostensibly mundane urban spaces become infused with political meaning—and what this symbolism holds for the conduct of urban fighting—is vital to understanding the future character of warfare.[5] Moreover, as professionalized Western armies continue to diminish in size, particularly in Europe, so the strategic importance attached to the costs of urban operations—both human and political—seems likely to become only more acute. This chapter argues that while the symbolic importance of Donetsk Airport was deeply rooted in the high costs and lived experience of the battle, the character of fighting itself was determined by the wider context of the internationalized civil war in the Donbass, and specifically, the principal-agent politics of proxy conflict.

Reexamining the Symbolic Perils of Urban Warfare

Combat in cities has traditionally been seen as a costly and difficult endeavor, leading armies to eschew urban fighting wherever possible. The ancient strategist Sun Tzu held that the "worst policy is to attack cities," while Napoleon repeatedly cautioned his lieutenants "not [to] engage in the streets" for "the greatest catastrophes [occur] as a result of armies rushing and diving into the narrow streets of towns."[6] Although armies have frequently struggled for control of small villages or isolated farmsteads, fighting inside large cities was historically rare. Instead, cities were typically defended by fixed fortifications at the periphery, which when breached, usually led to the collapse of the defense.[7] Deliberate fighting within urban centers only became commonplace in the middle of the twentieth century, as changes in technology rendered perimeter defenses obsolete. By then, the size of armies had expanded to the point where fighting for control of urban centers typically extended along fronts well beyond the city, with combat outside the suburbs every bit as intense as that within the city proper.[8] During the battle for Brest in 1944, for example, US divisions fighting in the surrounding countryside saw comparable rates of battle fatigue to those engaged within the city itself; a reality that still prevailed during the Vietnam War, despite the lived experience of the battles such as Hue City as "a black and white madness of destruction and death."[9]

In recent years, however, urban fighting has become an increasingly prominent feature of warfare. Rapid urbanization and population growth have seen cities become larger, denser, and more important than ever before. Simultaneously, downsized professional armies now lack the mass to maintain extended fronts, focusing fighting within the city itself.[10] Nonetheless, urban warfare continues to present distinct challenges for modern armies. The built environment impedes movement and communication, constrains the employment of weaponry, channels maneuver, and hinders command and control. Dense networks of buildings likewise provide defenders with a plethora of cover and concealment across three dimensions, providing ample opportunity for ambush and infiltration from above, below, and behind an attacker. Consequently, urban engagements often occur suddenly and at extremely close quarters, increasing the propensity for confusion and organizational atrophy.[11] As a result, urban spaces are considered to provide an inherent defensive advantage. Whereas doctrinal estimates require a concentration of three attackers for every one defender in open country, this ratio rises to 8:1 or higher in urban terrain.[12] In conventional military thought, therefore, the city is a potential graveyard for attacking armies—and the smaller the army, the greater the potential danger.

Yet, urban centers are more than simply an acute tactical problem. As centers of population, commerce, industry, and governance, cities also provide the locus for a nation's social and political life, reflecting "the values, social perceptions, and interactions" of their inhabitants.[13] This political quality can serve to limit the conduct of urban operations, exacerbating the practical challenges of city fighting. Historically, for instance, sieges represented "the most highly structured, ritualized and rule-bound forms of military conflict of the early modern era," and a besieged city could reasonably expect the opportunity to surrender honorably as soon as its walls were breached.[14] Nonetheless, a city taken by storm was liable to sack, and the protection of civilian populations and symbolic places remains difficult even today. As Alice Hills has argued, the tactical challenges of street fighting mean that "high levels of violence are inherent in urban war, regardless of the nationality of those involved or the scale of operations," creating profound tensions between the vision of restrained "liberal" warfare and the reality of urban fighting—as recent operations in Mosul and elsewhere attest.[15] Consequently, the city has become a refuge of choice for guerrillas and insurgents, prompting Western armies to pursue more-precise urban tactics in response.[16] Close-quarters battle techniques, once the preserve of small elite units, are now the stock-in-trade of the ordinary infantryman, becoming a de facto measure of modern military professionalism via a process sometimes described as a "special forcification."[17]

Conversely, though, these social and political meanings also sometimes intensify the conduct of urban fighting, generating disproportionately high costs. During World War I, General Falkenhayn's efforts to "bleed France white" at Verdun relied in no small part on the city's symbolic importance to French politics. In 1870, Verdun was the last fortress to fall to the Prussians before Paris, and according to legend, its ill-fated commandant committed suicide rather than surrender to the besieging Austro-Prussian army in 1792. Hence, in 1916, abandoning the Verdun salient was as politically unacceptable as it was militarily inadvisable, despite the strain this placed on the French Army. In the words of Alistair Horne, Verdun has subsequently become "a sacred national legend" for the French, but simultaneously also "a modern synonym for a Pyrrhic Victory."[18] Perhaps the most infamous example of this phenomenon is the Battle of Stalingrad. Although initially irrelevant to German military planning on the Eastern Front, the city's symbolic name transformed its stubborn Soviet defense into a personal affront to "Hitler's image as a military genius and the Social Darwinist concept of German racial superiority."[19] More recently, the destruction of Dubrovnik's Old Town and Mostar's Stari Most during the breakup of Yugoslavia owed much to their symbolic association with particular ethnonational communities, and their rebuilding has likewise become a potent totem of postconflict reconstruction.[20]

However, there is little consensus as to why cities sometimes assume a particular symbolic value, or how these meanings affect the conduct of urban fighting. Max Neiman, for example, has argued that some urban centers are intrinsically more important than others, concluding that places that inherently "hold special religious or historical meaning are likely to affect combatants in ways that cities without such intensely held symbolic or nationalistic implications are not."[21] Yet, the innate value of an urban space is not self-evident; nor do such characteristics always generate the same symbolic effect. In 1944, for example, the Allies were careful to preserve historic Paris intact, but extended no such courtesy to Saint Malo's historic intra muros, which was destroyed during liberation.[22] Consequently, Posen has argued that the political value of urban spaces—and by extension their symbolic potential—depends not on the intrinsic properties of the place itself, but on its strategic context. Accordingly, the meaning and importance of city can be fluid, depending on its relationship to the wider strategic peculiarities of the conflict. Hence, while Stalingrad assumed an immense symbolic importance in 1942–1943, Moscow did not demonstrate the same strategic resonance in 1812—as Napoleon discovered to his cost. Nonetheless, Posen concluded that where "urban battles occur for high-level strategic reasons, whether those ideas are right or wrong, they are very likely to take on the maximalist character" seen at Verdun or Stalingrad. Moreover, although the symbolic imperatives behind battles such as Stalingrad "may seem stupid in retrospect . . . a repetition of such thinking is not impossible."[23]

Recently, though, David Betz and Hugo Stanford-Tuck have dismissed the symbolic influence of urban spaces altogether. They argue that a scholarly and doctrinal focus on Stalingrad—"one titanic and highly peculiar battle" considered "extremely unusual in the strength of its political symbolism"—effectively "distorts perceptions of the problem at hand."[24] Instead, they ascribe elevated political interest in urban warfare as a product of the high human and material costs traditionally associated with fighting in such environments, leading to something akin to a phobia of urban combat among commanders and policymakers alike. As Betz and Stanford-Tuck have argued, for "the contemporary Western politician, conflict in the urban environment . . . is beyond the public's tolerance in terms of expenditure of 'blood and treasure.'"[25] More broadly, this concern with the political implications of battlefield cost reflects the so-called Wootton Bassett phenomenon, in which grassroots commemoration of British war dead exacerbated political opposition to the conflict in Afghanistan.[26] The pair conclude that a renewed focus on the generation of effective urban tactical capabilities will prevent political intrusions into the conduct of urban warfare, irrespective of a city's political status or strategic value, by lowering the costs of operational success.

In principle, of course, high costs can conceivably also produce the very inverse of this urban aversion, by inflating the political importance of particular urban spaces through a symbolic logic of sunk costs. Certainly, high costs in war can develop distinct political meanings of their own, even when their accumulation is otherwise accepted as worthwhile.[27] Yet, the political significance of any casualties or costs incurred in urban fighting, as in all warfare, can only be understood in relation to the perceived importance of the goals pursued at the time. Indeed, modest costs incurred in a discretionary or unimportant operation might be unacceptable, while far greater costs would be tolerated in pursuit of a more existential goal.[28] A city may likewise be destroyed in heavy fighting because the landscape was unimportant to risk soldiers' lives protecting, or paradoxically also because it was too important to give up without a fight. In fact, the willingness to incur costs in urban fighting *beyond those required by pure military necessity* is the very sine qua non of urban symbolism as a phenomenon.[29] Consequently, this chapter argues that the symbolic quality of urban fighting is a product both of strategic context *and* elevated casualties, and the two interact with each other in important ways. Bloody attrition is the product of strategic context, which in turn imbues these urban sites with distinct symbolic meanings engendered by the high costs of fighting.

Donetsk Airport as a Case Study in Urban Symbolism

The case of Donetsk Airport provides an important opportunity to understand the causes and consequences of urban symbolism in recent warfare, for a number of reasons. First, the fighting there displayed an abundance of exactly the kind of symbolic meanings in question. The battles for control of the airport have acquired a legendary, almost mythical status in Ukraine, imbuing both the airport and its defenders with politically symbolic meanings. As the fighting progressed, the airport's Ukrainian defenders became known as "cyborgs"—resolute, terminator-like part-man, part-machine. The sobriquet apparently originated among the separatist ranks, perhaps as a derogatory dismissal of the defenders' endurance as a marker of something less-than-human, but was rapidly adopted in the Ukrainian press as a reference to their purportedly superhuman tenacity.[30] However, the symbology of the airport's cyborg defender was also propagandistically exploited by the separatists after the airport's fall, as a means to reinforce their own claims to military prowess. As one separatist commander disingenuously told journalists, "Those on the Ukrainian side, who survived this *took a massive hit from us*. They deserve respect."[31] Moreover, the airport does not appear to have enjoyed any particularly special status or inherently political significance prior to the conflict. Admittedly, the airport represented one of Ukraine's more modern and internationally recognized pieces of infrastructure, having been substantially renovated ahead of the Euro 2012 football championships at a cost of £537 million (approximately US$701 million).[32] Yet, the airport did not possess any particularly unique historic, religious, or cultural connotations prior to the fighting, and prestige alone cannot account for the totemic importance the airport subsequently assumed. Indeed, this symbolic meaning only emerged during the fighting itself.

Second, while control of the airport did provide important tactical advantages to each side, these do not appear objectively significant enough to justify the escalating intensity of the fighting there, reinforcing the focus on urban symbolism. Situated on high ground at the northern edge of the city, close to a series of major road and rail routes, the airport provided Ukrainian troops with a vantage point overlooking Donetsk from which to project force into the separatist-controlled city. Moreover, government possession denied the separatists use of the airport's long runway and modern logistical handling facilities, precluding direct resupply from Russia by air. For the separatists, by extension, the airport represented "the most important gate into the city," and the government presence there constituted an "advance intrusion into the heart of the rebellion" that impeded rebel consolidation in and around Donetsk.[33] Yet, the strategic importance of the airport *as an airport* also declined in inverse proportion to the intensity of the fighting, as the airport's facilities were progressively destroyed. Moreover, either side could effectively deny the other use of its runway without having to physically occupy the airport itself, via the siting of ground-based anti-air systems on adjacent territory within their possession. Indeed, conventional military wisdom generally holds that an "inferior port or airfield, away from a large urban environment, which can be taken and held with moderate casualties, and improved with engineering" is usually "preferable to theoretically better facilities that can be had only through hard fighting, and which would be badly damaged by such fighting in any case."[34]

Last, but by no means least given the theme of this book, the scale of fighting at Donetsk Airport was comparatively modest—at least in numerical terms—when compared with previous examples of symbolic urban battles such as Stalingrad or Grozny. The airport was initially defended by a single Ukrainian volunteer battalion, rising to the 93rd Mechanized Brigade supported by the 3rd Spetsnaz Regiment by the autumn of 2014. These were subsequently replaced by elements of the 79th and 95th Airborne Brigades, supported by various militia and volunteer battalions, before the 93rd Mechanized Brigade resumed the defense in January 2015, alongside the 81st Airborne Brigade and elements of the 80th and 95th Air Assault Brigades. Nonetheless, these formations were typically understrength and undertrained, and the forces actually present in the terminals can seldom have amounted to more than a few companies of infantry at any given time. The 81st Airborne Brigade, for example, was composed of two battalion-sized maneuver units, which despite its elite status, were both largely manned by mobilized reservists and wartime volunteers.[35] If anything, the airport's separatist attackers were an even more ad hoc affair; one rebel militia even called itself the "Somali Battalion" after its rag-tag, pirate-like appearance.[36] Yet, while both sides may have lacked the pronounced capabilities Betz and Stanford-Tuck advocate, they certainly did not lack the will to absorb casualties. Indeed, the fighting for control of the airport's terminal buildings developed exactly the maximalist characteristics highlighted by Posen, despite the otherwise quite limited scale of the battle. At times, forward positions were so close that Ukrainian paratroopers could hear the separatists "baiting us from behind the walls," shouting "time to surrender, Ukies, we're coming to cut your throats,"[37] or pretending to be feared Chechen foreign fighters in an effort to psychologically intimidate.[38]

If the fighting at Donetsk Airport thus provides an ideal case study into the emergence and influence of urban symbolism on warfare, this chapter argues that the airport's symbolic meaning was intimately connected with, and derived from, the intensity of the fighting experienced there. However, this intensity did not emerge as the escalatory product of symbolic sunk costs, or because of the airport's inherent societal value, but was instead the product of wider political imperatives rooted in the fighting's strategic context as a proxy conflict. Proxy warfare can be understood as a principal-agent problem, in which the principal employs an agent to conduct activity on its behalf, typically in pursuit of some shared goal. For the principal, proxies provide a means to pursue objectives that would otherwise be too costly or risky to be worthwhile by providing an agent with assistance or reward in return for the conduct of the desired activity on its behalf. However, the interests of each party rarely align exactly. This can create incentives for the agent to shirk particular activities, or misappropriate the patron's resources for its own ends, in turn necessitating a degree of carrot-and-stick supervision by the principal to ensure it gets what it wants.[39] In a proxy conflict, these "interest asymmetries" can lead the agent to distort the aims and objectives of the patron, while the principal's efforts to secure its own discrete interests (often at the expense of the agent) likewise affect the conduct of the fighting.[40] Importantly, the Donbass conflict has been widely described as a proxy war, in which the policy goals of the DPR's Russian backer shaped the conduct of the fighting as much as the aims and ideas of the cyborgs and separatists themselves.[41]

Methodologically, a range of sources have been used to reconstruct the fighting at Donetsk Airport. The Organization for Security and Co-operation in Europe (OSCE) Special Monitoring Mission (SMM) maintained a team in Donetsk City throughout the battle. Established in March 2014 to act as a humanitarian observer in the Donbass, the SMM subsequently gained a formal role in cease-fire monitoring, providing a daily "seismograph on the actual intensity of the conflict." However, the mission's neutral civilian status meant that, by its own admission, OSCE monitoring sometimes offered little more than "a good account of the intensive and continuing shelling around Donetsk Airport."[42] Consequently, SMM reporting has been augmented with media accounts to provide further detail on the perspectives and experiences of combatants themselves. A number of investigative journalists gained impressive access to belligerents on both sides, sometimes even reporting on the fighting firsthand in close to real time. That said, such accounts are typically filtered through the journalistic lenses applied by reporter and editor, as well as the self-censorship of the original subjects, and must be treated with caution. Moreover, neither the Russian nor the Ukrainian press can be considered impartial, while Western media typically sympathized with the Ukrainian cause—although the Kremlin's deliberate exploitation of Russian media for information operations was particularly problematic.[43] Even so, propaganda created by the belligerents themselves is particularly useful for understanding how each side sought to frame the fighting. Finally, existing scholarly research has been used to situate the battle in its wider strategic context, while the handful of published US military assessments provide valuable tactical benchmarks. Critically, the use of multiple different types of sources has enabled information to be cross-referenced for veracity, offsetting the limitations of each while underpinning analytical rigor.

The chapter now provides a brief chronology of the first and second battles for Donetsk Airport. I then examine the emergence of the airport's symbolic cyborg narrative, concluding that this symbolic image alone cannot explain the escalation of the fighting at the airport. I then place this symbolism in its wider strategic context, examining the principal-agent politics behind the separatist offensives, together with the wider strategic dynamics of the fighting at Donetsk Airport. In the final section, I conclude that the strategy and politics of proxy warfare best account for the unusual intensity of the fighting at Donetsk Airport and, in so doing, significantly explain the emergence of its totemic status.

The First and Second Battles for Donetsk Airport

The war in the Donbass region of eastern Ukraine began in the spring of 2014, precipitated by tumultuous events earlier that year. In late 2013, President Viktor Yanukovych's decision not to sign an association agreement with the European Union sparked mass protests in Kiev. By February 2014, this "Euromaidan" movement had spiraled into revolution, leading to the overthrow of Yanukovych and the collapse of the government. Russia viewed these events with alarm, fearing a shift in Ukrainian policy that might threaten its strategic warm-water naval base at Sevastopol on the Black Sea. Almost immediately, undisclosed Russian troops seized key instillations in Crimea, rapidly assuming de facto control of the peninsula in a political and military fait accompli. Russia formally annexed Crimea in March, though de jure ownership remains disputed. Simultaneously, anti-Kiev protests began in eastern Ukraine, in the Donbass oblasts of Donetsk and Luhansk. Assisted by Russian intelligence, pro-Russian activists took over government offices and local administration buildings, in many cases with the tacit support of local police and security agencies. The following month, separatist leaders proclaimed two breakaway "People's Republics," centered in Donetsk and Luhansk, respectively. However, with Kiev still engulfed in political turmoil, the new Ukrainian government did not immediately respond. Instead, nationalist militias self-organized and took it upon themselves to resist, leading to a series of tit-for-tat skirmishes as the separatists attempted to consolidate their territory. Then, in mid-April, the interim Ukrainian administration initiated an "Anti-Terrorist Operation" in the Donbass in an effort to reassert Ukrainian sovereignty. While the euphemistic rhetoric was intended to preclude full-scale hostilities with Russia, it ultimately failed to prevent a further escalation of the conflict.[44]

The fighting for control of Donetsk Airport began that spring, as insurrection descended into full-blown civil war. In the early hours of May 26, 2014, between 100 and 200 separatist militiamen from the DPR arrived unexpectedly at Sergei Prokofiev International Airport, having deployed from their makeshift barracks in the city in commandeered civilian buses. Many had only recently joined the cause, having been smuggled across the Russian border during the preceding week.[45] The separatists established a hasty defense of the main terminal buildings, driving out government sentries and placing support weapons and snipers on the New Terminal roof. They also evacuated travelers from the terminals, though the airport received two arrivals as normal that morning before flights were suspended. However, Kiev refused to cede control of the airport and instead demanded the separatists withdraw. Then, at around 2 p.m., the Ukrainian military launched a combined arms attack to recapture the airport. This was spearheaded by several companies of air assault infantry, supported by Mi-24 helicopter gunships and ground attack aircraft, and made skillful use of the airport's exposed terrain to fix the separatists in place.[46] In the words of one militiaman, government troops "forced us tightly into the building and were bombing from all sides. They had missile launchers around the perimeter of the airport and were firing on the terminal."[47]

By late afternoon, government troops were threatening to cut off the terminal complex, forcing the separatists to abandon the airport altogether. Having already been driven from the roof and surrounding areas, the remaining militiamen boarded a series of KamAz trucks inside the New Terminal and drove at speed for the city. A cover party was left behind to facilitate the withdrawal, later slipping away on foot under cover of darkness, though the convoy itself was ambushed as it passed through Donetsk's northern suburbs. The DPR later claimed it had been fired on by government forces, but it appears that separatist militiamen actually mistook their retreating comrades for Ukrainian troops.[48] Although skirmishing continued around the airport into the following day, it was evident by then that the DPR had lost the first battle for Donetsk Airport. In all, the separatists suffered an estimated thirty to thirty-five killed and a further fifteen to twenty injured, provoking both desertions and renewed determination in equal measure.[49] In early June, DPR leaders called for additional volunteers to help prepare for an expected government attack, and checkpoints along access routes into the city were strengthened.[50] With government troops also digging in at the airport, the battle lines were now drawn for a further confrontation.

The second battle for Donetsk Airport effectively began in early September, when sporadic fighting at the airport developed into a new separatist offensive. DPR troops used "technicals" mounted with antiaircraft autocannons, together with recoilless rifle, mortar, and artillery fire spotted from apartment blocks in the adjoining districts, to advance along access roads in the airport's eastern service area toward the terminal complex. They progressively seized the airport's control center, police station, and hotel, along with a series of hangars, but stalled briefly in the open ground surrounding the Old Terminal building.[51] Then, on October 3, the separatists used main battle tanks and a smoke screen to break into the Old Terminal, although the assaulting Oplot Battalion reportedly suffered 40 percent casualties in the process.[52] The offensive petered out again toward the end of October, but not before the DPR gained a further foothold in the New Terminal as well. Simultaneously, the separatists worked to surround the airport, attacking the ring of villages along the northern side of the runway, beginning in Spartak and progressing west toward Opytne by the following month.[53] Meanwhile, the Ukrainian defenders retained a tenuous line of communication west from the terminals via the village of Pisky, partially covered by support weapons sited in the government-occupied control tower. Nonetheless, government resupply became increasingly difficult even under armor.

By early November, the top two floors of the New Terminal had collapsed under shellfire. The DPR occupied the New Terminal's remaining third floor and parts of its basement, while the Ukrainian Army retained an enclave in the Old Terminal, together with the lower stories of the New Terminal.[54] A temporary cease-fire was agreed overnight on November 16, allowing the separatists to recover their dead and wounded, although fighting resumed promptly the next morning.[55] A further separatist push in late November failed to dislodge the Old Terminal's remaining defenders, but left the building so badly damaged that the Ukrainian Army decided to abandon it overnight of December 5.[56] This was followed by a more extensive cease-fire, providing safe passage for the Ukrainian military to rotate their forces at the airport. On December 12, fifty-one Ukrainian soldiers were driven out through DPR-controlled territory and replaced by forty-eight incoming soldiers that day and a further thirty-six on December 15, driven in by separatist drivers who then handed over the transports to government troops at the south side of the New Terminal. Further roulements were completed on December 20 and again on Christmas Day, though skirmishing continued elsewhere, as separatist troops advanced to the edge of Pisky.[57]

These arrangements held into the new year but were not to last. Around January 8, the separatists launched a renewed assault on the New Terminal, preceded by an intense artillery bombardment. The control tower finally collapsed under shell fire on January 13, and separatist troops seized around a third of the New Terminal soon after.[58] Here, main battle tanks were again used in close support, reducing two government positions in the New Terminal and forcing the Ukrainian Army to blow up a third in order to create an obstacle to slow the DPR's advance. According to Colonel Yevgeny Moysyuk, then commanding the Ukrainian 81st Airmobile Brigade, the separatists "drove up two tanks right in front of the terminal and shot it point-blank a few times and went away before our artillery could get them, and they would do it again and again, completely turning the terminal into a sieve."[59] This fire was deliberately concentrated on corners of the building in an effort to collapse parts of the structure, enabling the separatists to use the rubble to infiltrate above the defenders. By mid-January, the defenders had become progressively confined to the mezzanine balconies on the second story of the New Terminal, while the separatists controlled the floors above them as well as parts of the ground floor and basement. Moreover, Ukrainian resupply was becoming a major challenge. As one Ukrainian officer explained, "We couldn't get our tanks or armored vehicles there anymore because they had used the truce time to fortify all the side approaches and deploy all kinds of heavy weapons."[60]

Confronted with an increasingly bleak situation, the Ukrainian military mounted a series of ever more risky operations to relieve their stranded troops. On January 17, efforts to extract wounded from the New Terminal failed, prompting a larger-scale operation the following day. This time, Ukrainian armor managed to force a passage along the southern perimeter of the airport, briefly allowing a relief convoy to reach the New Terminal.[61] Even so, the defense remained precarious, with the separatists throwing first grenades and later tear gas down onto the defenders in the terminal in an effort to flush them out.[62] Then, on January 19, DPR troops used repurposed antiship mines to blow out the New Terminal's remaining interior walls. The explosion was sufficiently powerful that Ukrainian officers orchestrating the defense felt the blast from their basement command post 6 kilometers away.[63] A further attempt to relieve the terminal was made by the 81st Brigade that night, using borrowed armored vehicles crewed with their own men, after the original drivers refused to go. Some of the vehicles got lost in heavy fog and drove to a separatist-held building by mistake, but the operation did manage to evacuate the most severely wounded from the New Terminal.[64] An additional battalion of Ukrainian paratroops arrived early the next morning and was immediately thrown into action, but without success. The separatists then used explosives to collapse the upper floor of the New Terminal onto the remaining defenders below. By then, command and control in the terminal was beginning to erode, and some of the survivors decided to try and escape under cover of nightfall. The following morning, only about sixteen Ukrainian soldiers were left in the New Terminal. After a brief discussion, the remaining defenders surrendered on January 21, led by a sergeant.[65]

Emergence of Symbolism at Donetsk Airport

Donetsk Airport's symbolic importance emerged during the course of this fighting, with the image and potency of the airport's "cyborg" narrative intimately derived from the intense combat experienced there. Indeed, at the start of the conflict, the airport did not appear to possess any particularly noteworthy symbolic meanings. Admittedly, the DPR's initial attempt to seize Donetsk Airport betrayed a certain propagandistic quality and may have been intended as a show of separatist political strength. The day before, as the rest of Ukraine went to the polls to elect a new president, the DPR had defiantly paraded its militias through Donetsk city center, to the cheers of many (but by no means all) local residents.[66] Yet, this timing also reflected a high degree of opportunism, with Ukrainian elections likely seen as useful cover for a DPR land-grab while Kiev's back was turned. Certainly, the separatist militiamen sent to the airport in May did not expect to meet much resistance. One militiaman recalled how his superiors had "told us no one would fire at us. Just pose for the cameras and that's all. They would see us, get scared, give up"; the DPR's Vostok Battalion even left its antiaircraft weapons behind in the mistaken belief that they wouldn't be needed.[67] Although no Ukrainian aircraft were lost in the first battle for Donetsk Airport, separatist forces did shoot down a Ukrainian transport aircraft at Luhansk Airport just a few weeks later, then representing the single greatest loss of government life in the conflict to date.[68] Accordingly, the relative ease with which the Ukrainian military retook the airport in May reflected not only the tactical importance of government air power, but also the DPR's significant underestimation of Ukrainian will to fight for it. Hence, while the airport may have been militarily and politically important from the start, it was not initially all that symbolic in and of itself.

Rather, the airport's symbolic value developed as a product of the intense fighting experienced during the second battle for Donetsk Airport. As with civil conflicts elsewhere, fighting across the Donbass frequently assumed a brutal, zero-sum dynamic, rooted in the contested identity politics of the region. At Donetsk, for example, Ukrainian artillery began targeting cultural sites with perceived associations to the separatist cause from late August, first shelling a number of Orthodox churches belonging to the Russian Patriarchy, and later also the city's regional history museum.[69]

However, the narrative of the cyborg owed more to the lived experience of defending the terminals than anything else, reflecting the arduous conditions experienced by Ukrainian soldiers—especially in the terminals. In October, for example, one Ukrainian soldier described how the "holes in the walls account for more space than the rest of the structure," such that there was "not a single place where bullets or shrapnel cannot reach you at any given time of the day."[70] Moreover, these conditions only deteriorated as the fighting progressed. Come winter, Ukrainian troops in the terminals had to switch off their generators and even some of their radios at night, as the separatists were so close that any noise or light drew immediate fire in the darkness.[71] As one defender told journalists, "It's cold. It's dark. It's dangerous all the time. All this affects how you think. You get used to shots being fired pretty quickly."[72] By January, defenders described separatist infantry "crawling all over the place like rats—above, below, and on either side," commenting that "the worst thing was this sense of phantoms flying around you . . . people writhing in agony, moaning, crying for help."[73]

Importantly, the longevity of the second battle served to reinforce the exceptionalism of the airport's defense in the public imagination, transforming the precarity endured by the defenders into reputation for active heroism rather than passive victimhood. Here, media accounts typically juxtaposed the ruin and decay of the terminal buildings with the stoicism of the combatants' daily struggle to survive and fight in them. In one interview, for example, a Ukrainian officer described how the "terminals we are holding on to are weaker than the Three Little Pigs' houses, and it is a miracle that they are still standing,"[74] while another related the everyday techniques soldiers employed to survive:

> When they're firing tank shells, you simply lie on the floor as flat as you can because there's no real cover, only drywall around you. We were in what was the airport's customs department, where there were conveyors that moved luggage. These conveyors are metal constructions that go up to your waist. So you could hide behind those. That was the one thing that could save you—these three millimeters of metal running along the conveyor. So we lie down, we sleep, we work, and we take our breaks all behind that space.[75]

To a significant extent, even the progressive dilapidation of the airport—so striking to outside observers—became a central feature of the cyborg legend, highlighting how separatist forces were obliged to destroy more and more of the complex in an effort to dislodge the defenders. As one Ukrainian officer asserted after the airport finally fell, "The cyborgs withstood the final attack; the concrete didn't."[76] Here, the image of the cyborg at Donetsk Airport is reminiscent of the symbolism that emerged from Verdun in World War I, where the endurance of the French *pilou* returning glassy-eyed from the maelstrom along the Voie Sacrée provided a central component of Verdun's public *on ne passe pas* (they shall not get past) narrative.[77]

Yet, while the intensity of the fighting undoubtedly gave shape and resonance to the image of the cyborg, it is not clear that this symbolic meaning reciprocally affected the character of the fighting. Certainly, some Ukrainian commanders felt that the operations mounted to relieve pressure on the terminals during the final days of the defense were less than militarily sound. Major Ruslan Prusov, for example, publicly described his battalion's mission to relieve the terminal as "just sheer idiocy," complaining that to "really storm the airport and win it back, we need 10,000 men, tanks and armored vehicles, but first of all one or two hours of serious artillery or missile bombardments." Instead, he "lost two men dead, seven wounded and seven vehicles before even reaching the tarmac."[78] A brigade commander even told one journalist that "we should have evacuated our men a few days earlier, then waited for the separs [separatists] to converge on the premises and bury them under the ruins," but by then, the Ukrainian Army had "fallen hostage to this beautiful cyborg legend."[79] At the same time, however, the airport's symbolic importance does not seem to have been universally accepted, even among Ukrainian troops. As one defender remarked, "Not everyone likes the [cyborg] name. . . . We are humans. Almost every one of us is married with children. Our thoughts are human too—we want this to be over soon."[80] Moreover, the airport's loss was not uniformly recognized as militarily or psychologically significant by combatants themselves. One Ukrainian soldier, for example, later asserted that "defending the terminal was a pointless endeavour, there was nowhere from which to defend," claiming instead that abandoning the terminals had "only strengthened our morale."[81]

Importantly, the potency of the cyborg legend owed as much to the activities of journalists and politicians *after* the battle as it did to the actions and experiences of combatants themselves. The public resonance of the cyborg image is largely a product of media accounts of the fighting—including dedicated blogs that tracked the course of the battle day-by-day in near real-time—which drew public attention to the battle while simultaneously building sympathy for the Ukrainian cause.[82] This contemporary reporting was subsequently reinforced by accounts released after the fighting was over, including a spate of books, documentaries, and at least one novel written by journalists who had witnessed the fighting firsthand.[83] The Ukrainian state has subsequently sought to control this process, seeking to direct and appropriate the image of the cyborg as a national symbol. In 2017, a feature film lionizing the defense of the airport was released, funded by the Ukrainian government, entitled *Cyborgs: Heroes Never Die*.[84] The Ukrainian government has also designated January 16 as "Cyborg Commemoration Day," and the Ukrainian postal service released a set of commemorative postage stamps to coincide with the fifth anniversary of the fall of Donetsk Airport, emblazoned with the motto "they withstood, the concrete didn't," as military bands played requiem concerts in airport terminals across the country.[85] Veterans organizations have likewise been active in curating the image of the cyborg. In 2017, for example, an Apple promotional video describing a Russian swimmer with a prosthetic leg as a "cyborg" drew particular ire.[86] Yet, the relationship between the Ukrainian government and veterans groups has not been universally harmonious, and these parallel efforts to shape the identity and legacy of the cyborg reflect what Ilmari Käihkö has termed "a nation-in-the-making during a state-in-breaking."[87]

Critically, while the airport's cyborg symbolism is undoubtedly rooted in the intensity of the fighting there, this narrative alone cannot account for that intensity. The airport's symbolic meaning does not explain the escalation of the fighting from the first battle to the second and can only partially account for its perpetuation during the second. Despite the political capital invested in the legend of the cyborg defender, this narrative actively survived the airport's fall; if anything, the Ukrainian defeat at Donetsk Airport has actively reinforced the tragic heroism of the cyborg image. Moreover, this symbolic Ukrainian narrative cannot explain the separatists' repeated efforts to storm the airport complex. Indeed, without both high levels of Ukrainian *and* separatist resolve, the battle could not have assumed the intensity inherent in the image of the cyborg defender. In order to understand why the airport became so fiercely contested, we must instead look to the political imperatives behind each side's military strategy in the Donbass. If the airport's totemic status was a product of the intense fighting rather than its cause, what motivated the DPR's repeated assaults, and what alternate strategic rationales explain continued Ukrainian resistance?

Patriots and Proxies: Politics and Strategy in the Donbass

Although the conflict in the Donbass was a logical extension of Russia's annexation of Crimea, Russian aims and commitment there were very different. Russia's annexation of Crimea was intended to secure its strategic Sevastopol naval base, preempting an expected shift in Ukrainian policy in the aftermath of the Euromaidan revolution. In contrast, control of the Donbass was nowhere near as vital to Russian national interests, and it is debatable whether Russia ever seriously expected to annex the region. Indeed, Russian support for separatism in the Donbass was largely opportunistic. In the words of one former Russian general staff officer, "had the Ukrainian's fought for Crimea, we would not now be fighting in the Donbas."[88] The region certainly offered tangible benefits for Russia. Donetsk and Luhansk represented some of Ukraine's richest industrial provinces, while control of the Black Sea coastline could have provided the first step in a land corridor from Russia to Crimea. In its absence, Russia has been forced to construct an expensive bridge across the Kerch Strait in order to resupply the peninsula. Equally, though, Moscow continues to view Ukraine as a buffer with the West, and seeks to maintain influence over Ukrainian policy. Traditionally, this had been accomplished through economic levers, as with the post-Soviet provision of natural gas discounts in return for Russian basing rights in Crimea.[89] However, Russia's actions in Crimea hardened attitudes in Kiev, limiting the utility of such an approach, while Ukrainian resistance made outright annexation more difficult. Consequently, insurgency provided a new source of potential leverage, enabling Russia to hold Ukrainian policy to ransom by stoking or subduing the violence—but it was a leverage that relied on the threat of succession rather than its actual accomplishment.[90]

For the separatists, in contrast, the conflict was essentially zero-sum. Donetsk has historically displayed a greater concentration of ethnic Russians than elsewhere in Ukraine, and Russian remains the predominant first language. Although ethnicity is a crude indicator of identity, the gradual ascendancy of Ukrainian as the country's official language after the breakup of the Soviet Union helped to isolate and marginalize many Russian-speakers in the Donbass, who concomitantly struggled to gain access to state jobs and services. However, these Russophile tendencies were assuaged by the Yanukovych regime, which supported the adoption of Russian as a second official language and sought to balance Westernization with cordial relations with Russia. Yanukovych himself came from the Donbass, where he had previously served as governor of Donetsk oblast, and his Party of the Regions drew significant electoral support from the region. Yanukovych's ouster thus raised concerns about the future status of Donetsk's Russian majority, itself a minority at the national level, while simultaneously removing access to important sources of informal government largesse. With politics in Kiev shifting toward a more Ukrainian nationalist agenda, public sentiment in the Donbass began to coalesce around an alternate sense of Russian cultural identity.[91]

Consequently, the separatist leadership aspired either to full annexation by Russia or else the formation of a Russian-backed breakaway enclave. Early in the rebellion, the two breakaway "People's Republics" declared an intent to confederate as Novorossiya, or "New Russia," appropriating the historic name for the area when first incorporated into the Russian Empire in the mid-eighteenth century.[92] DPR militias made extensive use of the Novorossiyan saltire as a combat identification symbol, alongside Russian tricolors and the black-and-orange Ribbon of Saint George, another traditional Russian military emblem. Some units even styled themselves as "Cossacks," replete with furry papakha hats. Moreover, many separatist militia commanders—in addition to their strong Russian ties—were political nonentities prior to the conflict and owed their position and notoriety to the separatist cause. For example, the DPR's self-declared defense minister and head of its Slovyansk militia, Igor Girkin (aka, Strelkov or "gunman"), was both an avid proponent of Novorossiya and a former Russian colonel. Similarly, the DPR's Vostok Battalion was led by Alexander Khodakovsky, a former Ukrainian major who had defected from Yanukovych's "Alfa" counterterrorism unit, while the Sparta Battalion was commanded by Arsen "Motorola" Pavlov, a Russian national who had spent some time as a signaller in the Russian marine infantry.[93] The campaign at Donetsk Airport itself appears to have been partly directed by Mikhail Tolstykh, call sign "Givi," a Ukrainian of Georgian descent who used the battle to cultivate his media image as a warlord.[94]

However, if the separatists' goals differed from Russian policy in important ways, they nonetheless became increasingly reliant on Russian military aid as the Ukrainian Anti-Terrorist Organization gathered pace. Indeed, the DPR itself was less a breakaway statelet than a fragile coalition of independent militias sharing a somewhat similar political ideology. By late July, for example, the Ukrainian Army was able to conduct a major sally into separatist territory, punching through DPR positions near Debaltseve before turning north to relieve besieged Ukrainian troops at Luhansk Airport.[95] In response, Russia was forced to commit increasing numbers of its own troops to prevent separatist collapse; first as advisers (or "vacationers") and then as formed units, shifting from what Tetyana Malyarenko and David Galbreath have described as "nomadic" or "creeping" occupation by proxy to overt military "consolidation."[96] In all, Russia committed at least six reconnaissance groups and ten battalion tactical groups to the Donbass during the summer of 2014, supported by indirect fire from the 90,000-odd Russian troops massed along Ukraine's border. However, this support remained bounded by Russia's strategic ambitions in the Donbass, as well as the reciprocal pressures generated by the conflict on Russia. Russian battalion tactical groups, for example, were composite formations drawn from a parent regular army brigade, and sustaining their deployment at scale placed a significant strain on the Russian military. Equally, the rising number of Russian casualties began to draw unwelcome domestic criticism, exacerbated by the effect of international sanctions on Russia's economy.[97]

This underlying disagreement between Russian and separatist aims was exposed by the Minsk I cease-fire protocol, agreed to by Russia and Ukraine in early September. Brokered by European nations eager to prevent a wider escalation of the conflict, Minsk I provided for an immediate cease-fire and the tacit withdrawal of Russian troops, together with the longer-term reintegration of the separatist regions into a federalized Ukraine under a new semiautonomous special status. Consequently, Minsk I essentially recognized the key policy goals Moscow had sought to protect through force of arms that summer and represented a positive development for Russia. For Kiev, the idea of ceding authority and legitimacy to Russian-backed separatists in the Donbass was less than palatable, but the agreement did offer the Ukrainian military some much-needed respite. Critically, the agreement was not acceptable for the separatists, who still aspired to a Russian protectorate rather than devolution within Ukraine and had to be coerced into acquiescence by Russia.[98] Worse still, Russian-formed units were withdrawn from the front line shortly after the agreement, representing a further nail in the coffin of the separatists' Novorossiyan project.[99] Thus, while Russia may have doubted the Ukrainian government's ability to make good on the political vision put forward in the Minsk I protocol, the tenets of that agreement—and the cease-fire which these provided for—were viewed very differently in Donetsk than in Moscow.

The DPR appears to have initiated the second battle for Donetsk Airport in direct response to the Minsk I protocol, deliberately undermining the cease-fire in an effort to force further Russian intervention. Fighting at the airport had largely subsided during the summer of 2014, as Russian forces focused on checking Ukrainian advances further east. However, conditions for residents in Donetsk city had continued to deteriorate, undermining the credibility of separatist claims to authority. Fighting elsewhere regularly damaged pumping stations, interrupting the city's water supply and leading to sporadic water rationing in June and bucket-chain queues for water tankers by August.[100] Artillery exchanges also began to cause significant collateral damage, exacerbated by the DPR's frequent siting of artillery in residential areas, drawing government counterbattery fire, and the use of inaccurate rocket artillery to target the airport, falling short in adjacent suburbs.[101] On August 7, for example, a hospital and residential high-rise were both struck by government shellfire, apparently intended for the nearby DPR-controlled Security Service of Ukraine (SBU) building. Two days later, part of the Donetsk Prison was destroyed by shelling, precipitating the escape of some prisoners.[102] Against this backdrop, the Minsk I agreement transformed the government presence at the airport from an irritant into an acid test for the separatist cause, providing a means for the DPR to demonstrate its own agency by undermining the Minsk cease-fire while attempting to compel further Russian intervention. Indeed, these political imperatives are apparent in the ebb and flow of the battle. In November, for example, the DPR offensive halted briefly while separatist elections were conducted, in contravention of the Minsk protocol. The DPR likewise exploited every opportunity to claim victory in the media, even prematurely hoisting flags over the New Terminal while the Ukrainian Army still held out below.[103]

For Ukraine, meanwhile, continued resistance at the airport served a series of similarly political functions. Although Ukraine maintained 130,000 troops on paper at the start of the conflict, these were significantly unprepared to mount a major military campaign. Much of the force was composed of undertrained and underequipped cadres, with deployable capability limited to a joint rapid reaction force of just 24,000 troops.[104] Consequently, Ukraine relied heavily on volunteer militias early in the fighting, variously raised by political organizations, local authorities, and regional oligarchs. Many owed little direct loyalty to the Ukrainian state and had to be later coaxed into the auspices of the army or national guard in exchange for artillery and armored support during the summer of 2014.[105] However, the official Ukrainian military continued to struggle with recruitment and retention, notwithstanding the reintroduction of conscription. Thus, while the Minsk cease-fire bought Ukraine time to consolidate, continued fighting at Donetsk Airport also provided a useful rallying cry. In the words of one Ukrainian militiaman, "the future of our country depends on whether we will be able to hold on to this airport or not. . . . That is why I am here."[106] The battle likewise served as an outlet for the more truculent militia groups such as Right Sector, which had refused to affiliate with either the army or national guard but agreed to cooperate at the airport. Its leader was later made an adviser to Ukraine's chief of the general staff, having himself been wounded at the airport, in an apparent effort to incorporate the group.[107] Moreover, as the defender, Ukraine could reasonably claim to be upholding the spirit of the Minsk cease-fire while simultaneously challenging the legitimacy of its political settlement, in a bid to elicit greater Western military support. As an article entitled "The Death of Ukraine's Cyborg Army" in the US magazine *Foreign Policy* asked, "If the United States is really supporting Ukraine, as President Obama claims, then why are Kiev's forces getting hammered?"[108]

Importantly, Russia's response to wayward separatist belligerence at Donetsk Airport only served to exacerbate the attritional character of the fighting, as Russia sought to exert greater control over the separatists. Russia continued to provide a degree of military support to the DPR but limited direct militia access to battle-winning systems, including heavy weapons and artillery—a process already underway since the downing of Malaysian Airlines Flight MH17.[109] Russia also acted to remove separatist leaders who refused to tow its line. Strelkov was removed from office in late 2014, having become increasingly critical of Russian policy. Moreover, a series of militia commanders were mysteriously assassinated shortly after the airport fell, initially in Luhansk, but also subsequently in the DPR. Although officially blamed on separatist infighting or Ukrainian special forces, the fact that a number of separatist leaders or their families were killed while in Russia lends credence to the idea that these assassinations represented the culmination of Russian efforts to control proxy behavior in the Donbass.[110] Russia also actively contributed to the Joint Centre for Control and Coordination (JCCC), a bilateral contact group established to negotiate the implementation of the Minsk agreements, which facilitated the enactment of a local cease-fire at the airport in December 2014, enabling the rotation of Ukrainian troops in the terminals.

Yet, as time went on, Russia also increased its covert military assistance to the DPR at the airport, in an effort to pressure Ukraine into compliance with the Minsk settlement. Indeed, Ukrainian officers have largely attributed their defeat there to Russian intervention. Toward the end of the battle, the defenders perceived an influx of DPR "advisers" replete with Russian accents and military terminology.[111] As one defender observed, "the separatists changed their tactics at Donetsk Airport": "If, at the beginning, they had 'stupidly stormed,' then all that changed. They concentrated a lot of artillery. They supressed our artillery before attacks, and then stormed. It felt like a regularly planned attack."[112]

Immediately after the fall of the airport, Russia launched a major offensive to capture Debaltseve, a strategic road and rail intersection midway between Donetsk and Luhansk, leading one US military analyst to describe Donetsk Airport "as the sinew between Russia's summer and winter offensives."[113] This fighting produced a further cease-fire agreement known as Minsk II, which obliged Ukraine to enshrine the Donbass's special autonomous status in the Ukrainian constitution, though with no more political will than before. Indeed, as Laurence Freedman has argued, Russia's salami-slicing approach has ultimately proved self-defeating, retrenching Ukrainian public opinion against reintegration of the Donbass while perpetuating separatist sentiment.[114] Certainly, fighting around the airport continued largely unabated, with the Ukrainian infantry company manning trenches in Pisky expending an average of 20,000 rounds of ammunition a week during the summer of 2015.[115] Consequently, Russia has sought to maintain a mutually hurting stalemate. It continues to provide layered antiair cover to the separatists, preventing the kind of Ukrainian overmatch seen during the initial fighting at Donetsk Airport, while simultaneously denying the rebels the ability to defeat the Ukrainian military outright. By early 2017, a visiting (retired) US general described the fighting as "World War One with technology"; when a Ukrainian commander was asked what US support he needed, he purportedly replied "concrete."[116]

Conclusion

At Donetsk Airport, the creation of symbolic meaning went hand-in-hand with intense, costly, and extremely brutal urban fighting. The airport's totemic status was manifested in the narrative of the "cyborg" Ukrainian defender, holding out in the ruins of the shattered terminals against the odds. In fact, the very concept of cyborg-like Ukrainian resilience at Donetsk Airport was inherently connected to the ferocity of the fighting rather than the fabric of the airport itself, and the visceral destruction wrought by the conflict on the terminal buildings and the minds and bodies of the individual soldiers defending them only added to the potency of this symbolism. However, although the symbolism of the airport was intrinsically tied to the costs of its defense, this cannot account for the development of the fighting there. Moreover, the airport's totemic status does not seem to have been universally accepted as a justification for maintaining the defense in the eyes of many defenders; nor can it explain the separatists' repeated attempts to seize the complex. Instead, the perpetuation and escalation of the fighting for control of Donetsk Airport was a product of strategic context and, in particular, the principal-agent politics of proxy conflict and internationalized civil war.

Here, the unsatisfactory nature of the Minsk I settlement for both Ukraine and the DPR transformed the airport from a minor military objective into a strategic political tool. For the separatists, capturing the airport became a means to undermine the Minsk cease-fire, reassert their vision for the Donbass over and above the policy goals of their Russian patron, and compel Moscow into providing further military and political support. For Ukraine, its continued defense garnered much-needed international support for continued struggle without sacrificing the political high-ground, while simultaneously providing a focal point for domestic mobilization and military consolidation. Importantly, the drawn-out character of the second battle for Donetsk Airport reflected not only the belligerents' strength of will—both viewing the conflict (if not the battle) as existential—but also their evenly matched capabilities. Indeed, while Russian intervention ultimately facilitated separatist victory at Donetsk Airport, the Russian response to its proxy agent's divergent goals in the Donbass generally served to prolong rather than hasten the conflict, ensuring neither side had the ability to decisively overmatch the other.

The construction of urban symbolism at Donetsk Airport thus raises a number of important implications for future urban conflict. First, the symbolic meanings that frequently emerge from major urban battles are not intrinsic, but the product of prevailing political or strategic imperatives *and* high tactical costs. These costs are themselves shaped both by tactical capability and by the intersection of belligerents' will, such that the two ingredients of urban symbolism are themselves directly related. Second, this relationship calls into question our ability to prevent such battles in the future. If urban symbolism emerges from the interaction between the will and capability of belligerents in context, as the experience of Donetsk Airport suggests, then improvements in tactical practice alone are unlikely to prevent a repeat performance. Indeed, because military capability is a subjective, situationally defined characteristic, determined by relative strength compared to a given opponent and each party's aims, then urban symbolism is likely to emerge whenever the will to fight outstrips conventional tactical overmatch. Yet, despite the advent of precision, "special forcification," and similar advances in tactical urban praxis, it remains possible to envisage future scenarios in which Western forces will be pitted against urban enemies with similar or matched capabilities—and the will to use them. In fact, as Western military capability becomes ever more concentrated in smaller and more exquisite armies, and as cities become bigger and more important, such a prospect seems increasingly likely. When such a situation does occur, urban fighting is very likely to assume the same symbolic importance of a Stalingrad or Donetsk Airport.

Finally, the centrality of principal-agent politics to the escalation and continuation of fighting at Donetsk Airport stands as a salutary lesson in the potential pitfalls of proxy warfare. Both Moscow and Kiev made use of various surrogates and militias at Donetsk Airport, and both exploited the fighting there as an opportunity to exert greater control over these groups. Yet, despite these efforts, both principals appear to have struggled to maintain complete control over their local agents—and with them, the pace of events on the ground. That agents can and do exercise their own independent agency, potentially frustrating (or at the least complicating) their patron's agendas, should come as no surprise. It does, however, provide a cautionary corrective to the notion that proxies and auxiliaries can be relied upon to augment their patron's lack of military mass in future urban operations without extracting their own costs in turn.

Notes

1. Fox, "Little Stalingrad"; Beckhusen, "Airport Siege 'Stalingrad.'"
2. Posen, "Urban Operations," 157.
3. Betz and Stanford-Tuck, "City Is Neutral."
4. Stryzhova, "Battle for Luhansk Airport"; Shramovych, "Ukraine's Deadliest Day."
5. Harris et al., "Megacities and the United States Army."
6. Tzu, *Art of War*, 78; Colson, *Napoleon on War*, 333.
7. Duffy, *Siege Warfare*; Duffy, *Fortress in the Age of Vauban*.
8. King, *Combat Soldier*, 237–238.
9. Helmus and Glenn, *Steeling the Mind*, 39–55.
10. King, *Urban Warfare*; King, *Combat Soldier*, 237–238; Smith, *Utility of Force*; Evans, "Lethal Genes," 515–552.
11. See British Army Review, *Special Report*, vol. I; British Army Review, *Special Report,* vol. II.
12. Hills, *Future War in Cities*, 30, 54.
13. Neiman, "Urban Operations," 143.
14. Daly, "Siege, Sack and Violence," 163.
15. Hills, *Future War in Cities*, 220; Mosul Study Group, *What the Battle for Mosul Teaches the Force*.
16. Kilcullen, *Out of the Mountains*; Rosenau, "Lessons of Modern Urban Warfare," 371–394.
17. King, *Combat Soldier*, 237–241, 305–322; King, "Close Quarters Battle," 276–300; Ben-Ari et al., *Rethinking the Sociology of Combat*.
18. Horne, *Price of Glory*, 1–2.

19. Jukes, *Hitler's Stalingrad Decisions*, 5, 19; see also Garner, *Myth of Stalingrad*.
20. Mahečić, "Case of Dubrovnik," 27–41; Forde, "Bridge on the Neretva," 467–483.
21. Neiman, "Urban Operations," 143.
22. Clout, "Place Annihilation," 165–180.
23. Posen, "Urban Operations," 157.
24. Betz and Stanford-Tuck, "City Is Neutral."
25. Ibid.; Glenn, *Combat in Hell*, 5.
26. See Jenkings et al., "Wootton Bassett," 356–363.
27. On the politics of meanings generated by the costs of war, see Carden-Coyne, *Politics of Wounds*; Mosse, *Fallen Soldiers*; Fussell, *Great War and Modern Memory*.
28. Spencer, "City Is Not Neutral."
29. Weinberg, "Stalingrad and Berlin," 16; Posen, "Urban Operations," 157.
30. Grytsenko, "Cyborgs"; Shevchenko, "Ukraine Conflict"; "Battles Rage in Donetsk," *BBC News*.
31. "Dispatch 95," *Vice News* (emphasis added). Footage also emerged of the same individual abusing Ukrainian prisoners of war. See Kates and Removska, "Disturbing Videos."
32. "Ukraine's Coveted Prize," *BBC News*.
33. Ibid.; "Dispatch 95," *Vice News*.
34. Posen, "Urban Operations," 158.
35. Grytsenko, "Cyborgs"; Galeotti, *Russia's War in Ukraine*, 54–56; Goncharova, "1,373 Soldiers Killed"; Kovalenko and Titish, "Moysyuk About Airport"; Fox, "Little Stalingrad," 7.
36. Matveeva, "No Moscow Stooges", 33–34; Galeotti, *Russia's War in Ukraine*, 23.
37. Carroll, "Bloody Battle."
38. Kovalenko and Titish, "Moysyuk About Airport"; Carroll, "Trapped in Donetsk Airport."
39. Berman and Lake, *Proxy Wars*.
40. Marshall, "Civil War to Proxy War," 183–195; Brown, "War by Proxy," 243–257; Hughes, "Perils of Proxy Warfare," 522–538.
41. Malyarenko and Galbreath, "Paramilitary Motivation in Ukraine," 113–138; Rauta, "Towards a Typology of Non-State Actors," 1–20.
42. Neukirch, "Operational Challenges and New Horizons," 183–197; Neukirch, "OSCE Conflict Management in Ukraine," 229–239.
43. While Ukraine has improved in the World Press Freedom Index, moving from 127 to 102 between 2014 and 2019, Russia has declined from 148 to 149. Reporters Without Borders, "2014 World Press Freedom Index"; Reporters Without Borders, "2019 World Press Freedom Index."
44. "Ukraine Profile," *BBC News*; "Turchynov Announces Anti-Terror Operation," *BBC News*.
45. Shakirov, "Interview."
46. Ibid.; Walker, "Fighting Leaves Dozens Dead"; Salem, "Let Them Eat Cake"; OSCE, "Information Received Until 26 May."
47. Shakirov, "Interview."
48. Langston, "Dispatch 44"; Shakirov, "Interview."
49. OSCE, "Information Received until 18:00 hrs, 27 May."
50. OSCE, "Information Received until 18:00 hrs, 4 June."
51. Thorp and Gamio, "Cease-Fire War in Donetsk"; Associated Press, "Attempt to Seize Donetsk Airport."
52. "Heavy Fighting for Donetsk Airport," *BBC News*; Grytsenko, "Cyborgs."

53. OSCE, "Spot Report . . . 7 September 2014."
54. Loiko, "Ukraine Fighters."
55. Zverev, "Artillery Explosions."
56. "New Battle Rages at Donetsk Airport," *BBC News*; "Defenders to Leave Old Terminal," *Ukrainian Mirror Weekly*.
57. OSCE, "Information Received as of 12 December"; OSCE, "Spot Report . . . 16 December 2014"; OSCE, "Information Received as of 18:00 (Kyiv time), 21 December 2014"; OSCE, "Information Received as of 18:00 (Kyiv time) 26 December"; OSCE, "Information Received as of 18:00 (Kyiv time), 30 December"; OSCE, "Information Received as of 18:00 (Kyiv time), 2 January 2015."
58. OSCE, "Information Received as of 18:00 (Kyiv time), 9 January 2015"; Harress, "Pro-Russian Forces."
59. Loiko, "Outgunned 'Cyborgs.'"
60. Ibid.; Kovalenko and Titish, "Moysyuk About Airport."
61. Carroll, "Bloody Battle"; Kudrytski, "Ukraine Forces Retake Most of Donetsk Airport."
62. Kovalenko and Titish, "Moysyuk About Airport"; Carroll, "Trapped in Donetsk Airport"; Carroll, "Bloody Battle"; OSCE, "Information Received as of 18:00 (Kyiv time), 20 January 2015"; Goldsmith, "Were Chemical Weapons Used."
63. Carroll, "Bloody Battle"; Carroll, "Trapped in Donetsk Airport"; Kovalenko and Titish, "Moysyuk About Airport."
64. Carroll, "Bloody Battle"; Loiko, "Outgunned 'Cyborgs.'"
65. Carroll, "Trapped in Donetsk Airport"; Walker and Grysenko, "Ukraine Forces Admit Loss."
66. Ostrovsky, "Dispatch 43."
67. Shakirov, "Interview."
68. "Military Plane Shot Down in Luhansk," *BBC News*.
69. OSCE, "Information Received as of 18:00 (Kyiv time), 25 August 2014"; OSCE, "Information Received as of 18:00 (Kyiv time), 27 August 2014"; OSCE, "Information Received as of 18:00 (Kyiv time), 22 August 2014"; Ostrovsky, "Dispatch 84."
70. Loiko, "Ukraine Fighters."
71. Ibid.; Shevchenko, "Ukraine Conflict"; Fox, "Little Stalingrad," 7.
72. Shevchenko, "Ukraine Conflict."
73. Carroll, "Bloody Battle."
74. Loiko, "Ukraine Fighters."
75. Radio Free Europe/Radio Liberty, "Nine Days Defending Donetsk Airport."
76. Loiko, "Outgunned 'Cyborgs.'"
77. Ousby, *Road to Verdun*.
78. Loiko, "Outgunned 'Cyborgs.'"
79. Ibid.
80. Shevchenko, "Ukraine Conflict."
81. Langston, "Dispatch 91."
82. See, for example, "Ukraine Live," *Interpretermag*.
83. Tulaganova, "Airport Donetsk"; Shtohrin and Loiko, *AD 424*.
84. Miller, "Art of War."
85. Ministry of Defence of Ukraine, "'Cyborgs' Commemoration Day"; Racurs, "Ukrainian Post Service"; "International Airport," *Dnepropetrovsk International Airport*.
86. Kupfer, "Ukrainian Veterans."
87. Käihkö, "Nation-in-the-Making," 147; Puglisi, "General Zhukov and the Cyborgs," 1–22.

88. Galeotti, *Russia's War in Ukraine*, 12.

89. Biersack and O'Lear, "Geopolitics of Russia's Annexation"; Sherr, "Russia-Ukraine *Rapprochement*?" 33–50.

90. Charap and Darden, "Russia and Ukraine," 7–14; Freedman, "Crisis Management," 7–42; Bowen, "Coercive Diplomacy and the Donbas," 312–334.

91. Sotiriou, "The Irreversibility of History," 51–70; Matveeva, "No Moscow Stooges," 25–50; Loshkariov and Sushentsov, "Radicalization of Russians in Ukraine," 71–90.

92. Matveeva, "No Moscow Stooges," 37.

93. Galeotti, *Russia's War in Ukraine*, 20–24; Malyarenko and Galbreath, "Paramilitary Motivation in Ukraine," 126–128; Matveeva, "No Moscow Stooges," 33–34.

94. See Anon, "One Day with Givi."

95. Ministry of Defence of Ukraine, "Raid Documentary"; Kovalenko and Titish, "Moysyuk About Airport"; Fox and Rossow, "Making Sense of Russian Hybrid Warfare," 11.

96. Malyarenko and Galbreath, "Paramilitary Motivation in Ukraine," 126.

97. Sutyagin, "Russian Forces in Ukraine," 1–3; Bowen, "Coercive Diplomacy and the Donbas," 331–333; Freedman, "Limited War," 15–17.

98. International Crisis Group, "Rebels Without a Cause"; Wittke, "The Minsk Agreements," 264–290.

99. Sutyagin, "Russian Forces in Ukraine," 6–7.

100. OSCE, "Spot Report . . . 15 June 2014"; OSCE, "Information Received until 18:00 hrs, 19 June"; OSCE, "Information Received as of 18:00 (Kyiv time), 19 August 2014"; OSCE, "Information Received as of 18:00 (Kyiv time), 20 August 2014"; OSCE, "Information Received as of 18:00 (Kyiv time), 22 August 2014."

101. OSCE, "Spot Report . . . SMM Observes Civilian Fatalities"; Walker, "Ukrainian Forces and Pro-Russia Rebels Clash."

102. OSCE, "Latest Report . . . on Alleged Shelling of Donetsk Hospital"; OSCE, "Information Received as of 18:00 (Kyiv time), 12 August 2014."

103. OSCE, "Information Received as of 18:00 (Kyiv time), 6 November 2014"; Associated Press, "Russia-Backed Separatists Seize Donetsk Airport"; Carroll, "Residents Flee Savagery."

104. Sanders, "The War We Want," 33–36; Klein, "Ukraine's Volunteer Battalions," 1.

105. Malyarenko and Galbreath, "Paramilitary Motivation in Ukraine," 119–121; Karagiannis, "Ukrainian Volunteer Fighters," 139–147; Käihkö, "A Nation-in-the-Making," 147–156.

106. Loiko, "Ukraine Fighters"; Prykhodko et al., "Servicemen's Motivation," 347–366; see also, Zinets, "Ukraine Struggles to Recruit."

107. Ministry of Defence of Ukraine, "Dmytro Yarosh."

108. Miller and Vaux, "Death of Ukraine's Cyborg Army."

109. Galeotti, *Russia's War in Ukraine*, passim; Ferguson and Jenzen-Jones, "Raising Red Flags."

110. Galeotti, *Russia's War in Ukraine*, 24; "Givi Dies in Rocket Attack," *BBC News*.

111. Loiko, "Outgunned 'Cyborgs.'"

112. Kovalenko and Titish, "Moysyuk About Airport."

113. Fox, "Little Stalingrad," v.

114. Freedman, "Limited War," 18–20.

115. Gibbons-Neff, "Point 18 in Eastern Ukraine."

116. Hix, "Future Combined Arms Operations."

5

Operational Advantages and Limitations

Louise A. Tumchewics

FOR FIVE MONTHS IN 2017, A FIERCE URBAN BATTLE RAGED IN MARAWI, Philippines, the largest military operation in that country since World War II.[1] At the time, however, events in Marawi were largely overshadowed as the global media spotlight shone on US and Iraqi security forces efforts to liberate the northern Iraqi city of Mosul from Islamic State (IS) occupation. Yet, like its Iraqi counterpart, the Battle of Marawi offers many pertinent lessons for contemporary armies contemplating the challenges of urban operations, and particularly counterinsurgency operations in urban centers.

The Battle of Marawi demonstrates how a city can offer haven for well-resourced and determined defenders. Marawi demonstrates the challenge of thwarting a violent extremist organization while minimizing casualties, both military and civilian, and the need to honor human rights and provide for civilian safety and welfare, both during the battle and in its aftermath. The battle is also an instructive example of the centrality of influence operations to the urban battlefield, and how control of the narrative and understanding of the social terrain are as vital to success as controlling specific ground.

The struggle for Marawi was documented at the time by major international news outlets, including the *Washington Post*, the *New York Times*, BBC, Al Jazeera, and National Public Radio, as well as Southeast Asian regional publications such as *The Diplomat* and the *South China Morning Post*. Philippines-based digital media company Rappler, noted for its critical reporting of the administration of Rodrigo Duterte, covered the battle extensively. Such media reporting provides a detailed chronology of events in Marawi.

Further analysis, with a more academic focus, can be found in the pages of *Foreign Affairs* and *Foreign Policy*. The United States and Australia sent troops in training and advisory capacities to assist their Filipino counterparts, and practitioner reflections from Australian Defence Force members on those experiences provide insight into the tactical lessons learned over the course of the battle. In addition, human rights organizations such as Amnesty International and International Crisis Group have chronicled the human rights failures of both insurgent forces and the government and armed forces of the Philippines throughout the battle and the subsequent reconstruction.

Southeast Asian terrorism and security scholars Rohan Gunaratna, Joseph Franco, Jasminder Singh, and Muhammad Haziq Jani have chronicled the Battle of Marawi and the rise of violent extremist organizations in the Philippines in reports for regional security think tanks. In 2020, Rommel C. Banlaoi and fellow scholars produced an anthology entitled *The Marawi Siege and Its Aftermath: The Continuing Terrorist Threat*, which explores various aspects of insurgency and violence in Mindanao. Australian academics Charles Knight and Katja Theodorakis have produced the sole study of information operations in Marawi, as well as additional analyses of the battle, and their articles provide instructive analysis on the role of information and influence operations in urban operations. The urban terrorism in Marawi is the subject of a Master of Military Art and Sciences thesis by Major Ervin C. Divinagracia of the AFP. The Modern War Institute at West Point has also highlighted the Battle of Marawi, and relevant lessons learned, in its podcasts and articles.

In this chapter, I build on the information provided in these publications to trace the long running history of the violence in Mindanao, of which Marawi was the latest, and largest, episode before shifting to the battle itself. I discuss how the defenders were able to present such a strong resistance over so many months and how the urban landscape enabled a strong defense, along with the response of the Philippine government and Armed Forces of the Philippines (AFP), the use of influence operations by both sides, and the effect of widespread destruction on the peace and reconciliation process.

Marawi Geography and History

The city of Marawi is situated on the shores of Lake Lanao, in the northwest of the island of Mindanao, the most southerly island of the Philippine archipelago. The capital of Lanao del Sur Province, one of five provinces in the Autonomous Region of Muslim Mindanao (ARMM), Marawi was the educational and economic center of the province, an urban center in a province dominated by dense jungle.[2]

Spanish explorers first landed on Mindanao in 1543. By this time, Islam was well established among the local population, as Taurig preachers from the Malay peninsula had brought their faith to the island in the previous century. Although Roman Catholicism was adopted by some on Mindanao, a significant population retained their Islamic faith. The city of Marawi was established as Dansalan in 1639 by Francis Atenza, over a century after the discovery of the island. The conquistador's stay was short-lived, however, as he and his compatriots were forced to retreat from their new city by Moro warriors.[3]

Clashing identities and religions lay at the heart of conflict that plagued Mindanao for centuries. Colonizing Spanish forces arrived in Southeast Asia with a heritage of battling the Muslim Moors in Spain. They discovered the co-religionists of their long-time opponent in the new South Pacific colony of the Philippines and attempted to eliminate them through force or conversion. However, fierce Moro defense, and the ineffective governance of the Spanish, meant Mindanao remained largely unconquered and unconverted, developing an identity apart from the rest of the Philippines. The Spanish animosity toward the Muslim population was inherited by the Catholic majority of the Philippines, and the Muslim minority reciprocated the hostility, creating a culture of intercommunal unease and tension that persisted from generation to generation.[4]

A lone Muslim enclave in the overwhelmingly Catholic archipelago, Mindanao resisted Spanish rule for centuries, remaining geographically, ethnically, and religiously distinct from the rest of the Philippines. When Spanish control ceded to US authority in the late nineteenth century, US forces succeeded in subduing Muslim opposition. During World War II, Moro fighters fought an insurgency against occupying Japanese forces. This long heritage of resistance shaped a distinct Moro identity.[5]

Mindanao was governed separately from the rest of the Philippines until 1946.[6] Tensions persisted long after the two entities were formally joined, with the Muslim population continuing to view itself as separate and distinct from the rest of the Philippines. Government-sponsored Christian migration into traditionally Muslim areas of Mindanao fueled separatist sentiment, as well as the alleged killing of Moro army recruits. These feelings swelled into open conflict between militant Muslim separatists and government forces in 1970.[7] The government of Ferdinand Marcos imposed martial law on Mindanao, enforced by the AFP in a harsh and repressive manner.[8]

The Muslim separatist cause gained the support of Muammar Gaddafi's Libya, and Libyan sponsorship fostered the growth of the Muslim National Liberation Front (MNLF). The MNLF claimed its political ideology of egalitarianism, rather than religious motivation. After several years, in which thousands of people were killed or displaced, conflict between MNLF fighters and government forces ground into a bloody stalemate, and a cease-fire was negotiated, with the assistance of Libya, in 1976.[9]

Within two years, however, conflict reignited, and a new iteration of separatist violence began. In 1978, disenchanted former MNLF members regrouped and formed the Moro Islamic Liberation Front (MILF), under the leadership of the Egyptian-born cleric Hashim Salamat. Salamat's Islamism, combined with a clear goal of establishing an independent Islamic state in Mindanao, proved attractive, recruiting over 12,000 members, mostly young men from Mindanao.[10] Throughout the 1980s, MNLF and MILF fighters engaged in cycles of conflict and cease-fires against government forces.

Violence was not limited to Moro independence organizations. The New People's Army (NPA), the armed wing of the Communist Party of the Philippines, maintained armed opposition to the AFP, waging a guerrilla war in seventy-three of the country's eighty-one provinces, including Mindanao.[11] In addition to the separatist movements, clan-based violence, known as *rido*, compounded the instability of Mindanao. As Wilfredo Torres writes, these disputes often arose over political rivalries and land ownership, and were influenced by notions of family honour.[12]

In 1990, a Filipino militant, Abdulrajah Janjalani, established a new, more radical Muslim separatist group, Abu Sayyaf, based on the islands of Vole and Basilan, to the west of Mindanao. Once a MNLF member, Janjalani had become radicalized during his education abroad. Abu Sayyaf initially gained notoriety with grenade attacks on Christian targets and similar acts of violent terrorism. Later the group shifted to kidnapping for ransom, which proved to be a far more profitable enterprise. Profit from nefarious activity surpassed the appeal of religious ideals, and members of the criminal underground moved into key leadership positions within the organization. After Janjalani's death in 1998, Abu Sayyaf devolved into a collection of underworld organizations aligned out of convenience rather than conviction, operating throughout the Philippines.[13]

By the late 1990s, the character of the mainstream Moro independence movements had changed. In 1996, the MNLF signed a landmark peace agreement with the Philippines that saw the creation of ARMM, an area comprising the five predominantly Muslim provinces of the Mindanao island group: Basilan (except Isabela City), Lanao del Sur, Maguindanao, Sulu, and Tawi-Tawi. Under the terms of the ARMM, the predominantly Muslim population of the region enjoyed a measure of self-rule. The MILF signed a peace deal with Manila in 2001, though violence from MILF splinter groups continued in the following decades.

In 2014, a faction of Abu Sayyaf, led by Isnilon Hapilon, pledged allegiance to Abu Bakr al-Baghdadi, the leader of the IS. Hapilon, already on the FBI's Most Wanted Terrorist List with a $5 million bounty on his head, established himself as the emir of Southeast Asia. The alliance between IS and Abu Sayyaf had benefits for both parties. Aligning with IS gave Abu Sayyaf greater moral credibility and a coherent narrative. For IS, having an ally in Southeast Asia increased its attractiveness to a global audience of potential recruits. Moreover, Abu Sayyaf Group (ASG) had the ability to generate a steady stream of (criminally acquired) revenue. After a government offensive against its stronghold on the island of Basilon in 2016, the Hapilon faction fled to Mindanao, taking refuge in Lanao del Sur Province, where thick jungles and hills offered a more secure, less accessible base for the group.[14]

Mindanao was home to the most recently formed insurgent organization, the Maute group, a clan-based enterprise headquartered in the rural town of Butig, 30 miles from Marawi. The Maute clan had roots in Lanao del Sur and was locally influential and relatively wealthy. They had links to other jihadist organizations in Southeast Asia and the Middle East, as both Maute brothers had been educated in Egypt. The whole Maute family (parents and five sons) played key roles in the group, originally formed as a private militia against rival clans.[15] Called Dailah Islamiyah Fi Ranao (Islamic State of Ranao) by followers, the group formally pledged allegiance to the IS leader Baghdadi in 2014. The Maute group trained two armed groups, Khalifah sa Javbul Uhod and Khilafah sa Ranao, headed by brothers Omar Maute and Abdullah Maute, respectively. The family organized a headquarters in their hometown of Butig and operated satellite camps in neighboring villages and towns, including Marawi.[16]

Cooperation between the Maute and Hapilon groups increased between 2014 and 2016, a significant development in the coordination of radical Islamic fighters in Southeast Asia. Although fighting between militant groups and government forces was a longstanding, periodic occurrence on Mindanao, the IS links of the emerging groups were a new and disturbing element to the Muslim separatist movement. With the IS gaining territory in the Middle East and committing acts of horrific brutality, the prospect of the Southeast Asian expansion of the violent extremist organization brought a worrying possibility that IS-directed or -inspired violence could spread among the region's Muslim populations.[17]

In addition to the Maute group, there was a constellation of other armed groups operating on Mindanao, including Anshar Khalifa Philippines, Khilafa Islamiyah Mindanao, Bangsa Mora Islamic Freedom Fighters, and Rajah Solaiman Movement.[18] Of these organizations, the Maute group and Abu Sayyaf were the most significant in the battle of Marawi, and as such, I refer to the insurgents active in the battle as the ASG-Maute Group, following the example of Knight and Theodorakis.

More radical in direction than their predecessors, Islamist groups cleverly employed social media platforms to increase their notoriety. Maute-Hapilon propaganda began circulating well before the Battle of Marawi on the secure messaging app Telegram. From 2016 onward, videos were shared that portrayed the group as an active IS division, training recruits and documenting their leaders' activities. Insurgents copied the aesthetics of IS's Amaq News agency, to draw in Southeast Asia–based IS supporters. Videos depicted victims dressed in orange jumpsuits, in the same style as IS prisoners in Syria and Iraq, and an Arabic script overlay was added to images.[19]

Not only did they threaten the security of Mindanao, the Maute-Hapilon Coalition presented a threat to the established Moro separatist organizations on Mindanao. The Maute group had strong local ties throughout Lanao del Sur Province, making it possible to build a support base and draw membership away from the MILF and MNLF. The comparatively less radical, more mainstream, and political autonomy–focused MILF and MNLF had cooperated with successive government attempts to reach peace agreements and also reconciliation between the two separatist factions.[20]

In 2014, the government of Corazon Aquino signed the Comprehensive Agreement on the Bangsamoro Peace agreement, which would have granted increased autonomy to the Moro people. However, in 2015, after an outburst of separatist-linked violence, the Bangsamoro Basic Law failed to pass in the Philippines Congress.[21]

While the failed passage of the Basic Law did not spark the insurgent uprising in Marawi, the failure of agreements to address the Muslim minority's desire for autonomy contributed to continued disbelief and cynicism toward the Philippine state. Fifty years of conflict on Mindanao had killed thousands and caused the island's economy to stagnate, despite its abundance of natural resources.

A heritage of violence, the inability of the more mainstream and comparatively moderate separatist movements to reach a peace deal with the government, combined with poverty, instability, and underdevelopment, created ideal conditions for radical ideologies and extremist groups such as the Maute and ASG to gain traction, particularly among disenfranchised young men who were disillusioned with the more mainstream groups.[22]

In late 2015, several months after the failed passage of the Bangsamoro Basic Law, Duterte was elected president. The first president of the Philippines from Mindanao, he came to power pledging to resolve the long-standing conflict and division on the island. Once in office, however, he refused to revive the stalled peace process and instead suggested an alternative governance arrangement for Mindanao. Little substantive progress was made on a peace process in the first year of Duterte's term, and the Duterte government focused its attentions on other priorities.[23]

The Battle of Marawi Begins

Marawi in early 2017 was a busy mercantile city, with a population of approximately 201,785 and its own distinctive, clan-based subculture.[24] Beneath many buildings lay concrete bunkers, colloquially known as *buhos*, used for food storage and shelter in times of interclan conflict. Arms caches were also found throughout most of the city.[25] Above ground, commercial buildings and houses were made from poured concrete rather than the cinder block construction used in other communities on Mindanao and throughout the Philippines. Reinforced concrete can withstand airdropped weapons and indirect fires, as the AFP would soon experience.[26]

Most of the city's population were Muslim Moros, although there was a substantial Catholic minority. Testament to the ubiquity of Islam in the city, the skyline of Marawi was dotted with domes and minarets, including the the golden-topped Capitol Mosque and the largest mosque in Marawi City, the King Faisal Mosque, donated by the former king of Saudi Arabia to Mindanao State University.[27]

As a result of successive peace agreements, the AFP had a minimal presence within the city. The headquarters of the 103rd Infantry Brigade was just outside the city, but military personnel existed apart from community life and were not fully trusted by the local community. AFP forces were aware of the Maute-Hapilon Coalition and had been engaged in clashes with Maute fighters in and around their home base of Butig several times throughout 2016.[28]

In May of 2017, Isnilon Hapilon, leader of the Hapilon faction of the ASG, made his way into Marawi, possibly to meet with members of the Maute family, sneaking in among crowds gathering to celebrate a local religious holiday on the eve of Ramadan. Hapilon's intention was to meet with the leadership of the Maute group to plan an uprising for the first day of Ramadan, May 26.[29] This date had obvious religious significance and had practical advantage, too: most Muslim families would store additional food for Ramadan, so there would be more supplies available for the militants.[30] News of Hapilon's arrival in Marawi reached security forces, who were concerned about his presence in the city and possible militant activity in the forthcoming period of Ramadan.

On May 23, 2017, as daily life in the city moved at its usual bustling pace, a company-sized all-arms AFP team set out to arrest Hapilon and others from his group believed to be hiding at a safe house in Marawi. The team consisted of special forces, rangers, reconnaissance troops, and an armored vehicle unit. The raiding team was well prepared for its mission but did not anticipate the situation that unfolded. It expected to encounter a handful of insurgents surrounding Hapilon, at a safe house, but as the team approached the suspected safe house, around 100 militants surged out from nearby buildings, catching the arrest team in a sustained firefight while Hapilon and other insurgent leaders escaped through a hole in the wall.[31]

The AFP raid on the safe house and resulting firefight precipitated the premature launch of the militants' Ramadan operation. Across the city, militants emerged from their prepared positions and joined the fray. Equipped with high-powered rifles, rocket-propelled grenades, and Molotov cocktails, insurgent fighters rampaged throughout the city, seizing the hospital, police station, and prison. Police officers and guards were killed, and prisoners set free and given weapons. A group of insurgents occupied the city hall, while others set the Catholic college alight, burned the Catholic cathedral of Maria Auxiliadora, and raced through the streets hoisting black flags, symbolizing their allegiance to IS. Camp Ranao, the headquarters of the 103rd Infantry Brigade, came under attack, reportedly by a large group of up to 500 insurgents. Catholics attempting to flee the city were held up and killed at insurgent checkpoints.[32] A Catholic priest and his parishioners were taken hostage.[33] Police and army personnel in armored vehicles who rushed into the city to provide rescue and reinforcements to the trapped arrest raid team were ambushed. Two armored vehicles were destroyed in the ambush, and the surviving crews were rescued five days later by reinforcing light infantry troops.[34]

As the situation in Marawi unfolded, President Duterte was out of the country on an official visit to the Russian Federation. When news of the militant seizure of Marawi reached him in Moscow, Duterte cut short his visit and returned to Manila. He declared martial law and the suspension of habeas corpus on Mindanao, and a large-scale, comprehensive response to the insurgent uprising began.[35] It would prove to be a lengthy, multistage effort—a day of chaotic violence across the city, followed by several weeks of evacuating and securing the perimeter, then months of clearance efforts and years of reconstruction.

Evacuation and Rescue

Residents of Marawi began to leave within the first few days of conflict, shocked by the rapid escalation of brutality and destruction, and fearful of the potential repression that might accompany the imposition of martial law. By May 27, four days after the initial burst of violence, 90 percent of the population had fled.[36] Some took refuge with relatives and friends in other towns and villages, while thousands of others headed for Illigan and Cagayan de Oro, the nearest large cities. Had this voluntary evacuation not occurred, large numbers of civilians would have been killed or injured in the battle for the city, with severe popular and political consequences for the government of the Philippines and partner nations that later joined the fight.

Whereas the majority of Maranaos managed to flee to relative safety, several thousand residents remained trapped in the city, either held as hostages, caught in the crossfire between militants and AFP, or hiding in fear of militant brutality. Civilians with access to mobile phones or two-way radios called for help from inside the city, fearing for their lives.[37] Community members, alongside police and military personnel, attempted to rescue hostages, venturing back into the city to negotiate for their release. The AFP coordinated with the MILF and MNLF members on rescue operations, recognizing the role of these more moderate organizations in local authority.[38]

The AFP had only a brigade-sized formation based on the outskirts of Marawi. Of that brigade, the arrest team was trapped in the city, and others were in the jungle, fighting insurgents. When the uprising occurred, others were rapidly recalled from the jungle, where they had been engaged fighting NPA militants there. Within 24 hours, additional troops landed on Mindanao. Within a week of the initial uprising, still more troops arrived by sea from other parts of the country.[39] By the end of the battle, 12,000 AFP troops were involved in the fighting for Marawi, vastly outnumbering the insurgent forces.

A joint task force (JTF Marawi) was established to reclaim the city from the militants. The task force comprised three brigade-sized formations, two army and one marine, and a task organized for urban warfare, with attachments of engineers, artillery, and medics. Additionally, two AFP brigades were assigned to managing the welfare of the displaced civilian population, with one brigade dedicated to civil-military relationships and the other to information and the needs of internally displaced persons (IDPs).[40]

Though the city had effectively fallen to the insurgents, the AFP accorded priority to meeting the needs of the displaced population, aware that the Maranaos' perceptions and attitudes were key terrain in the battle. Army engineers were tasked with building emergency shelters in IDP camps well outside the city. Efforts were made to deliver culturally sensitive assistance, and respect for Islamic cultural practices was an important part of countering the militant anti-government rhetoric. The AFP consulted Moro community leaders on IDP-related issues.

International Assistance

President Duterte had come to power less than two years prior to the uprising in Marawi, advocating for an independent foreign policy, with far less reliance on the United States than previous administrations. He sought to establish new relationships with China and Russia and create greater strategic distance from the United States.[41]

When the fighting began, President Duterte was initially reluctant to request external assistance, seeking to retain his independent foreign policy. A month into the fighting, however, the AFP, a formidable political force in its own right, formally requested the assistance of the United States, the Philippines' sole treaty ally. The AFP had a history of cooperation with US forces on countering violent extremism. For nearly 15 years, under Operation Enduring Freedom–Philippines, American Special Forces units had provided training and support for counterterrorism operations. With an outbreak of violent extremism in their own territory, the AFP needed the support of their US ally.

In early June, Washington provided Manila with a new cache of advanced weapons and tactical intelligence gathered by unmanned aerial vehicles. By the middle of that month, a contingent of US Special Forces arrived to provide specialized urban warfare training to its Filipino counterparts.[42]

Nearby Australia deployed two AP-30 Orion surveillance aircraft, and an Australian Defence Force Land Mentoring and Training Team to assist Filipino forces in filling urban warfare capability gaps.[43] Member countries of the Association of Southeast Asian Nations were quick to provide humanitarian and security assistance when the fighting erupted in Marawi. Singapore was one of the first countries to send in humanitarian supplies to the beleaguered city. Indonesia, Malaysia, and the Philippines put into place joint, trilateral border patrols to prevent the spread of armed conflict.[44] Aside from dealing with the consequences of the Marawi siege, Southeast Asian countries cooperated closely in sharing knowledge and information to confront violent extremist groups.[45]

The fight against an IS offshoot prompted a greater response because of the globally recognized threat of IS in northern Iraq and Syria. Containing IS and preventing its particular brand of extremism from taking root in other Islamic communities was a national security priority for the United States and Australia, as well as many other countries around the world. Without this priority, a long-running domestic insurgency might not have attracted the same degree of international attention.

The Defensive Battle

Initially the AFP set a deadline of June 2 for reclaiming the city. In previous episodes of violence, including the battle of Zamboanga, cease-fires were reached within a matter of days or weeks. A cease-fire agreement, facilitated by the MILF, was agreed to by government forces and insurgents on June 4. However, frontline AFP troops fighting within the city refused to observe the cease-fire terms, only allowing the evacuation of civilians from the outskirts of insurgent-controlled territory. It was apparent that this battle would not be resolved in a matter of weeks, but rather months. Insurgents massed in the city center, where extensive preparations had already been made for a defensive battle.[46] The militants were well supplied and well armed, numbering around 1,000 fighters. Most militants were local to Mindanao, though there were a small number of foreign fighters and a few recent recruits from the raid on the prison in the heady opening days of conflict.[47]

Insurgents did not muster a groundswell of popular support from Maranaos as most of the population fled in the early days of the fighting. Maranaos may have been dissatisfied or fatigued with the status quo, but they were not prepared to support or join an extremist uprising that was actively damaging their own city, taking hostages from among the population, and prompting a potentially severe government response.

Marawi's geography was well suited to defense, with limited access routes and the natural boundary provided by Lake Lanao. The city was bisected by the Argus River into east and west areas, with a limited number of bridges for access between the sides. It was difficult for any offensive action to infiltrate the city and easier for the defending insurgents to barricade entry to the city. The main battle area of the downtown core was not large, measuring approximately 0.8 kilometer by 1 kilometer, a reasonably condensed area that insurgents could move about and control on foot.[48] A nearly empty city, with most of the population displaced, meant buildings could be entered without resistance or obstruction and properties looted for food and valuables.[49]

The built environment offered a haven for defenders. Concrete homes and businesses offered ready-made, fortified fighting positions. The thick walls of many buildings offered protection from small arms fire, and small openings could be cut in interior walls to provide protected sniper hides. The *buhos* beneath the city provided a network of spaces for storing weapons and ammunition, food, and other supplies as well as providing shelter and space for concealed communications with external supporters via computers, cell phones, or two-way radios. Some of the few civilians remaining in Marawi were held as hostages, used as forced labor to enhance the network of underground bunkers, and threatened with violence if they refused to cooperate with insurgents' orders.[50]

The AFP used close air support to locate insurgents and to target sniper positions. Insurgents were observant and resourceful, adapting their movements to the pattern of aircraft activity. They quickly learned to hide when airplanes were overhead, moving only when aircraft had withdrawn. To confuse aircrews, insurgents used smoke and dummy markers, making it difficult to distinguish between friend and foe.[51] Air strikes were effective at eliminating insurgent positions yet engendered significant risk to their own forces, causing a number of AFP friendly-fire casualties.

Commercial unmanned aerial systems—the sort of consumer-grade, off-the-shelf drones that are readily available from online retailers—gave militants an improvised intelligence, surveillance, and reconnaissance (ISR) capability. Drones were flown regularly around the battlespace to provide militants with information on potential AFP countersniper locations and activity.[52] On the ground, insurgents benefited from their local knowledge, as many lived in or around the city and knew shortcuts and hiding places ideal for launching sniper attacks and ambushes. The combination of local knowledge and drone oversight initially gave the insurgents situational dominance.[53]

Marawi initially offered a stronger defensive position. However, the ASG-Maute group had to take up this defensive position because, although more capable than other previous separatist factions, they were not strong enough to launch a sustained offensive against the AFP.[54] The attack on the arrest team and subsequent seizure of the city owed more to surprise, chaos, and fear than offensive capabilities. Holding urban terrain, ambushing AFP troops, and then disappearing back into the morass of homes, buildings, and underground bunkers allowed ASG-Maute to antagonize the AFP while preserving forces and supplies.

As attackers, the AFP had a relative disadvantage in the city. Though a competent and capable army, the AFP had limited experience with urban warfare because much of their training and equipment was intended for jungle warfare. The AFP had fought a breakaway MILF faction in an uprising in the nearby town of Zamboanga in 2013, but that battle had been relatively short-lived, lasting only two weeks before ending in cease-fire. Furthermore, Zamboanga was a small town, with buildings made of cinder blocks and wood, easily destroyed by explosive firepower.[55] Marawi's size and concrete structures presented a new challenge, and the AFP had to rapidly adapt to this new and complex combat environment.

Whereas the ASG-Maute insurgents could slip through, around, and beneath buildings, the AFP had to move along streets, bridges, and alleyways, where they were visible and vulnerable to attack and ambush. Hiding behind or within buildings, using murder holes cut in interior walls or the turns of staircases for hides, snipers were able to ambush dismounted troops at close range. As James Lewis describes, these insurgent "snipers" were not technically snipers, as they were mostly firing from short- or middle-range distances and did not have the expert training or equipment of their conventional-force counterparts.[56] Yet the cover provided by buildings compensated for militants' relative lack of sniping skill. Militants also ambushed armored vehicles with petrol bombs (Molotov cocktails) and RPG-2s, killing or injuring the crews inside. In a clash on June 9, such an ambush led to the near loss of an entire Marine company. The incident led to a theater-wide initiative by the AFP to outfit its armored personnel carriers and fire support vehicles with improvised wooden armor.[57]

Wooden armor, often decorated with morale-boosting graffiti, offered limited protection against projectiles but provided psychological reinforcement for AFP troops. Armored vehicles, mostly M-113 armored personnel carriers, played essential roles in the battle, providing close support to dismounted infantry units advancing through the city and breaching capabilities for engineers. Armored vehicles also aided with casualty extraction.[58] However, armored vehicles' mobility was impaired by the narrow streets and alleyways of Marawi, piles of rubble from air and artillery strikes, and obstacles deliberately placed by militants, such as parked cars. Later in the battle, armored bulldozers, as seen in Ramadi and Mosul, were brought into the city to clear debris and assist with restoring mobility, another tactical lesson learned from the recent US experience in Iraq.

By mid-June, the AFP announced they had begun "normalization" operations, having regained control of the majority of Marawi. However, the city was far from normal. Thousands of civilians remained displaced, and insurgents clung to the central core of the city, resolutely fighting on. On June 24, both sides observed a 24-hour cease-fire for Eid al-Fitr, the end of Ramadan. MILF members used this brief break in the hostilities to rescue civilians still trapped in the fighting area.[59]

Despite the destruction wrought by close air support, indirect fires, and armored vehicles, the buildings and streets of Marawi still needed to be cleared by dismounted units to achieve a decisive victory. Small, dismounted teams had to enter and clear buildings, one by one, to ensure the city was fully clear.[60] Technology undoubtedly provided advantages to both insurgents and AFP troops, but as with so many other urban battles throughout history, it was ultimately an intense close-quarters fight.

Sniping and ambushing were limited to daylight hours, as insurgents had no night-fighting capability. Once darkness fell, militants withdrew to defensive positions. In the night, militants often indulged in *shabu*, a locally popular form of methamphetamine.[61] The drug enhanced insurgents' ability to endure long hours of fighting. Insurgents' drug habits enabled AFP special forces units to take advantage of their distraction to move about the city center and occupy sniper positions undetected. Special forces units had night-fighting equipment, unlike their adversary and green army counterparts, who were restricted to daylight fighting.[62]

To recover the city, the AFP developed an approach that was systematic, if slow-moving. Known as SLICE (strategize, locate, isolate, constrict, and eliminate), it involved dividing the city into three sectors, each allocated to one of the marine and army task-force groups with armored vehicles and artillery attached. In each sector, units first located and isolated insurgent positions, constricting their mobility.[63] Finally, insurgent positions were eliminated with explosive firepower and infantry assaults. Each attack was carefully planned and practiced to reduce the likelihood of friendly-fire casualties.[64]

The SLICE method of deliberate attack was similar to techniques used in recent battles against IS in the Middle East. The similarity in methods was not surprising as the AFP had the support and advice of US forces who had fought IS in Mosul just weeks before and shared their experience with Filipino counterparts.[65] Although SLICE sought to minimize AFP casualties, it came at the expense of the urban landscape. To clear the city without sustaining unbearable casualties, AFP used heavy firepower against the structures within Marawi, in the form of close air support, artillery (105 millimeter field guns) in a direct fire role, and thousands of mortar rounds. One infantry company noted it had used 10,000 mortar rounds in three months.[66]

Deliberately exempt from fires—direct and indirect—were religious sites. President Duterte ordered explosive weapons not be used against mosques, to demonstrate respect for Islamic sacred spaces and to avoid antagonizing the Moro population. Insurgents took advantage of this restraint and used mosques to hold hostages, food, and weapons and for fighting positions. To clear the buildings without inflicting structural damage to the mosques, AFP troops used CS gas (a tear gas) to force militants from the buildings.[67]

As AFP troops cleared building after building, insurgents used large numbers of improvised explosive devices (IEDs), hiding their devices in doorways, hallways, and ceiling cavities. As a result, the AFP suffered many IED casualties in building entranceways. To avoid entering a building through doorways or windows and risk triggering an IED, troops opted instead to breach a hole in a wall, often destroying much of the building in the process.[68] Once inside, the danger of improvised explosives was not over. Militants wearing suicide belts would wait until troops had entered a room before detonating their weapons, the explosion magnified by the confined space, causing extensive damage, serious injuries, and deaths.[69]

The extensive tactical use of IEDs and suicide bombing was a hallmark of IS insurgency, used to stunning and deadly effect in combat in Iraq and Syria. The transfer of technology may have come with foreign fighters from other conflicts, study of IS videos and other propaganda material, and access to IS technical and tactical instruction in the construction and deployment of these weapons. This was a marked departure from earlier outbursts of separatist violence on Mindanao, where other militants had relied on conventional weapons and petrol bombs.[70]

Beyond their arsenal, what set the Maute-ASG insurgents apart from other separatist fighters was their determination to fight. Compared to the more secular motives of the MNLF rebels, the ASG-Maute group intended to fight the AFP to the death. Like IS counterparts in Mosul and Raqqa, ASG-Maute fighters saw themselves as engaged in an existential struggle rather than merely a political one, a cause that required and justified martyrdom.[71] This mindset gave the defenders an even greater determination, a willingness to die for their cause, and an intention to take as many AFP casualties as possible.

The AFP's progress through the city was painstaking. It took two months for marines to clear one key location, the Mapandi (Baloi) bridge, and over five months for the AFP to clear the city center. The slow pace was due to the AFP's underdeveloped urban warfighting capability and the challenge of fighting a well-entrenched and resourceful adversary in a dense urban environment.[72] By August, after months of grueling fighting, the AFP had recaptured the police station, the Grand Mosque, and the Cathedral, cultural sites for the city's Muslim and Catholic populations. Within weeks, the AFP had secured all the bridges linking the fighting area to the rest of Marawi.

Influence Operations

Beyond the urban battlefield, insurgents and the AFP battled to shape public opinion and convince the local population—as well as a wider national and international audience—of the righteousness of their respective causes.[73] Both sides understood that shaping popular perceptions was key to winning the longer-term battle for popular support and changing the political will of the target population.

Insurgents' messaging sought to depict them as the protectors and liberators of the Maranaos. They emphasized the lack of economic opportunity, welfare provision, and social justice for Maranaos, framing the battle of Marawi as the AFP's deliberate destruction of Muslim heritage, culture, and society, another chapter in a chronicle of historic disregard for the Muslim population. Militants engaged in direct offline influence activities, conducting door-to-door visits to inform locals of intended actions, and forced local clerics to repeal earlier counter-IS statements. Militants drew upon existing relationships and clan ties to further their messaging and expand their influence.[74] In the opening days of the battle, insurgents used social media platforms to communicate with the outside world. The Philippine government asked the Facebook office in Singapore to remove accounts and content associated with the ASG-Maute group, and cooperation between government and the private sector deprived insurgents of a broadcast channel.[75]

A benefit of affiliation with IS was access and attention from the organization's Amaq News agency, who produced far more sophisticated propaganda materials for a much wider audience. In June 2017, the IS magazine *Rumiyah* published a special issue titled "The Jihad in East Asia" that featured stories from Marawi.[76]

The magazine feature was followed by propaganda videos in the "Inside the Khilafah" series, depicting ASG and Maute fighters attacking the AFP, emphasizing how the jihad in Marawi was a defensive campaign to protect a part of the global caliphate from a hostile crusader army. IS video footage featured graphic images of the desecration of religious inconography in a Catholic church, a deliberately provocative gesture in the majority Catholic Philippines.[77] IS messaging incorporated themes of colonial oppression, the unfair occupation of Mindanao by Catholic incomers, and the purity of Islamic governance in contrast with the repeated failures and violent actions of successive secular governments and American involvement in the region, in the same vein as its messaging about the Middle East.[78] Videos also encouraged potential IS recruits to consider traveling to the Philippines should Syria or Iraq prove to be a difficult destination. The goal of the IS information campaign was to herald IS successful takeover of the city, creating an image of global reach and continued power, despite the organization's recent significant losses of territory in the Middle East.[79]

To counter insurgent narratives and propaganda material, JTF Marawi created a dedicated information operations cell, the Civil-Military Operation Coordination Center, with several local task groups located throughout Mindanao. The cell's purpose was to actively counter militant messaging and influence public opinion, for both the affected local population and the wider national audience over the immediate and longer-term. A 24-hour media center was formed to assist the messaging campaign, and an Influence Operation cell made use of online social media platforms, as well as more traditional means—leaflet drops, banners, radio, and loud-speaker broadcasts—to spread their messages.[80]

For the population of Marawi and wider Mindanao, JTF Marawi's messaging sought to prevent civilian casualties, encourage safe civilian evacuation from the city, and inhibit further recruitment of local people into insurgent organizations.[81] AFP leadership recognized that the Maranao population had negative perceptions of the armed forces and, by extension, the Philippine government, derived from years of repressive military occupation of Mindanao. The commander of the JTF Marawi, Lieutenant General Rolando Bautista, openly acknowledged the historic Muslim distrust of the military and the continuing tensions between the Muslim minority and Christian majority.[82] Changing this perception was essential to winning the cooperation and support of the local population. Messaging, therefore, sought to present the AFP as a friendly, caring force, committed to ensuring the safety and well-being of Maranaos.[83]

For a national audience, the conflict was deliberately presented as a struggle against a foreign insurgency, an IS offshoot rather than another Moro separatist movement, and not the start of a civil war between Muslim minority and Christian majority populations.[84] To rally popular support throughout the rest of the country, government counternarratives presented heroic combat footage, accompanied by emotive music. Patriotic hashtags, such as #supportourtroops, accompanied online messaging and imagery to rally support for the struggle in Marawi.[85]

AFP narratives received little press scrutiny as the military largely controlled the flow of information out of Marawi, ostensibly for operational security reasons. These restrictions also included limitations on the kind of reports allowed on the progress of events in Marawi, as only positive media coverage of the battle was permitted. These restrictions afforded the military a significant degree of control over the narrative and shaped national and international audiences' understanding of the situation in Marawi. Restricted media access to the battle area mirrored the wider press restrictions in the Philippines, as the Duterte administration sought to exert greater control over the media.[86]

But who won the battle of the narrative? As Steve Tatham explains further in Chapter 8, the outcome of influence operations cannot be controlled, and influence operations to alter a population's perceptions may not result in rapid, measurable changes of behaviors. Determining a victor is therefore challenging given that the insurgents did not gain widespread support among Maranaos for their caliphate or encourage similar outbursts of violence elsewhere in the region. The widespread violence, hostage-taking, and destruction of the city did not endear the insurgents to the local population.[87]

The countering influence operation by the AFP was a dedicated effort to present a three-part opposing narrative. Their information campaign needed to intimidate their adversary, to portray the military as more humane to win support of Maranaos, and to promote a more inclusive national identity among the wider population of the Philippines. However, opposition and distrust was deep-rooted and unlikely to be swayed by a few months of messaging. And the ground gained with the displaced Maranao population was threatened by how the government handled the post-battle reconstruction process.[88]

Both militant and military influence campaigns were central features of the urban operations; both sides appreciated that winning the battle of the narrative was a key element of the battle. Insurgents and the AFP alike understood prevailing local narratives and attitudes and could adapt their messages accordingly. As Knight and Tatham concur, this sort of cultural intelligence is vital to influence operations.[89]

Rebuilding Marawi

The AFP declared victory on October 23, 2017. After five months of war, there was an estimated 925 insurgent casualties, 165 army casualties, and 45 civilian casualties, though this was a conservative estimate. Among the dead were Omar Maute and Isnilon Hapilon, leader of the Hapilon faction, though his three sons escaped capture.[90]

President Duterte heralded the victory and pledged to rebuild Marawi and restore it to its former glory. It was quite an undertaking, as entire city blocks had been reduced to rubble and iconic mosques pockmarked with holes from firefights and air and artillery bombardments. No homes remained in the main battle area. Unexploded ordnance, both conventional weapons and IEDs, rendered the city unsafe for displaced Maranaos wishing to return home.[91] Rebuilding the city and restoring it to habitable conditions would take considerable time and investment.

The rebuilding plan favored by the Duterte administration featured an extensive modernization plan for Marawi, with the main battle area in the downtown core turned into an "economic zone," although it was uncertain what this zone would entail.[92] In addition, a new military base would occupy a prime location in the heart of the city, establishing a permanent

AFP presence—and by extension, a symbol of Manila's control and oversight, the opposite of the autonomy sought by the Maranaos.[93]

Duterte's ambitious reconstruction plan failed to address the complex landownership status of Marawi, where competing deeds and informal claim to land had been a perennial source of conflict. Many among the displaced population did not own the deeds to their homes and, without these deeds, were unable to receive compensation for the destruction of their home or the redevelopment of the land. Pouring fresh concrete and erecting new buildings over the disputed deeds and destroyed homes would not solve the underlying land-claim issues that had long plagued Marawi, nor provide adequate long-term housing for displaced Maranaos.[94]

Opposition leaders criticized Duterte's lofty rebuilding plans and the awarding of contracts to a consortium of Chinese construction firms blacklisted by the World Bank for unscrupulous business practices. Maranao leaders were dismayed at the lack of local consultation on the reconstruction project and the sluggish pace of reconstruction.[95] Steps were made, however, toward political recognition, as the Bangsamoro Basic Law passed into legislation in July 2018, coming into effect the following spring.[96]

For Maranaos, displacement continued not for months but for years, as deadlines for rebuilding and rehabilitation came and went, and martial law remained in effect. By 2019, two years after the cessation of hostilities, over 100,000 Maranaos remained in displaced persons camps, unable to return home.[97] The camps, hastily constructed to provide emergency shelter at the beginning of the conflict, lacked adequate sanitation and clean water facilities and were never intended for long-term habitation. Temporary schools were provided for displaced children, and some IDPs established ad hoc businesses: selling food or basic items along the roadside. IDPs in organized camps received a government subsidy equivalent to US$1,400, but this did not alleviate their poverty nor replace the homes, buildings, and livelihoods lost to the conflict. Those who had sought refuge with friends or relatives did not receive such a subsidy. The delayed reconstruction and resettlement risked legitimizing popular grievances and prolonging instability and poverty, conditions that extremist groups could readily exploit.

Two years after Duterte had declared the city "free from terrorist influence," the desolate conditions of the displaced persons camps provided optimal conditions for the reconstituted Maute group to recruit and regroup.[98] Preying on unaccompanied teenage boys, group members sought to reestablish a presence by capitalizing on disenfranchisement and despair. They promised a joining bonus and regular monthly salary to prospective members, an attractive proposition to those living in poverty with few economic or educational prospects. The recruitment of another generation to the remnants of extremist movements suggests that the battle of Marawi may not be the last episode of violence on Mindanao.[99]

Conclusion

Marawi, with its limited allegiance to Manila, historic distrust of the AFP, and alternative local power structures, bears a resemblance to the urban and peri-urban areas of cities, not just in the developing world but also in large cities in Western Europe. There are lessons to be learned in engaging and communicating with a disaffected, historically disenfranchised population and incorporating local leadership into crisis response and resolution.

The Battle of Marawi demonstrates the advantages the urban environment offers a defender—even a small, unconventional force such as the ASG-Maute group. The cityscape of Marawi provided cover and concealment, concrete buildings were ready-made fortifications, and the network of underground bunkers and tunnels beneath the city served as storage facilities and access routes.[100] Maute-ASG fighters could determine when and where to engage AFP forces and had abundant hiding places for IEDs. Their intimate knowledge of the city and improvised ISR capabilities gave them the ability to observe AFP activity and move to avoid or invite contact. They had the advantage of surprise.

Violent jihadist activity may have tried to capitalize on the historic divisions between Muslim and Christian, but the Maute and Hapilon groups failed to galvanize popular support. The underlying frustration they sought to exploit was political rather than ideological—a long-standing desire for greater autonomy, recognition, and self-governance rather than adherence to an extreme fundamentalist interpretation of Islam. Over 90 percent of the population fled, fearing for their lives, property, and livelihoods.[101] Violent jihadist activity had concentrated considerable military attention on Marawi, but it was not going to deliver the political settlement and increased economic prosperity that would measurably improve quality of life for Maranaos.

Insurgents misread the social terrain as the local population feared the insurgents' violence and the potential army and government reprisal. There was little appeal or perceived benefit to supporting the ASG-Maute group. Insurgents were unable, despite an influence campaign that played on historic grievances, to win widespread local support. Nor, despite their allegiance to IS, could insurgents rely on material support from IS central leadership.

AFP forces went into the city at a tactical disadvantage, attempting to unseat a radicalized adversary that had embedded itself within the city. Though a competent and professional army, their limited preparation for the complex context of Marawi underscores the importance of combined arms training for urban environments and the importance of studying history.[102] Moreover, study and training in the principles of positional warfare, such as the siege warfare described in Chapter 3, and modern siege techniques detailed in Chapter 9, could offer an alternative to hugely destructive clearance practices such as the SLICE approach.

"Destroying a city to save it" has become a typography of urban warfare in the past twenty years, an established best practice for removing violent extremist organizations from entrenched defensive positions in urban areas. Large quantities of high explosives, in the form of air-dropped weapons and direct and indirect fires, are used against concrete buildings. The practice has become commonplace in recent decades in part because of a lack of precision weapons and a distaste for long, drawn-out overseas engagements and high casualty rates among small, professional Western armies.[103] Destroying the physical domain of the city is deemed preferable to sustaining a large number of military or civilian casualties. Insurgent groups such as al-Qaeda in Iraq and IS in Iraq and Syria have been quelled in this way, but the destruction of the city addresses only the immediate tactical and operational objectives. In Marawi, the destruction of the physical environs of the city conflicted with the long-term strategic objectives.

The widespread and indiscriminate destruction in urban areas such as Marawi extracted a high personal and public cost, as homes, schools, shops, and places of worship and cultural significance are turned to rubble. The physical destruction of Marawi fundamentally disturbed what Wendy Pullan described in Chapter 2 as the "urban order" of the city. The relationship between people and place was irrevocably damaged. Gone were places of social and economic importance, Moro historical and religious sites, and the rituals and habits that had previously shaped life in the city. There were relatively few civilian casualties in Marawi because of the rapid evacuation from the city in the opening days of the conflict. However, the destruction of the city and of the daily life that took place within it, compounded by the slow process of reconstruction, risked creating new grievances, aggravating existing ones, and losing the progress made in gaining the support of the Moro population. Winning the urban battle did not end the struggle against insurgency.

Notes

1. International Crisis Group, "Philippines: Addressing Islamist Militancy."
2. Ibid.
3. Tarafder and Custodio, "Microfinance in a Conflict Zone."
4. Steven Rogers, "Beyond Abu Sayyaf: The Lessons of Failure in the Philippines," *Foreign Affairs* 84, no. 1 (2004), https://www.foreignaffairs.com/articles/philippines/2004-01-01/beyond-abu-sayyaf.
5. Rogers, "Beyond Abu Sayyaf"; Knight and Theodorakis, *Marawi Crisis*.
6. Rogers, "Beyond Abu Sayyaf."
7. Rogers, "Beyond Abu Sayyaf."
8. Centre for International Security and Cooperation, "Moro National Liberation Front," Stanford University, https://cisac.fsi.stanford.edu/mappingmilitants/profiles/moro-national-liberation-front#text_block_20212.

9. Ibid.
10. Rogers, "Beyond Abu Sayyaf."
11. Centre for International Security and Cooperation, "Communist Party of the Philippines—New People's Army," Stanford University, https://cisac.fsi.stanford.edu/mappingmilitants/profiles/communist-party-philippines-new-peoples-army#highlight_text_13192.
12. Torres, *Rido: Clan Feuding and Conflict Management in Mindanao*, 16.
13. Rogers, "Beyond Abu Sayyaf."
14. Franco, "Assessing the Feasibility of 'Wilayah Mindanao,'" 30.
15. Gunaratna, "Siege of Marawi," 1.
16. Ibid, 1.
17. Franco, "Assessing the Feasibility of 'Wilayah Mindanao,'" 34.
18. Major E. Divinagracia, "Urban Terrorism: The Siege of Marawi City," 56.
19. Franco, "Assessing the Feasibility of 'Wilayah Mindanao,'" 32.
20. Heydarian, "Crisis in Mindanao."
21. Hartman, "Marawi Battle Highlights the Perils."
22. Fealy, "Battle for Marawi."
23. Heydarian, "Crisis in Mindanao."
24. Tarafder and Custodio, "Microfinance in a Conflict Zone."
25. Franco, *Battle for Mawari*.
26. Ibid.
27. Tarafder, and Custodio, "Microfinance in a Conflict Zone."
28. Franco, *Battle for Mawari*.
29. Franco, "Assessing the Feasibility of 'Wilayah Mindanao,'" 32.
30. Spencer and Knight, "Battle of Marawi."
31. Tarafder and Custodio, "Microfinance in a Conflict Zone."
32. Ibid.
33. Gunaratna, "Ending the Fight in Marawi," 2.
34. Franco, *Battle for Mawari*.
35. Heydarian, "Crisis in Mindanao."
36. Amnesty International, "'Battle of Marawi' Leaves Trail."
37. Knight and Theodorakis, *Marawi Crisis*.
38. Spencer and Knight, "Battle of Marawi."
39. Ibid.
40. Ibid.
41. Heydarian, "Crisis in Mindanao."
42. Ibid.
43. Knight and Theodorakis, *Marawi Crisis*.
44. Franco, "Assessing the Feasibility of 'Wilayat Mindanao,'" 35.
45. Gunaratna, "Ending the Fight in Marawi," 5.
46. Knight and Theodorakis, *Marawi Crisis*.
47. Tarafder and Custodio, "Microfinance in a Conflict Zone."
48. Knight and Theodorakis, *Marawi Crisis*.
49. Gunaratna, "Ending the Fight in Marawi," 1.
50. Ibid., 2.
51. Lewis, "Battle of Marawi."
52. Ibid.
53. Spencer, "Eight Rules of Urban Warfare."
54. Ibid.
55. Franco, *Battle for Mawari*.
56. Lewis, "Battle of Marawi."

57. Franco, *Battle for Mawari.*
58. Lewis, "Battle of Marawi."
59. Knight and Theodorakis, *Marawi Crisis.*
60. Lewis, "Battle of Marawi."
61. Ibid.
62. Ibid.
63. Knight and Theodorakis, *Marawi Crisis.*
64. Spencer and Knight, "Battle of Marawi."
65. Knight and Theodorakis, *Marawi Crisis.*
66. Lewis, "Battle of Marawi."
67. Knight and Theodorakis, *Marawi Crisis.*
68. Ibid.
69. Ibid.
70. Ibid.
71. Franco, "Assessing the Feasibility of 'Wilayah Mindanao,'" 32.
72. Ibid.
73. Knight and Theodorakis, *Marawi Crisis.*
74. Ibid.
75. Ibid.
76. Franco, "Assessing the Feasibility of 'Wilayat Mindanao,'" 32.
77. Ibid., 33.
78. Knight and Theodorakis, *Marawi Crisis.*
79. Franco, "Assessing the Feasibility of 'Wilayat Mindanao.'"
80. Spencer and Knight, "Battle of Marawi."
81. Ibid.
82. Knight and Theodorakis, *Marawi Crisis.*
83. Ibid.
84. Ibid.
85. Ibid.
86. Ibid.
87. Spencer and Knight, "Battle of Marawi."
88. International Crisis Group, "Philippines: Addressing Islamist Militancy."
89. Ibid.
90. Amnesty International, "'Battle of Marawi' Leaves Trail."
91. Ibid.
92. International Crisis Group, "Philippines: Addressing Islamist Militancy."
93. Ibid.
94. Ibid.
95. Ibid.
96. International Crisis Group, "Philippines: Militancy and the New Bangasamoro."
97. Auditor, "Marawi Rehabilitation Delays."
98. Ibid.
99. Regencio, "Delay in Return."
100. Spencer and Knight, "Battle of Marawi."
101. Franco, "Assessing the Feasibility of 'Wilayat Mindanao,'" 29.
102. Spencer and Knight, "Battle of Marawi."
103. Spencer, "Eight Rules of Urban Warfare."

6

The Power of "Place"

Patrick Finnegan

CROSSMAGLEN, A SMALL VILLAGE IN THE NORTHERN IRISH COUNTY OF Armagh, was not the scene of extended gun battles or of close-quarter, high-attrition battles like those found in other chapters of this volume, but in a way that is the point. It was the spiritual home of Irish republicanism in the area. Part of the wider South Armagh area, Crossmaglen became both the scene of many acts of violence but also the symbol for local resistance to British rule on the one hand and the focal point of British efforts to stamp out militant republicanism on the other. The common perception of urban warfare is of rubble and soldiers fighting over it or of highly demarcated areas fought over by large bodies of troops. The combat witnessed by Crossmaglen was different, but it is important nonetheless. In this case, conflict was often at the squad or platoon level, and it was based on surprise and precision rather than sledgehammer blows or saturation occupation, it was above all else a tactical contest that had higher ramifications. The violence witnessed here was intimate and protracted but occurred within a "normal" situation where civilians went about their daily business.[1] It was fought between two learning organizations who sought to control the environment they operated in. The military altered the physical environment to suit their needs, while their opponents dominated the social plain and used the locality as it stood to their advantage. By examining Crossmaglen, we can highlight important lessons for any urban operation, such as the problems associated with physical and social isolation and the difficulty of operating within a liminal geographic space, especially the physical cost of controlling territory in the face of local, popular opposition.

Cross' in Context

The prolonged period of violence in Northern Ireland known as the Troubles lasted from 1969 to 1998, or from 1996 to the later 2000s if we wanted to take a longer view. Various attempts have been made to examine this conflict and its root causes; for the purposes of this work, it is enough to understand that the Troubles were fought between Irish republicans with the reunification of Ireland as their ultimate aim and those associated with the continuance of British rule. This included both official forces of the state as well as local loyalist forces who acted outside of the law, engaging in their own (counter)terror campaign. Throughout that period, the British military sought first to "aid the civil power" in restoring order after widespread rioting, then they fought a larger-scale insurgency between the years 1970 and 1976–1977, before having to tackle a smaller, but more directed terror campaign as the Irish Republican Army (IRA) reorganized. The efforts of the British military during the Troubles—Operation Banner—have been described as bringing about stability and room for governance rather than achieving outright victory.[2] Throughout this, the army and marines in particular had to operate in urban centers such as Belfast and Derry during one deployment, then perhaps travel to Germany before again returning to patrol the country lanes of Fermanagh and Antrim. At times, violence extended to England and Europe, but for the most part it was concentrated in Northern Ireland, and most violence occurred in larger urban centers, such as Belfast, or in distinctly rural settings, such as East Tyrone. Units would deploy for a particular mission in a particular place and would be trained as such, learning how to lay covert observation posts in hedgerows or abandoned buildings or learning how to conduct urban patrols with multiple "bricks" to ensure that all avenues of fire and escape were covered. But there were other deployments, ones that blended these two sides of the coin into one and that became correspondingly more difficult. These were deployments to areas that stood on the boundary between urban and rural. These deployments shared the difficulties of both urban and rural deployments and offered none of the benefits. This chapter will discuss one of these deployments within this "liminal space" between urban and rural.

Known to locals as Cross' and to veterans of the Troubles as XMG, the village of Crossmaglen was the scene of significant levels of violence throughout Operation Banner, far beyond anything its meager size would suggest. The home of a platoon-sized garrison, this area formed the focal point of some of the most intense militarization within Northern Ireland because those operating within it had to contend with the problems of urban terrorism and rural insurgency at the same time. Any discussion solely of Crossmaglen excludes important information about the wider activity of the Provisional Irish Republican Army (PIRA) within Northern Ireland and within South Armagh as well as the neighboring town of Dundalk more specifically, which were all seen as parts of a larger whole from the outset of the conflict.[3] Where possible, information and events concerning Crossmaglen will be contextualized with comparison to other areas in the vicinity, such as Forkhill and Newtownhamilton. Much has been written about the difficulties of operating in areas such as Belfast and Derry when it comes to histories of the British Army in Northern Ireland, but this chapter aims to address the gap of urban operations within smaller environments.[4] This chapter is not a discussion about the impact the South Armagh PIRA had on the wider campaign; it is an investigation into dealing with the threat "where they lived" rather than the violence they were able to export. For the first stage of contextualization, we can consider conflict data and then the historical context. While there are other figures available, Conflict Archive on the Internet (CAIN) is the most widely used and will be discussed here as a baseline.[5]

With a total of 3,532 fatalities, security forces (police and military) represent roughly one-third of those killed during the period (1,114). Almost all of these were related to terrorism (98 percent) and the security forces accounted for 35 percent of all those lost to terrorist violence (see Table 6.1). It should be noted, however, that these figures do not include those who were injured, which were significantly higher. County Armagh has been included as a subset to provide the first contextualization for Crossmaglen. It should be noted that these figures are only those who are recorded in relation to County Armagh, not those who died in other areas as a consequence of Armagh-based Active Service Units (ASUs). Armagh as a subset accounts for 15 percent of the total fatalities in the conflict and 22

Table 6.1 Fatality Breakdowns According to the CAIN (Conflict Archive on the Internet) Database

	Total	Terrorist Action	Percentage of Total That Were Terrorism-Related	Terrorist Action in County Armagh	Percentage of Terrorist Action from County Armagh
Fatalities	3,532	3,084[a]	87%	477	15%
Security fatalities	1,114	1,094	98%	238	22%
Security as percentage of total	32%	35%		50%	

Source: All figures taken from the Sutton Index of Deaths, CAIN Index, University of Ulster, https://cain.ulster.ac.uk/sutton/index.html, and rounded to the nearest decimal point.

Note: a. Includes Republican and Loyalist.

percent for security casualties specifically. Half of those who died in the area were members of the security forces.

Belfast and Derry, the largest urban centers, were the main areas of conflict during the Troubles and saw the most activity and casualties. Studies of the Troubles, especially of urban operations, should acknowledge this, but here, as we are assessing smaller urban centers, it is worth looking at the figures without the skewing involved through including these areas (see Table 6.2). Some of the information remains the same, while other elements undergo interesting changes. In both cases, there is a roughly 10 percent rise in the percentages relating to security forces, suggesting better terrorist targeting or increased vulnerability outside of larger cities. Alternatively, the lower number of civilians may have resulted in less collateral damage. In any case, County Armagh accounted for more than 30 percent of Troubles-related fatalities if we exclude the largest cities. However, these representations do not provide a fair picture, as most fatalities occurred in distinctly rural areas. As such, we need to delve into Armagh in more depth to really place Crossmaglen within its proper context.

Crossmaglen, Forkhill, and Newtownhamilton are three urbanizations within County Armagh.[6] Forkhill and Newtownhamilton represent a "similar" and a "different" comparison to Crossmaglen. Forkhill, as a similar comparison to Crossmaglen, was also predominantly nationalist, it was a nodal point, and the army maintained a fortified barracks reachable mainly by helicopter during normal operations. Meanwhile, Newtownhamilton was farther from the border and mirrored the demographics of North Armagh, being predominantly unionist and Protestant. When we look at County Armagh by itself, there are some interesting findings. Within Armagh, a total of 477 people died as a result of the Troubles, with 50 percent of them belonging to the

Table 6.2 Fatality Breakdowns Excluding Belfast and Derry Cities

	Total	Terrorist Action	Percentage of Total That Were Terrorism-Related	Terrorist Action in County Armagh	Percentage of Terrorist Action from County Armagh
Fatalities	1,764	1,466	83%	477	33%
Security fatalities	716	707	99%	238	34%
Security as percentage of total	41%	48%		50%	

Source: All figures taken from the Sutton Index of Deaths, CAIN Index, University of Ulster, https://cain.ulster.ac.uk/sutton/index.html, and rounded to the nearest decimal point.

security forces. Throughout all operational areas, Armagh had the highest number of security forces fatalities with a total of 238 (see Table 6.3).

Within this we can also see variation between the three selected urban areas. Forkhill and Newtownhamilton accounted for between 5 and 6 percent of fatalities within Armagh. Crossmaglen, on the other hand, was an outlier on all accounts. It accounted for 17 percent of the total casualties for Armagh but 26 percent for the security forces, with 75 percent of all those killed in and around the village a member of the security forces. Another, perhaps unusual, sign of its outlier status was that Crossmaglen was also one of the few areas that Margaret Thatcher visited on her day-long visit on August 29, 1979.[7]

Regardless of whether we include or exclude Belfast and Derry, Crossmaglen is associated with a disproportionately high number of security force fatalities. Including the major urban centers, Crossmaglen accounted for 6 percent of the total fatalities during Operation Banner, and if we exclude them it accounted for nearly 10 percent. While West Belfast and Derry may have been the locations of more violence, we need to think of the scale involved. Belfast accounted for an average of 20 percent of Northern Ireland's population during the Troubles while simultaneously accounting for 37 percent of security personnel killed by terrorist groups. Crossmaglen,

Table 6.3 Armagh Specific Casualty Breakdown

Area	Armagh	Crossmaglen	Forkhill	Newtownhamilton
Total fatalities	477	81	27	26
Percentage of Armagh total		17%	6%	5%
Security fatalities	238	61	17	15
Percentage of Armagh security		26%	7%	6%
Security as percentage of total	50%	75%	63%	58%
Percentage of total security including Belfast and Derry (1,094)	22%	6%	2%	1%
Percentage of total security excluding Belfast and Derry (716)	33%	9%	2%	2%

Source: All figures taken from the Sutton Index of Deaths, CAIN Index, University of Ulster, https://cain.ulster.ac.uk/sutton/index.html, and rounded to the nearest decimal point.

on the other hand, accounted for an average population of .09 percent of Northern Ireland's total, while also responsible for 6 or 9 percent of the fatalities, depending on whether we include the two largest cities (see Table 6.4). In other words, for every 1,077 residents of Belfast, there was 1 security casualty, whereas for Crossmaglen the equivalent was 22 residents. Everything else being equal, if we project the casualty rate from Crossmaglen to Belfast, it would have resulted in 13,930 casualties. While this kind of number crunching may seem odd, it does highlight the outlying nature of Crossmaglen. For such a small place, it is associated with too many casualties to be ignored in a discussion of urban operations in Northern Ireland (see Figure 6.1).

These figures represent an important part of understanding this area within the wider Troubles, but to suggest that this was always going to be so is a mistake. To fully contextualize Crossmaglen and the violence that occurred there, we also need to include some historical information. The most common understanding of Crossmaglen is its role at the heart of South Armagh's "Bandit Country." This is the area that saw the Kingsmill Massacre, the Warrenpoint/Narrow Water bombing, and the disappearance of Robert Nairaic, and was the source of the lorry bombs destined for London and Manchester in the 1990s. The most widely known work on the area, Toby Harnden's *Bandit Country*, states: "Nowhere else in Ireland was the commitment to the IRA's armed struggle so strong."[8] He goes on to claim the following:

> History and landscape are rooted deep in the psyche of the people of South Armagh. . . . For republicans, Crossmaglen is their stronghold, the capital of the *de facto* independent republic of South Armagh. Proximity

Table 6.4 Population According to Northern Ireland Census Data

	Northern Ireland Population	Belfast	Percentage Northern Ireland	Crossmaglen	Percentage Northern Ireland
1971	1,536,065	360,150	23.45	1,257	0.08
1981	1,532,619	314,270	20.51	1,333	0.09
1991	1,573,282	279,237	17.75	1,586	0.10

	Forkhill	%NI	Newtown-hamilton	Percentage Northern Ireland
1971	403	0.03	670	0.04
1981	263	0.02	552	0.04
1991	343	0.02	761	0.05

Source: All figures taken from the Sutton Index of Deaths, CAIN Index, University of Ulster, https://cain.ulster.ac.uk/sutton/index.html, and rounded to the nearest decimal point.

> to the border, the absence of a protestant community, the undulating terrain and the powerful sense of rebellion throughout history have all combined to make South Armagh the ideal operating ground for the IRA. . . . Modern-day republicanism in South Armagh is simply the latest manifestation of its independence and refusal to submit to the Queen's Writ or any other rule from beyond its hinterland.[9]

The ultimate lesson from Harnden's description of the area is that "for republicans in South Armagh, the IRA's campaign . . . is simply the latest manifestation of a rebellion stretching back centuries."[10] This reputation for resistance and identification with the republican cause has been replicated in many formats, such as in this local saying: "From Carrickmacross to Crossmaglen, there are more rogues than honest men." Or in songs such as "The Fighting Men of Crossmaglen."[11] All of these present an image of proud and prolonged resistance to centralized authority, particularly British authority. This is a mythology that both sides drew upon. It provided the rhetorical justification for the militarization of the area, and, depending on the speaker, "bandit" could be a badge of honor. For others, the militarization was a sign that Britain was targeting the area for special treatment. This mythology is both useful and misleading.

It is useful as it provides a context wherein tradition is strong and provides a clear understanding of the area during and after the Troubles, but it is misleading as it suggests that it had always been so. This is not true. Crossmaglen and its surrounding area has always been a place apart. It did not recognize the "North" but did not belong to the "South." This began

Figure 6.1 Location of Fatalities

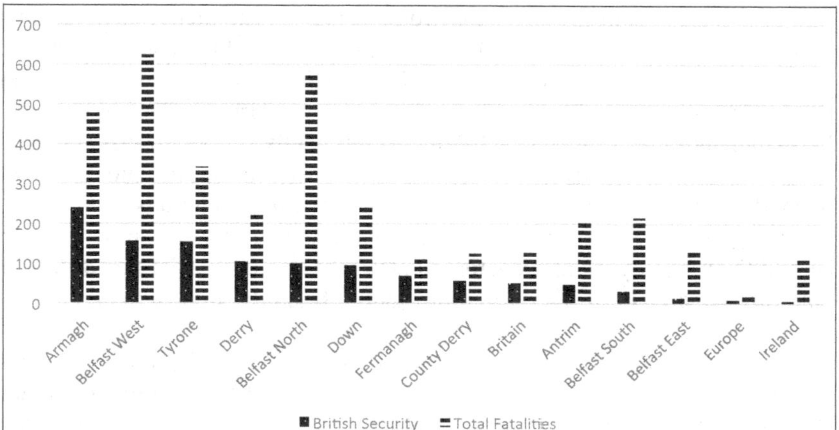

Source: All figures taken from the Sutton Index of Deaths, CAIN Index, University of Ulster, https://cain.ulster.ac.uk/sutton/index.html, and rounded to the nearest decimal point.

with the partition of the island: "the people of South Armagh were convinced that the Boundary Commission would transfer them from Northern Ireland to the Free State. The Northern Ireland authorities shared that view and regarded South Armagh as an unwelcome possession that could not be relinquished too soon."[12] While the Republic would have welcomed the transition of authority over many nationalist areas along the border, including South Armagh, the problems with the Boundary Commission of the 1920s meant that this transition never took place. Another account put it as follows: "Marooned by a border that was not of its choosing, Crossmaglen is a landlocked island cut off by the tide of history."[13] The border would have an important role in how the Troubles would play out in the area, but it was not the only factor. As one former member of the Irish Defence Forces states: "Not to understand the border was to understate its hazard, underestimate its risk and underplay its peril. What made the border dangerous were armed men with deadly intent and local knowledge. You might not always see them but they were always there."[14]

Resistance to the "Bandit Country" myth has grown in recent years, with texts such as Darach MacDonald's *The Chosen Fews* setting out counterarguments to a malignant label, but Patrick Mulroe provides perhaps the most concise challenge. Although South Armagh would become "the problem area": "To explain this transformation requires a degree of sophistication that has hitherto been absent from much of the discourse on the border and the 'Troubles.' The legacy of past republican campaigns certainly had a role in legitimising political violence. Likewise, the experience of generations partaking in the 'black economy' was significant."[15]

Mulroe goes on to show how tradition mattered, whether for involvement in violence or for illicit economies, which both provided two important aspects to PIRA's military capability: combatants and experienced logisticians. But what is just as important is his highlighting of the fact that the area was not actually as active in previous periods of conflict. The same is just as true for Crossmaglen as it was for other areas, such as Forkhill: "When the later IRA campaign was launched in 1970, Forkhill barely figured. It was not until the saturation British Army occupation of the area that the IRA mobilised its Second Battalion, South Armagh. . . . Despite several attempts to keep up, including spectacular raids and gun battles, the district's IRA units fell far behind their colleagues in the Crossmaglen area as the conflict raged. It was not until the later 1970s that Forkhill began to figure prominently."[16]

These relatively calm descriptions are at odds with the storied reputation of Crossmaglen and its surrounds, which may have been exaggerated on both sides for propaganda purposes, both as a justification for increased security measures along this area of the border or as a badge of honor for those engaging in PIRA-related violence. By scratching the surface on mythologies like this, we can also see what Kieran McConaghy means when he encourages us to view state efforts to deal with terrorist threats as their own version of "propaganda of the deed."[17] When the Troubles began in earnest in 1969–1970, South Armagh remained relatively quiet, including Crossmaglen. It was only when a local man was killed in Belfast that the area "awoke." This is a critical point about substate violence and the wider question about the interconnectedness of soft social systems. While South Armagh, and Crossmaglen in particular, would have likely become involved in the conflict eventually, the area was known for the mistrust of outsiders and for its independence from the wider PIRA organization. It took the death of a local, Harry Thornton, wrongly shot in Belfast by paratrooper Sergeant Allan McVitie after his van backfired, to bring "the war home."[18] Indeed, Mulroe makes a strong case to look elsewhere along the border in the earliest years of the Troubles for the real hotspots.[19]

The Operational Environment and Threat

Now that Crossmaglen has been contextualized, let us discuss the specifics of the operational environment and the threat faced by the security forces. As previously stated, Crossmaglen is a small village within a wider rural area close to the border with the Republic. The role of the border could take up the length of this contribution, but our focus must remain on the urban aspects of the topic. Additionally, the border stretched for hundreds of miles and had many larger urban centers along its route that did not represent similar threats. The border was not in and of itself a threat to the security forces; it is what those who lived on it did that presented the threat.[20]

In many ways Crossmaglen, or any other town in similar circumstances, was an operational nightmare. This was for two main reasons. First was its liminality. As with many rural villages and towns, the geography within this area can rapidly turn from urban to rural and back again. Troops may be operating in the alleys of a housing estate one minute and conducting extended, rural patrols the next. This required the mastery of both skill sets and an ability to mesh the two when needed. The training offered by Northern Ireland Tactics and Training provided a good foundation for those deploying within Operation Banner, but very often deployments were classed as urban or rural and did not account for the fluidity of the landscape. One former commander claimed that "we did a lot of training on riot control but none of that was needed in South Armagh at all."[21] Soldiers needed to be able to blend and substitute urban and rural training as needed. For some this was easier than for others. One veteran made this claim:

> You adapted rather than changed significantly. I don't remember it being particularly different. In the urban environment you had the clues to look for, lots of people all the time. In the rural environment you are, it's almost a bit like a real war because if it's going to happen they might have

> a go at you and you might have a chance to shoot whereas in the urban environment the chance is low and then if it does happen in a rural environment you can fire back willy-nilly. In an urban environment you have to think about what you are doing because there are people there that you could hurt.[22]

The second major difficulty was the isolation. This isolation can be seen in many ways. The first was physical. Although the army maintained other positions in South Armagh, Crossmaglen was particularly isolated. Not long into the conflict, overt road travel was declared too risky due to improvised explosive devices (IEDs), especially culvert bombs, along the narrow country roads in the area. As such, all travel was either covert or required helicopter assistance. This meant that support, should anything go wrong, would have to be in force or come from helicopter assistance, which was not always present.

A second form of isolation was social. The population of Crossmaglen was nationalist from the outset of the conflict, although not automatically republican, meaning that sympathies for deployed troops were low, and intimidation and mistakes by the security forces, including the killing of innocents, compounded these problems.[23] Perhaps the most persistent bone of contention between the locals and the army was the army's use of sections of the local sporting field as a helicopter landing zone. This was an issue that was discussed with surprising regularity by civil services of both countries and actually reacted to at the highest levels, and it was the topic of discussion between prime ministers and taoisigh.[24] If nothing else, this shows how the continuation of conflict and everyday life run parallel, rather than in isolation from one another. Tactical decisions in the area that negatively affected popular relations were compounded by the development of the Golf and Romeo towers that ringed the area to gather visual and signals intelligence.[25] This social isolation made it incredibly difficult to gather any kind of human intelligence (HUMINT), resulting in even more isolation than normal for a platoon base and the formation of a significant siege mentality. As a former Royal Marines intelligence officer stated in relation to this,

> South Armagh PIRA's reputation for self-discipline was pretty ferocious and do you need a cellular structure when everybody is either related to or drinks with or you know are so closely bonded in such a special place anyway? And then over the top of that you have this sort of harsh punishment system that brooks no suggestion of treachery, let alone actual treachery. I mean, yeah, I think South Armagh is a special place. . . . There are books written by people who claim to be high level sources but certainly down in our part of the world, as far as we were aware, there was nobody who was talking. . . . I am not sure about intelligence, I think that was part of the reason why we did operations as aggressively as we did, and we just didn't have the intelligence to suggest it should be done in a different way.[26]

Another account from the area during the 1980s paints a picture of troops who felt depressed and morbid because of their perceived and real physical, social, and operational isolation.[27] This isolation prevented any kind of "hearts and minds" operations from taking place; in reality, the troops and the population they were trying to protect were becoming increasingly hostile to one another. At times, with both sides knowing clearly who was who, this led to routine targeting while on patrol and almost ritualistic interactions: "I have to say that we reciprocated in kind. . . . I can remember to this day shaking down one of the [locals] for no other reason than we just wanted to shake him down just to make his life miserable because he made our lives miserable. There was little hearts and minds, there was very little . . . you know securing the population, we didn't try and secure the population in Crossmaglen, it would have been pointless they would've, you know, killed us."[28]

Within this context, the security forces had to operate out of limited space. The initial police presence in Crossmaglen relied on a small building that housed a few constables, as with many other villages of this kind. However, as the conflict progressed and this barracks came under regular attack, it became a hardened "Border Fort" first and then a full Forward Operating Base (one of the first installations to actually bear the name), complete with steel walls, mortar resistant housing, and an external sangar (Boruki Sangar), which directly abutted the village square.[29] Although the final state of these facilities would provide enough space for police and army personnel, for the majority of the conflict the buildings were cramped and vulnerable. Depending on the unit in question, this either provided motivation to increase patrolling to alleviate the threat and the overcrowding, or it hurt their morale and resulted in less activity as personnel lived under siege.[30] In many ways the barracks itself became the center of the struggle. The security forces had to operate out of it, and their opponents had to remove it to finalize their control over the area. This resulted in many attacks on the barracks and a resulting increase in fortification, which encouraged more elaborate attacks, and so on. In the end the fortification efforts won out, but in doing so increased the physical isolation of those deployed in the area and reinforced the siege mentality harbored by both sides. The soldiers and police were under siege by the PIRA and the locals suffered increased militarization. This need to hold territory will be discussed in more detail below. The initial figures presented in this chapter contained the information we have relating to casualties, but we can also assess Crossmaglen in other ways—specifically, the kinds of attacks that took place and who those attacks targeted. If we use the data contained in the Global Terrorism Database as another conservative estimate, we can trace the attacks that occurred directly in the village. In doing so, we can identify core trends that clearly outlined the threat faced by troops within this urban environment: the main perpetrator was the PIRA, the main target was the military, and the main threat was explosives based. This is important for several reasons, but its importance here is to draw out certain lessons.

The first of these is that even though they are the "hardest" target, the military can become the main target of violence. While other actors may have sought to achieve their aims by intimidating civilians or by targeting other nonstate forces, in areas dominated by one force, the military can and will become the primary target. This was PIRA's "home turf"; some other groups may have occasionally operated in the area, but 93 percent of attacks recorded within the town were accredited to PIRA (see Figure 6.2). The relative lack of other forces in the area meant that PIRA had increased operational freedom, which allowed increased time and space to conduct operations but also provided the context for increased military targeting (see Figure 6.3). The military were the primary security force in the area as the police were not resourced to tackle the threat; indeed unmarked cars are still used in the area today rather than overt policing. In contrast to areas such as Belfast, where the military and the police were confronted with a series of actors all competing for supremacy, here they confronted a single actor who was able to focus their efforts on them. This trend is perhaps best illustrated through a comparison to Belfast, where only 24 percent of targets were military in nature.

Moving to the kinds of attacks employed, we can see other trends present, particularly a reliance on explosives (see Figure 6.4). PIRA throughout its campaign relied more on explosives than firearms. This was due to

Figure 6.2 Attack Perpetrators

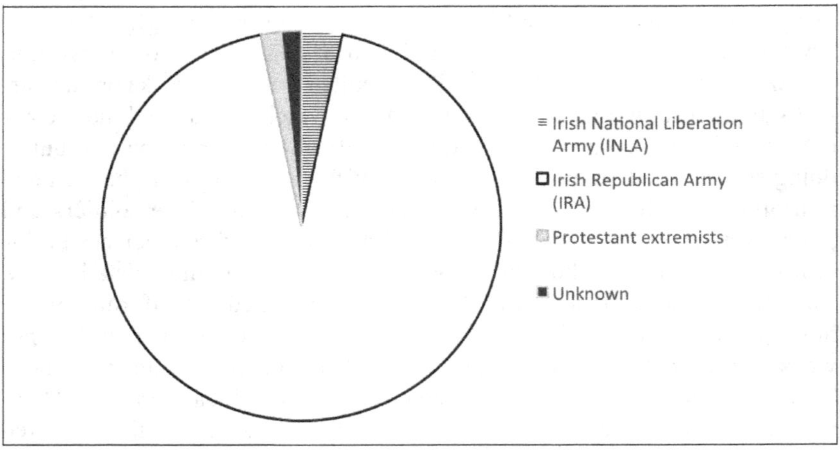

Source: All figures taken from the Sutton Index of Deaths, CAIN Index, University of Ulster, https://cain.ulster.ac.uk/sutton/index.html, and rounded to the nearest decimal point.

resources and training, but also PIRA eventually adopted a policy of caution and preference for remote attacks in order to reduce the risks suffered by its personnel.[31] There were other areas where the PIRA preferred to "come out and fight," but for the most part, and certainly in Crossmaglen, the preference was for IEDs and other explosives, such as mortars.[32] The threat from explosives in this area was also heightened due to the ready access to agricultural fertilizer and fuel, given the area's reliance on agriculture and fuel trading, which formed the basis of most of PIRA's explosives. This links to the problem of liminality mentioned before. The PIRA ASUs in the area had the resources and advantages of urban and rural guerrillas, with few of the disadvantages. This is part of the added difficulty in operating in between the urban and the rural; soldiers either tended to overreact to potential threats or they forgave what would be anomalies in any other urban setting as part of the flow of rural life. Sometimes this resulted in humorous instances, other times it resulted in casualties. A former Ulster Defence Regiment officer made the former point clear:

> I always remember the squadron leader . . . explaining to me "Don't laugh but as part of their pre–Northern Ireland training, I gave the boys cow and bull recognition courses." I had a vision of flipcharts of "points of note": udders and other such things. It probably wasn't quite like that but it was actually quite sensible of course . . . people were not familiar with rural life in general, including livestock and the normal pattern of rural life, which confused people and people would get worked up about something which was perfectly normal.[33]

Figure 6.3 Attack Targets

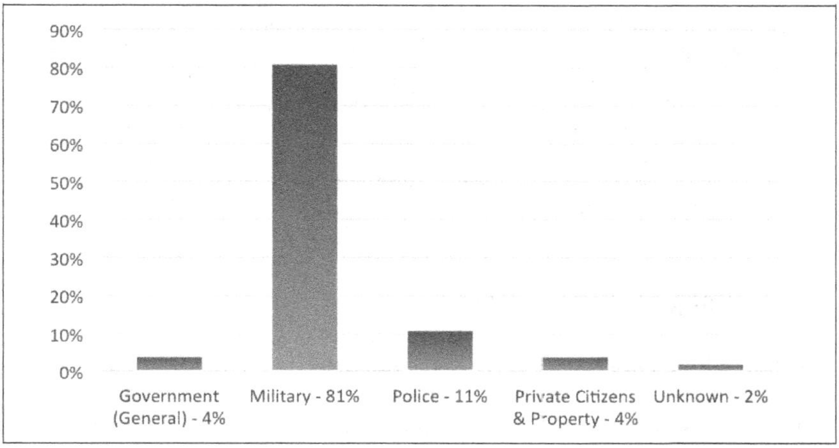

Source: All figures taken from the Sutton Index of Deaths, CAIN Index, University of Ulster, https://cain.ulster.ac.uk/sutton/index.html, and rounded to the nearest decimal point.

Figure 6.4 Type of Attack

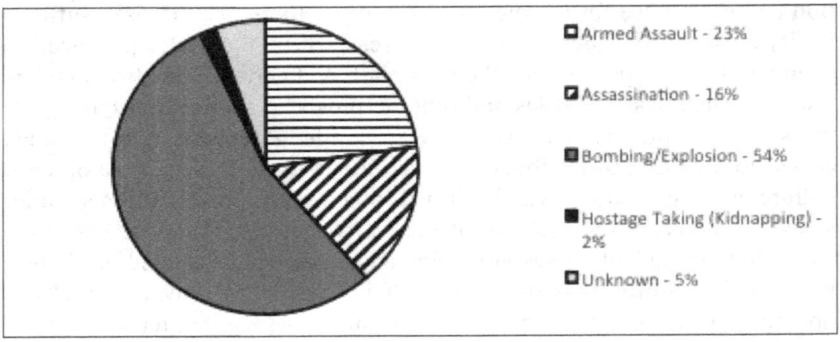

Source: All figures taken from the Sutton Index of Deaths, CAIN Index, University of Ulster, https://cain.ulster.ac.uk/sutton/index.html, and rounded to the nearest decimal point.

Edward Burke has made clear how soldiers struggle to operate effectively within "normal" social environments, but there is also the difficulty of operating within a fluid physical environment.[34] For instance, the appearance of a slurry tanker alongside Palace Barracks would have attracted substantially more attention in November 1992 than it did when one approached Boruki Sangar. The normality of this event in a rural village disguised the fact that it had been converted into a giant flamethrower, which set the sangar ablaze.[35] This was well summed up by one of Toby Harnden's interview participants:

> You normally find guerrilla armies are either urban or rural. South Armagh is rural but can operate very effectively in urban conurbations and that's what they did in England. Although there is farming there, there are also haulage companies, lorry companies and used car businesses. That sort of engineer-farmer thing becomes very interesting. These are people who can do welding and cutting and they've got bits of metal lying about and they've got a man who runs the metal works. All the ingredients, the scaffolding for the booster tubes, everything is there. Go round farmers over there you'll find all that stuff in their sheds.[36]

"The Fighting Men from Crossmaglen"

The preceding paragraphs have outlined why it is important to look at Crossmaglen and what the troops stationed there faced during their deployments. Let us now turn to an assessment of those who made up the PIRA units that these troops went up against. There are two main concerns here: (1) structure and practice and (2) levels of cohesion. Both interacted with one another to create a proficient terrorist threat, but one that was also highly resistant to interrogation and attrition in general. As will become clear, the military was aware of who they were fighting, but the policy of only engaging those directly involved in acts of terrorism meant that this knowledge could not be acted upon quickly. Targets needed to be found in the act, such as at Loughgall, where a PIRA ASU was wiped out by the Special Air Service (SAS) during a gun and bomb attack on the Royal Ulster Constabulary (RUC) barracks there. With more cautious units, however, this policy meant they operated for months if not years without being caught. This relative freedom allowed these units to develop certain practices that made them even more resistant and effective.

When assessing the PIRA ASUs in and around this area, there are important similarities with the wider organization but also some important differences. The first of these issues is that of structure. PIRA transitioned from a traditional ("brigade") style organization in 1977 and adopted a "cell structure." This involved a significant downsizing of the organization but also increased specialization. The reason behind this was a fear of infiltration and a desire to reduce attrition. However, South Armagh did not adopt this immediately. One British commander stated that "whilst the cell structure was adopted in most of Northern Ireland in response to the successes of the British Army and the intelligence services, the South Armagh Brigade continue to operate in the traditional way."[37] Some have argued that the failure of these units to adopt this structure was a symbol of its weakness and academic overemphasis. But this is a straw man argument.

South Armagh was a rural operational area with concentrated urban centers. In the PIRA policy paper that outlined its reorganization—the "Staff Report"—rural areas were separated from the towns they surrounded and were not to be included in the reorganization.[38] The cell structure was a response to urban infiltration, but because of the social isolation we mentioned earlier within rural and semirural areas, this was not a problem for this operational area. South Armagh was PIRA's most secure area. With this resilience to HUMINT and its rural nature, South Armagh was exempt from automatic inclusion in the reform process. However, as time went on, units in the area did transition to this structure. One former volunteer from the area stated that these units "did move into cells in a certain way in that smaller groups would operate and even if something happened in one part of the area this did not mean that everyone else knew about it."[39] This was the core rationale behind PIRA's reforms; increase the capability of each unit but also limit the amount of knowledge. These changes had significant impacts on the organization, but these changes were perhaps the most pronounced in this area. This is because the units here were some of the most socially dense and technically capable in PIRA. In other words, they had the highest levels of social and task cohesion.

If we begin with the idea of social cohesion—the levels of personal bonds and mutual support between unit members—we can look to the role of the family within this area. Harnden described those active in the area during the Troubles as "simply the latest manifestation of its independence and refusal to submit to the Queen's Writ or any other rule from beyond its hinterland."[40] While this is debatable, the idea is important. One former British commander described his time in the area as follows: "South Armagh was particularly dangerous for the Security Forces. Almost half of all the British soldiers murdered by the IRA in the first decade of the Troubles had been killed in this sparsely populated rural backwater, only about a hundred miles square. . . . What we were involved in here was counter insurgency, of the type the Army had conducted in Malaya and elsewhere, but fought in what should have been a sleepy corner of the British Isles."[41]

Building on this, Harnden claims that "while British Army units have come and gone . . . the same volunteers still form the backbone of the IRA. Some of the South Armagh Brigade's gunmen and bombers are in their forties or fifties; few in Belfast would be still active beyond their twenties."[42] Gill and Horgan have previously shown how the age of volunteers often identifies the length of time they have been active, as most generally become active at a young age, with the eldest members amassing years of experience.[43]

This prolonged career allowed for a different training approach to be taken than that used by most of the organization, who relied on formalized camps based in the Republic of Ireland and elsewhere. Brendan Burns, a notable PIRA volunteer from the area, who eventually blew himself up, put it thus:

> South Armagh developed a slow process of training the volunteer. They would have taken raw recruits and put them in with skilled people and took them on. They didn't push them too hard. And they got them used to what they were going to be faced with. And over a period of years . . . people working together in close proximity with each other, experiencing operations, came to trust and depend on each other. In fact they would have been loath to meet up with other people or to operate with volunteers outside their own immediate circle.[44]

Bonding between members was based on familiarity and a trust that "had to be earned" but was also based on "how a person's family is regarded. Betrayal has been rare because such an act is regarded as unnatural, cutting across blood ties as well as those of fealty."[45] This system, based on personal loyalty and familiarity, had its own costs and benefits. It produced a smaller number of highly proficient members, but it also provided opportunities for those same members to identify with themselves more than the wider organization. In some cases, this resulted in deviancy, such as when some of those in the area broke off to form the Real IRA. But the benefits ensured that they suffered the least from infiltration and conducted the most successful operations.

However, these developments were not automatic. Units in these areas were disproportionately active and had the benefits of operating within the urban and rural environments. What was a disadvantage for the military was an advantage for the ASUs. These units treated deviance harshly and judged others based upon their operational conduct. While they may not have mirrored wider developments within the organization, these units were effective operators and employed social cohesion to their benefit, rather than allowing it to dictate their actions, which was an important difference to other areas. Particularly toward the beginnings of the conflict, traditional republican families formed the basis of many PIRA units, and social connections and background could well determine the trajectory of one's career, but as PIRA reorganized this changed, and merit surpassed social connections in terms of importance.[46]

What made these units more effective was how they combined the effects of social connections with their commitment to achieving their goals; their commitment to expertise was combined with their commitment to one another. This is relatable to the relationships identified by Patrick Bury when he speaks about modern soldiers in Afghanistan.[47] Unit members are all capable, otherwise they would not be there, but they are also socially linked to one another. Their professionalism derived from their ability to operate effectively became embedded within their social relations. This was true for conducting terrorist violence as much as for other activities, including smuggling, which was extensively connected to PIRA units in this area:

> Certainly on the South Armagh side there was a lot more influence there from money because the smuggling was big business and you didn't do that unless you had the IRA on your side and most people were entirely synonymous, it's the same thing. We used the terms *black forces, red forces* and *white forces* in human terrain. Black forces are your criminal, white forces are just people, red forces are the enemy, well in South Armagh the red and black were the same people. They were criminal and in the IRA, not because he was in the IRA but because he was a criminal, because he smuggles stuff.[48]

Smuggling was conducted before and after the Troubles.[49] With the close-knit communities along the border, it is entirely likely that individuals were involved with smuggling before they were active with PIRA; the practice and their involvement in it continued after their recruitment. A former volunteer described the relationship between group membership and everyday life as follows:

> Areas like South Armagh were much more methodical. Just the way of life there, the outlook of the world, families that knew each other. There was no point in asking "if you have a couple of new recruits make sure that you get them vetted by the intelligence officer in Belfast." For a start, they would have known their family, would have seen them playing football, some of them probably would [have] worked together, they would have known what school he went to. They would've known what they were like; they didn't need to go into the IRA to know what these guys were like. If they had seen someone on the football pitch at 16 that he was fucking useless and cowardly, he wasn't going to get into the IRA. . . . A way of putting it is the difference between; you recruit say an 18-year-old. Say an 18-year-old guy from Turflodge joins the IRA and he is fine. People sort of know him, the way they would, there is a huge difference in that and the guy who grew up on the farm across the way from . . . who knows his dad, knew his uncles, knows his brother, knew his ma, went out with his sister.[50]

What all of this meant was that a normative "shield" developed wherein these social connections introduced an extra incentive to resist interrogation. With those involved so socially connected, any breaking under security-force pressure would have an effect on their wider social network, not just on those who operated alongside them.[51] A former British commander noted this:

> I think the family had an awful lot to do with it . . . if your father had been a nationalist or a republican and had been involved in the IRA, it was almost part of the culture to be involved in it and the whole family, there were families that quite obviously were republican in nature and which supported the IRA or the Provisional IRA, and there were families that didn't . . . I mean for somebody in a family in South Armagh to actually wish to speak about something to the security forces, not only is their life expectancy pretty short but also, they were ostracised by not only their friends but by their family too. So, no amount of money or no amount of motivation is going to separate somebody from their family or it will do if people have been thrown out of the family already but other than that. So, I think you have got two different things at play: the need for security takes you away from having a military structure but if you have a very tight family structure, as you do in rural areas of Northern Ireland, that's far more dominant than the military structure anyway. . . . I mean, operating against other terrorist organisations, the combination of a cell structure and a very tight family structure, make the life of the security services very difficult. That's why Belfast was more difficult for PIRA to keep secure, or one of the reasons for PIRA to keep more secure than the rural areas, particularly South Armagh of course.[52]

Another claimed that "we had no success in penetrating South Armagh PIRA whatsoever, even though we knew everybody who was involved in it, virtually everybody who was involved in it. Whether it was the fear of what was going to happen to them if they did come over to our side or just the fact that they hated us so much, we were the unifying purpose if you like."[53]

This normative shield acted as an additional defensive force for PIRA; however, it did cause additional problems. It enforced a level of mistrust within the organization in other areas or units that inhibited cooperation, particularly with Belfast.[54]

All this mattered because it robbed the military of most of their potential HUMINT. This reinforced the reliance on signals and visual intelligence gathered by installations such as the Golf and Romeo towers, which in turn reinforced the militarization of the area and further enhanced the distance between troops and the local population. At the same time, the British military could not resort to blanket occupation and harsh counterinsurgency (COIN) measures. This was because all of this violence and the measures to stop it were still occurring on British soil. The United Kingdom was pursuing its normalization/criminalization policy, where the police were supposed to be the leading force. This does not mean that the military was "operating with an arm tied behind its back," but it did mean that it had to operate under the law of the United Kingdom and could not resort to the same measures used in previous COIN campaigns in the likes of Aden or Malaya. What this meant was that the military needed to rely on patience rather than power, so to speak. As will be shown in the next section, careful reconnaissance and targeted interrogation provided the best counter to ASUs rather than undirected violence, which just served PIRA's recruitment propaganda.

Drawing Lessons

To reinforce the lessons of this chapter—the problems of liminality, isolation, and the effectiveness of the opponent—we can turn to a single case study to show how a PIRA unit benefited from operating in this semi-urban space, was proficient in its tasks, and was incredibly socially connected. It was also undone by a failure in this normative protection by including an "outsider." The unit in this case was the Caraher sniper team.

This ASU was part of a two-team campaign of single-shot sniper attacks throughout South Armagh, but mainly within the different villages in the area, particularly Crossmaglen. These groups conducted a number of attacks using Armalites and between two and three Barrett .50 caliber rifles. Although these weapons had substantial ranges, the average shooting distance was 2–300 meters. These shots were fired out of the back of a modified car in order to blend into the environment, but also to escape after the shot. During one period, these kinds of attacks resulted in the killing of seven security personnel between February and December 1993, with another soldier killed in 1992 and 1997, bringing the total to nine.[55] At the time, some believed these attacks were conducted by a hired mercenary or other international actor.[56]

The ASU was four strong: Michael Caraher, James McArdle, Martin Mines, and Bernard McGinn. The first three lived around Silverbridge, Cullyhanna, and Crossmaglen. McGinn was from the County Monaghan town of Castleblayney, itself a short drive from Crossmaglen in the other direction. The Caraher sniper team was a prime example of how local units adopted the cell system. It was a small, specialized task that was self-contained. It sourced its own equipment and conducted its own operations. On the day of the arrest, the team was preparing for another attack but were surprised by the SAS and captured. How this arrest was planned has been the subject of some debate. At its height, the fallout of the arrest was subject to a Police Ombudsman of Northern Ireland investigation after claims that attacks were being allowed to take place while the unit was under surveillance; this included the killing of Lance Corporal Stephen Restorick, who was the last soldier to be killed during the Troubles. The report concluded that the security forces were aware of the team's activities in general and which vehicles and properties they might have used to launch attacks, but the materials to be used in this attack were not confirmed in advance. Specifically, when talking about what was known about the ASU and its activities, the report stated

> the car used in the attack was one of a number of vehicles under surveillance in the area at the time and that police should have directed closer and more precise surveillance of it. . . . The Police Ombudsman investigators were given access to sensitive and secret intelligence documents held by Special Branch and the Army. They also searched a number of police premises for relevant documentation. The Police Ombudsman, Mrs Nuala O'Loan, said she is concerned about the level of information and original documentation which had been destroyed including "decision" logs, policy files and documentation about a surveillance operation which was underway in the area at the time. She said there was no "audit trail" as to who made the decisions to destroy this material and why. . . . Mrs O'Loan said she is concerned that police did have an intelligence assessment that the Mazda car was to be used in an attack and did not do more with this information. She said that in the absence of a "decision log" it is not possible to attribute those decisions to any officers.[57]

This suggests that the security forces were closing in on the ASU following a wider surveillance operation and were able to identify the vehicle and premises that would be used for this final attack. Following the ambush of the ASU, with some physical confrontation but no casualties, the team was arrested and interrogated.

What matters to us is that McGinn, the "outsider," broke during his interrogation. Harnden argues that this was due to fear about missing out on his son growing up.[58] In the hopes of immunity, McGinn provided the police with significant information but was given a 495-year sentence.

This was later reduced as part of the Good Friday Agreement, and he was released in 2000.[59] He later died in 2013 of a suspected heart attack. Although he would later retract his statement, McGinn's decision to provide information matters for many reasons. First, by breaking, he proved locals right in their decision to not trust those from outside their own community. Although all of the team were found in the same act and others were facing other charges, he was the only one who broke. Using McGinn's confession and other evidence, the team was sentenced and given lengthy prison terms before they were released early under the Good Friday Agreement as well. Second, because McGinn was alone in confessing, he shows the different primary groups that may be active within political violence at a micro level. While all of those involved had to consider the potential impact on their families, McGinn's family was sufficiently distant, even if only a few miles away, for him to disregard the stigma that would be placed upon them. As a local volunteer later stated, McGinn should not have been included in that team in the first place.[60] He was a liability because he was not under the same normative pressures as those he was operating with. In a unit that had shown its technical skill and had successfully targeted the security forces on several occasions beforehand, McGinn was the weakness and was properly pressured by the security forces and broke. This is an example of the kind of patient approach needed to unpick the threat in an operating area like this while operating within an otherwise "normal" existence.

These moves were facilitated by greater success across the rest of Northern Ireland, with the intelligence campaign securing greater victories in places such as Belfast much sooner. This was combined with the increasing political progress to eventually end the Troubles. South Armagh could not wage their campaign in isolation, and as the rest of the PIRA was contained elsewhere, the security forces could focus more and more resources on the area and eventually destabilize it beyond recovery. The military could not use extreme force as it had done in previous campaigns, nor could it win through hearts and minds because it had isolated itself from the local population and prioritized force protection over integration and cohabitation. It had to bide its time and wait. This shows the difficulty of operating in this kind of liminal space. The security forces had to contend with the difficulties of rural insurgency and urban terror and needed to blend their response to be effective. Relying solely on urban counterterror measures may have pacified the village but not the surrounds. Likewise, extensive rural COIN operations would not have sufficiently undermined the immediate urban threat that saw security installations attacked so often, and even if it did show some success, would have simultaneously reinforced PIRA's narrative of fighting a force of occupation.

The Forgotten Role of Territory Control

The official campaign analysis of Operation Banner provides several lessons supposedly learned from the Northern Irish Troubles:

- the need to take the long view as well as addressing immediate concerns;
- the importance of a central figurehead to ensure a cohesive approach across all actors;
- a coherent, high-level information operations campaign;
- the need for a sensitive application of force;
- the need to address the causes of conflict as well as the symptoms;
- the need for highly developed intelligence gathering and analysis systems;
- the importance of sufficient command arrangements;
- the need to amend doctrine as required; and
- the need for effective training.[61]

Other works have placed the importance of multiple patrolling, with enhanced intelligence operations and the use of special forces, as the core lessons to be learned.[62] One thing that these sources do not account for is the role played by a need to control territory at what seems to be all costs. This appears to have had an influence at all levels. To demonstrate this, we can look at strategic and tactical levels for evidence.

If we turn first to the tactical level, Burke provides a great example of how violence and territorial control ran right down to the lowest levels on all sides. As he says, "The micro-conflict, fought over a small 'patch' of street or fields, was all-consuming."[63] The desire for dominance and control of their territory saw Scottish soldiers clash with any other groups seeking to challenge their authority. The IRA was a clear example of this, but they also clashed with Loyalists, youth gangs, and even members of the US military based in Derry.[64] At times these challenges occurred within the wider political context, at other times it was fueled more by the masculine antagonisms of those involved, wherein groups of young men clashed to establish dominance over each other. At times control of territory was surrendered at a tactical level and at other times it was actively contested. This can be seen through referring back to how different units deployed to Crossmaglen responded to their threat and environment. Some patrolled more to alleviate overcrowding, while others remained within their barracks to avoid casualties. While establishing control is always important in conflict scenarios, in this case it came at a disproportionate cost. This is both in terms of casualties to *contest* a village and later in terms of overall troop levels to *control* the wider area. From an RUC barracks with a few constables, the British presence in the area developed into a forward operating base capable of housing roughly 100 troops, and by the 1980s the wider area required nearly 1,000 troops.[65]

This all stemmed from a wider political concern about ceding territory. Even in places where British control was only nominal, the political need to maintain at least nominal control was important. From the outset of the Troubles, we can see examples of real territorial contestation. The Battle of the Bogside and the establishment of no-go areas in Belfast and Derry are clear in this regard. Indeed, Operation Motorman and its Derry-specific component, Operation Carcan, saw over 20,000 troops deployed in force to British cities to retake control of barricaded areas from insurgent control. It was also the only time during the Troubles that armor was deployed, in the form of a Centurion AVRE battle tank. Similarly, the targeting of RUC barracks in Tyrone, which culminated with the Loughgall ambush, was a deliberate attempt at territorial control by the IRA in that area.

Beyond this there were also considerations surrounding the border and how much control needed to be enforced upon it. Aside from the problems resulting from cratering or otherwise blocking border roads, there were high-level discussions about whether it should be policed at all. The question was asked, if the border presented such an operational disadvantage, why not pull back from it? In this line of questioning we see some similarities with recent debates about the Irish border and Brexit planning. It was recently suggested that government facilities, and therefore influence, should be placed "five to ten miles 'back' from the actual land frontier."[66] Similar points, as well as their inverse (i.e., more control), were raised in the 1980s. During the security reviews of South Armagh, Fermanagh, and Tyrone conducted by Maurice Oldfield, the border was a frequent point of discussion along with how the PIRA members were able to use it to their advantage and how the southern security forces seemed unable to stop them. In one element, the opinions of the military are made clear: "There is no point in trying to establish an Iron Curtain along the border. . . . Nor is there any point establishing PVCPs [permanent vehicle checkpoints] in South Armagh. . . . There is a continuing need for the Security forces to *dominate* the roads in South Armagh, while remaining independent of them."[67] The report continues and directly addresses the point about surrendering territory, embodied in the presence of police stations: "The defence of police stations is a major commitment for the Army in South Armagh. But any abandonment of these stations would be interpreted as a major political and military defeat and is not recommended."[68] Territory could not be surrendered, even if this meant a continuation of disproportionate casualties.

Although the above information may seem obvious, given Britain's emphasis on hearts and minds–based COIN and information-led counterterrorism, this inability to consider surrendering local control even when there are no hearts and minds to win over or when attempts to gather information for counterterrorism lead to increased hostility seems counterintuitive. However, these priorities are still present in British doctrine: "The ability of land forces to seize and hold ground enables the physical control of territory and signifies the highest levels of political commitment."[69] In the case of Crossmaglen, we see the need to demonstrate "political commitment" surpassing concerns about force protection and local relations. Placing troops in this area made them a target, one that local forces consistently targeted, often successfully.

Conclusion

In this chapter, I sought to discuss the hazards found in a historical case study of urban combat. Rather than looking at an intense citywide conflict to demonstrate these aims, I chose to look at a longer, more drawn-out campaign centered on a small village. The conflict zone in question, Crossmaglen, was contextualized in relation to the wider Northern Ireland theater and its more immediate surrounds. This contextualization demonstrated its outlier status and the need to try and understand this episode in urban combat. In doing so it sought to highlight the influence of liminal operating environments and how identity can shape operational considerations on both sides of the conflict. Alongside the role of identity was also a discussion of the role of territory within British military and political thinking. Commitments to holding ground made the disproportionate levels of casualties suffered holding a relatively small area politically acceptable. The liminal nature of this deployment meant that it was not entirely urban, but it also meant that urban aspects could not be ignored in favor of a strictly "rural" skill set. This was a campaign that required resilience and patience rather than extreme firepower. Soldiers operating in this area were not transplanted to "far off lands of which we know little"; they were operating in the "British" countryside while those around them went about their daily lives. This was a conflict at the squad and platoon level, based on surprise and precision. As stated in the introduction, it was a tactical contest that had higher ramifications. Violence was often intimate and or highly demonstrative.

All that remains is to ask whether there are lessons that can be scaled up to larger urban settings. Some of these are obvious, such as trying to find the balance between justifiable force protection measures on the one hand and militarization of everyday life and structures that prevent the collection of local HUMINT on the other. We can also point to the problem of dealing with a nonstate actor that thrives on a tradition of resistance, real or imagined, and how the state's own messaging can reinforce this narrative. On a more nuanced level, the events covered in this case study highlight the need to understand how even distant locations can be socially connected and how security force action in one area can have consequences elsewhere. This case should also stand as a challenge to any notion that a "small" deployment is in any way safer than deploying to a larger, more populated area. One issue that requires further analysis is the influence of substate control over areas. Crossmaglen and its surrounds were the IRA's turf; they did not need to regularly compete for control with others aside from the security forces. They had one target and they committed their attention to it. Other areas, such as Belfast, may have been larger in every aspect, but they were also the scenes of increased substate competition, which meant that these actors could not dedicate as much time to targeting the security forces. Whether this is a positive or negative implication remains to be seen.

Notes

1. Burke, "Counter-Insurgency Against 'Kith and Kin'?"
2. DiMarco, *Concrete Hell*, 150; Ministry of Defence (UK), *Operation Banner*, 8–14.
3. FCO 87/3: Activities of Sinn Fein and Irish Republican Army (Foreign Commonwealth Office, National Archives).
4. See Urban, *Big Boys' Rules*; Hamill, *Pig in the Middle*; Dewar, *British Army in Northern Ireland*; Davies, *Ten-Thirty-Three*; Lewis, *Fishers of Men*; Sanders, "Operation Motorman"; Bennett, "From Direct Rule to Motorman"; Dixon, "'Hearts and Minds'?"; Kennedy-Pipe and McInnes, "British Army in Northern Ireland."
5. Such as "FOI Smyth 02-01-2013-160507-018 correspondence dated: 29 December 2012," DASA-Health Information, Ministry of Defence, response to Freedom of Information request, https://www.whatdotheyknow.com/request/operation_banner_soldiers_that_d#incoming-357035.
6. Newry has not been included here as it fell within the South Down ASU operational area, rather than the South Armagh.
7. PREM 19/385: Prime Minister's Engagements Wednesday 29 August 1979 (Prime Minister's Office Files, National Archives).
8. Harnden, *Bandit Country*, 11.
9. Ibid., 12–13.
10. Ibid., 70.
11. Williams, "Proverbs," footnote 25.
12. Harnden, *Bandit Country*, 106.
13. MacDonald, *The Chosen Fews*, 16.
14. Harvey, *Soldiering Against Subversion*, 74.
15. Patrick Mulroe, "Moving Away from the 'Bandit Country' Myth," *Writing the "Troubles"* (blog), https://writingthetroublesweb.wordpress.com/2018/05/21/moving-away-from-the-bandit-country-myth/#_edn2.
16. Harvey, *Soldiering Against Subversion*, 139.
17. McConaghy, *Terrorism and the State*, 14.
18. Harnden, *Bandit Country*, 37.
19. Mulroe, "The Most Notorious Trouble Spot Along the Entire Border," 566–584.
20. PIRA Operations Officer and South Armagh Volunteer, interview by Patrick Finnegan.

21. British Commander (2), interview by Patrick Finnegan. The second British commander interviewed.

22. British Platoon Commander, interview by Patrick Finnegan.

23. Harnden, *Bandit Country*, 67.

24. CJ 4/4572: Army property requirements at Crossmaglen, County Armagh (Records of the Northern Ireland Office, National Archive); CJ 4/3053: British Army use of sports pitch in Crossmaglen: relationship with the Crossmaglen Gaelic Athletic Association (Records of the Northern Ireland Office, National Archive); CJ 4/2522: Gaelic Athletic Association (GAA), Crossmaglen (Records of the Northern Ireland Office, National Archive); FCO 87/2153: Gaelic Athletic Association (GAA), Crossmaglen (Foreign Commonwealth Office, National Archive).

25. Harnden, *Bandit Country*, 187.

26. Royal Marines Intelligence Office, interview by Patrick Finnegan.

27. Hockey, *Squaddies*.

28. Royal Marines Intelligence Officer, interview by Patrick Finnegan.

29. CJ 4/3085: Area reviews of security: purpose and format: individual surveys on County Londonderry, South Armagh, Fermanagh and East Tyrone (Records of the Northern Ireland Office, National Archive).

30. Harnden, *Bandit Country*, 291.

31. Ibid., 15.

32. Collins, *Rules of Engagement*, 45.

33. UDR officer, interview by Patrick Finnegan.

34. Burke, "Counter-Insurgency Against 'Kith and Kin'?"

35. Harnden, *Bandit Country*, 92.

36. Ibid., 253.

37. British Commander (2), interview by Patrick Finnegan.

38. PIRA Staff Report in FCO 87/725: Provisional Sinn Fein and Irish Republican Army (IRA) in Republic of Ireland (Foreign Commonwealth Office, National Archive).

39. South Armagh Volunteer, interview by Patrick Finnegan.

40. Harnden, *Bandit Country*, 13.

41. Jackson, *Soldier*, 146.

42. Harnden, *Bandit Country*, 32.

43. Gill and Horgan, "Who Were the Volunteers?"

44. Harnden, *Bandit Country*, 34.

45. Ibid., 35.

46. For a discussion of these changes, see Finnegan, "Professionalization of a Nonstate Actor."

47. Bury and King, "Profession of Love."

48. British Platoon Commander, interview by Patrick Finnegan.

49. CJ 4/3085: Area reviews of security (National Archives).

50. Ibid.

51. Bloom, "Constructing Expertise."

52. British Commander, interview by Patrick Finnegan.

53. Royal Marines Intelligence Officer, interview by Patrick Finnegan.

54. Moloney, *Voices from the Grave*, 276.

55. Harnden, *Bandit Country*, 290.

56. "Officer B" believed "that the paramilitary activity under surveillance was not just of a local nature but involved the 'crème de la crème,' who were possibly acting on a national and international basis." Police Ombudsman for Northern Ireland, "Report into a Complaint," 22.

57. Police Ombudsman for Northern Ireland, "No Available Evidence that Terrorist Attack on Soldier Could Have Been Prevented," December 12, 2006, https://www.policeombudsman.org/Investigation-Reports/Historical-Reports/No-available-evidence-that-terrorist-attack-on-sol.

58. Harnden, *Bandit Country*, 228.

59. *BBC News,* "IRA Killer Laughs at Sentence," March 19, 1999, http://news.bbc.co.uk/1/hi/uk/299635.stm.

60. South Armagh Volunteer, interview by Patrick Finnegan.

61. Ministry of Defence, Chief of the General Staff, *An Analysis of Military Operations in Northern Ireland,* Army Code 71842.

62. DiMarco, *Concrete Hell*, 146–150.

63. Burke, *Army of Tribes*, 100.

64. Ibid., 107, 168–170.

65. CJ 4/3072: South Armagh Area Review: report by Security Co-ordinator's Planning Staff on coordination of efforts of security forces and of civil government (Records of the Northern Ireland Office, National Archive).

66. Connelly, "UK Proposes Customs Posts."

67. South Armagh review, 1–2, in CJ 4/3085: Area reviews of security (National Archives) (emphasis in original).

68. Annex A: Summary of findings, in CJ 4/3085: Area reviews of security (National Archives).

69. UK Ministry of Defence, *UK Land Power*, 11.

7

Civil-Military Relations and Ethics

Matthew S. Wiseman

ON THE NIGHT OF APRIL 13, 1945, ONE BATTALION OF 2ND CANADIAN Infantry Division entered the city of Groningen in the northwest Netherlands and attacked a defending force of German soldiers reinforced by Dutch and Belgian Schutzstaffel (SS) troops. The attack commenced an arduous four-day battle that cost 209 Canadian casualties and took the lives of 140 Germans and 106 Dutch civilians.[1] Meeting a defending force of 7,000 head-on in a deeply entrenched city, the heavily outnumbered and relatively small force of Canadian troops captured their objective and took 2,400 prisoners.[2] The Canadian Army had little experience in an urban warfare environment, and the defending forces had several advantages, including personnel strength, fortifications, and local terrain knowledge. With no opportunity to escape, the battle for Groningen exposed a large local population to the fighting that occurred, resulting in the mass civilian casualties and extensive damage to family homes, businesses, and farms in and near the city.

Groningen's liberation freed local residents, displaced persons, and refugees from the tyranny of enemy occupation and control. The city's population numbered approximately 115,000 when the war broke out in 1939, but fighting in the southern provinces of the Netherlands drove tens of thousands of Dutch refugees out of their homes and north, quickly swelling Groningen's population to well over 150,000 by war's end. The city thus represented a strategic location that required immediate intervention, for both military and humanitarian reasons. By attacking Groningen and recapturing the city, the Canadians engaged in urban warfare, saved thousands of civilians in the region, and made an important contribution to achieving the greater Allied cause of liberating the Netherlands during World War II.

This chapter recounts the battle for Groningen and highlights some of the daunting civil-military challenges encountered by the Canadian forces who liberated this region of the Netherlands in April and May 1945. From the outset, the soldiers committed to Groningen were severely disadvantaged. They represented a small army facing a big city, and their objective was heavily fortified and extremely well protected by a larger defending force that knew the terrain and had prepared to fight with relentless ferocity. The defenders held the numerical advantage and were equipped with powerful weaponry, while the city itself was overcrowded by a local and transient population, which posed unique tactical challenges for the smaller attacking force that needed to engage the enemy and take their objective with as few civilian casualties as possible. Indeed, the Canadian experience with urban warfare at Groningen was not restricted to the battlefield alone. Despite facing difficult circumstances and a potential humanitarian crisis, the Canadian Army successfully liberated the city and remained on location well after the fighting ended, assuming responsibility for reestablishing civil control and supporting local Dutch rehabilitation efforts during the final months of the war. This wider civil-military context of the fight for Groningen is useful for understanding a stark reality of urban warfare: the attentiveness and commitment owed to the local population before, during, and after conflict.

Groningen's Recorded History

Dutch accounts are useful for understanding the battle for Groningen and the civil-military context of the Canadian liberation in the northern Netherlands. Several Dutch books appeared within months of war's end, including *En T'och Staat de Martini* and the photo book *Groningen in Vuur en Puin, De Bevrijding van Groningen*.[3] Nearly four decades later, in 1980, author M. H. Huizinga published *Maple Leaf Up*, using reports and eyewitness accounts to describe the liberation of northern Holland, and Groningen, in particular.[4] Huizinga continued his research and produced a second book in 1999, *Vier Dagen in April*, detailing specific events as experienced by Dutch citizens on the ground.[5] Other descriptive picture books marking and commemorating the seventieth anniversary of the liberation emerged in 2015, including Beno Hofman's *Groningen in de Tweede Wereldoorlog* and Martin Hillenga's *Groningen 40-45*.[6]

English-language sources about the Dutch experience at Groningen are rare, but librarian and scholar Greg Sennema recently published portions of his family's wartime diary to examine the liberation as experienced through the eyes of his grandfather, Theo Polman.[7] Using his opa's *dagboek*, Sennema draws on the personal experiences of his grandfather to contextualize Groningen's liberation as witnessed by the Dutch citizens who lived and worked in the city.[8] Theo and his wife, Bep, were the proprietors of Polman's Tobacco, a store located at the northern edge of Groningen's inner city. Theo recorded the everyday experiences of his family, friends, neighbors, and colleagues, leaving behind a rich and descriptive wartime diary for his children and grandchildren.

While the story of Groningen's liberation has received comparatively less attention in the English literature, a selection of key historical works traces the route of the Canadian Army from Normandy through to northern Germany. In his official history of the Canadian Army in World War II, C. P. Stacey documents the liberation of Groningen by focusing on the operational context of the battle itself.[9] One of Canada's foremost military historians, Terry Copp, expands on Stacey's work in *The Brigade* and *Cinderella Army*, using terrain knowledge and war diaries to analyze the operational capabilities of the Canadians on the ground.[10] This is also done briefly by historian Tim Cook in his second volume on the history of Canadians in World War II, *Fight to the Finish*.[11]

Several other scholars have examined the military reasons for the Canadian attack on Groningen. Historian Ralph Dykstra argues that the Dutch city was a "military necessity" for the Allies, whereas US Army Major Jeffrey D. Noll suggests that Groningen's liberation was purely political and humanitarian in purpose.[12] Conversely, Gregory Ashworth contemplates the German decision to stay and fight at Groningen, especially considering the viability of withdrawal toward Germany.[13] All three authors agree that the Canadian Army limited its use of firepower to reduce civilian casualties and control collateral damages in the city, but none explore or investigate civil-military relations in the region. Regimental histories, military records, and other surviving documentation held at national repositories in both Canada and the United Kingdom add another layer of contextualization, enabling a historical reconstruction of the events at Groningen and analysis of the Canadian experience with urban warfare and civil affairs during the closing months of World War II.[14]

The Allied Advance

By April 1945, the Allies had liberated France, had crossed the River Rhine, and were prepared for a final offensive push into Germany. All Canadian forces in the European theater operated in concert with the British Army. Field Marshal Bernard Montgomery commanded 21st Army Group, which included First Canadian Army, under General Harry Crerar. Portions of I Canadian Corps had previously fought in Sicily and Italy, and the remainder of the Canadian forces had served under the British in Western Europe during the Normandy campaign. March of 1945 saw the Canadian soldiers in Italy moved north to unify the forces for the final phase of the war.

The arrival of the additional forces brought the entire overseas formations of the Canadian Army under one unified command in Northwest Europe, where they played an important role in the Allied operation to cross the Rhine. Shortly after crossing the Rhine and entering Germany, 2nd Canadian Infantry Division received orders to travel northwest for the campaign to liberate the Netherlands. With 3rd Division on their left flank moving toward the Dutch province of Friesland, and both 4th and 5th Canadian Armored Divisions on their right, 2nd Division headed for the northernmost province in the Netherlands, Groningen.

Hoping to lead the charge into the heart of Germany with 21st Army Group, Montgomery and the British command proposed a direct line of attack toward Berlin.[15] Supreme Allied Commander General Dwight Eisenhower had other plans, despite the intentions of Montgomery and British Prime Minister Winston Churchill. On March 29, Eisenhower ordered a US-led attack south of Berlin, toward Leipzig and Dresden, while Montgomery and the British attacked across the Elbe to seize Hamburg and Kiel. This protected the northern flank of the US attack and cut off the Danish Peninsula from Germany, allowing the Soviet Army to advance on the German capital. Under Eisenhower's plan, Montgomery and 21st Army Group received orders to liberate Denmark and the Netherlands. Montgomery directed Crerar to deploy two divisions to western Holland, while the remaining Canadian forces cleared northeastern Holland and enemy naval establishments along the coast. This opened an Allied supply route through Arnhem, a strategic location that the Allies had failed to capture seven months earlier during Operation Market Garden.[16]

Allied assessments from February and March of 1945 estimated the presence of approximately 200,000 German forces in the Netherlands, including air defense units and administrative personnel. The British and Canadian armies expected the occupying German forces to hold firm defensive strongpoints but also readied for their potential withdrawal. Even though shortages of equipment and training had deteriorated the German forces, Crerar and his officers predicted that challenging terrain and recent flooding in the northern provinces of the Netherlands would make liberating the country difficult.[17]

While the exact number of enemy troops occupying and defending the city of Groningen is unknown, estimates placed the German forces at approximately 7,000 strong. Wehrmacht Infanterie Division 480 departed the area by train on April 5, leaving behind a disjointed collection of military forces that was unable to take full advantage of the fortifications constructed in and near the city.[18] In the absence of a unified regiment or brigade-style formation, the German forces constituted a mix of all available service personnel, including the Heer, Luftwaffe, and Kriegsmarine operating in and near the city. The defending forces also included a strong contingent of SS troops, composed of ethnic Germans and Dutch nationals, as well as nonmilitary or paramilitary organizations, which included the German Security Service, Hitler Youth, and local railway operators. The "Dutch SS troops knew there was no future for them but death," recalled Canadian Harry Fox, who observed the resistance at Groningen. "It was the very young ones, sixteen or seventeen years old, that were the most fanatical, however, and they would die before giving up."[19] Unequipped with heavy armor, the defending forces at Groningen had in their arsenal twin- and quadruple-mounted 20-millimeter antiaircraft guns that caused extensive devastation to the attacking ground troops. The defenders also had the Panzerfaust, a rocket-propelled antitank grenade that caused considerable damage to both the hard cover of the attacking forces and soldiers in open ground.

Despite the relative limitations of equipment and weaponry, the defending forces at Groningen had several clear advantages. In addition to the complex network of roads and waterways in and near the city, the occupying forces had grown familiar with the terrain and had fortified any preexisting weak points well in advance of the Canadian attack. In September 1944, the Germans had forced local Dutch men to dig and construct trenches, antitank ditches, and weapons pits along the banks of Groningen's wide ring canal. This forced labor saw the construction of formidable bunkers and barriers, sited strategically near the main bridges and access points to the city. The German defenders also received advanced warning of the impending attack. "The endangering of the area Emden–Groningen and (small smudge) of the mouth of the Ems by enemy landing from the sea or from the air, which is possible within a very short while, demands the rapid reinforcement of all branches of the armed forces in this area, and a unified control of the defence," stated an intelligence cypher intercepted by the Allies in March 1945.[20] Having lost the advantage of surprise, First Canadian Army faced a disjointed but strong collection of defensive forces that had fortified their position and were well prepared for the impending attack.

First Canadian Army

When the Canadians from Italy arrived in March and April 1945, they joined their comrades in Northwest Europe, under General Crerar's reunited First Canadian Army. The entire force comprised thirteen national divisions, totaling over 470,000 troops (although considerably fewer for combat), as well as 1,200 guns and 3,400 tanks.[21] Major General Bruce Matthews's 2nd Division fought under Lieutenant General Guy Simmonds's II Canadian Corps. Matthews replaced Lieutenant General Charles Foulkes as division commander in November 1944.[22] Foulkes was unpopular among the senior command, who reassigned him to replace the fired Major General E. L. M. Burns in Italy, where he could do less harm to the Canadian war effort.[23] Matthews had accrued an impressive record in Italy, Normandy, and the Scheldt, obtaining experience in set-piece battles and demonstrating his ability to take a leading role as the Canadians punched forward into Northwest Europe.

To Crerar's reunited First Canadian Army went the immediate task of liberating the northern Netherlands and occupying parts of Germany as the Allies pushed toward Berlin. A desperate race against time, the liberation was both military and humanitarian in purpose. The Nazi occupiers had made drastic cuts to local food rations, forcing Dutch civilians in the southern provinces of the Netherlands to fewer than 900 calories a day.[24] "I had never seen anything that moved me like the hunger I saw in Holland," one shaken Canadian infantryman observed.[25] The dire circumstances placed the most vulnerable populations at severe risk, leading to a spike in death among the elderly and increasing fears that the young would follow. Many Dutch civilians ate flower bulbs to survive, according to Canadian military historian Cook.[26]

With tens of thousands of people on the verge of starvation, the Allies knew they had to act quickly and aggressively to stave off a humanitarian disaster. Ground forces used shell-carrying trucks to transport food while bombers dropped supplies rather than explosives. Soldiers at the front did what they could, often sacrificing meals to feed the impoverished civilians they encountered. Gunner J. P. Brady, a Métis soldier from Wetaskiwin, Alberta, recalled saving his rations to feed the desperate children, some of whom were "so hungry they howl like little wolf whelps, while others whine weakly from starvation."[27] As the Canadians pushed ahead engaging enemy positons at Emden, Arnhem, and Apeldoorn, the count of starving Dutch mounted and the stark reality of life under forced occupation became painfully clear to the liberating forces. Most of the Germans surrendered with little bloodshed, until the Canadians reached Groningen.

The Canadian advance into the Netherlands began on March 31. In early April, First Canadian Army was tasked with opening a supply route through Arnhem, clearing northern Holland and the German coast toward the Elbe, and beginning the liberation of the western regions of the Netherlands.[28] Crerar assigned Arnhem and the liberation of Holland's old provinces to the redeployed I Canadian Corps. The II Canadian Corps moved north quickly, with 4th Division maintaining contact with the British on the right, and 3rd Division coordinating actions with I Corps on the left. Meanwhile, 2nd Division received the important responsibility for advancing toward Groningen. The main enemy forces in the area had started withdrawing to the Twente Canal, where the Germans had planned to establish a new defensive line.

Battle for Groningen

Bearing the same name as its province, the capital city of Groningen is an old university town that dates from the Middle Ages. Originally constructed in the fourteenth and fifteenth centuries, Groningen's inner city has narrow streets suited for one-way traffic. Apartments and close-set brick buildings line the streets. Ranging from three to five stories in height and arranged curbside along the tight and winding city streets, the buildings in wartime Groningen helped to fortify the defensive position of the occupying German forces. Several waterways characterized the landscape, including a wide ring canal that enclosed the inner city itself. Thirteen moveable transportation bridges and one pedestrian footbridge crossed over the ring canal, allowing the German forces to establish and maintain formidable defenses on location. The wide canal completely encircled the inner city, and the evenly distributed bridges served as access to the city's downtown core. Several canals flowing from the south and the west also served as obstacles for Canadian soldiers, equipment, vehicles, and supplies.

Groningen's vast and interconnected system of waterways provided isolation from the countryside and created an easily defensible city for the occupying forces. Only two roads from the south provided direct land access, but the defenses were strong, and both routes were blocked. The Germans had either destroyed or rendered inoperative many of the canal bridges before the Canadian approach in April 1945. Much of the fighting occurred in the inner-city core, which contained brick apartment flats and buildings interspersed among a dense network of small streets. Three boulevards running west to east through the inner city were wide enough for heavy military equipment, but the remaining streets and pathways were too narrow for tanks and other large vehicles. Two large squares stood adjacent to each other in the core of the city center, acting as the primary battlefield and experiencing substantial fighting and destruction. A moat protected the old town, and canals encircled much of the newly constructed areas. Fighting to capture the city was thus a daunting prospect, but the Canadians knew that liberating Groningen and the port of Delfzijl, while 3rd Division cleared the enemy from Emden, would enable the Royal Navy to open a critical supply route and end the U-boat threat in the Ems Estuary.[29]

The battle for Groningen commenced during the night of April 13, when Matthews ordered 4th Brigade to advance and begin clearing the city. He lacked solid information about the garrison, but early results indicated that the occupying enemy forces were prepared to fight aggressively. Prisoners revealed that, in addition to several ad hoc battle groups of German soldiers, a battalion of Dutch SS operated in the area.[30] Soldiers in 4th Brigade approached the city from the south, preparing to attack the main defenses head-on. Upon reaching the southern outskirts of the city, they encountered enemy resistance along the banks of the outer ring canal. After penetrating the city's outer edge from this location, 4th Brigade engaged in close quarters combat and house-to-house fighting against German and Dutch SS units. The Canadians encountered a strong defensive resistance with hidden machine guns and SS snipers disguised in civilian clothing.[31] Fighting occurred through the night and into the next morning.

At 6:00 a.m. on April 14, the Canadian command outlined a plan to attack and capture old town in the city's downtown core. Major Ken MacIntyre's A Company of the Essex Scottish led an advance at 10:30 a.m., supported by 6-pounder guns that knocked out the enemy's 40-millimeter antiaircraft guns. Fierce hand-to-hand combat characterized the fight for the railway station area, which included all three battalions.[32] Soldiers in A Company reached the canal south of the city at 2:15 p.m., finding one of the few bridges that remained intact and serviceable. The Canadians immediately brought the bridge under fire from the windows of a house overlooking the canal, preventing German soldiers from withdrawing across the bridge and back into Groningen's inner city. War diaries and memoranda of the fighting indicate that 4th Brigade gained this bridgehead in the southern portion of the city by rushing in Kangaroos—armored personnel carriers—and using small arms and bazookas to overcome the heavily defended enemy forces. Accessing the intact bridge enabled 6th Brigade to cross and begin clearing the city.

Under orders against using artillery and airstrikes to capture Dutch cities, the Canadian forces that attacked Groningen exercised relative restraint during the battle. Exactly who issued the order for restraint is unclear, however.[33] Although Lieutenant General Simmonds, the commander of II Canadian Corps, and Major General Matthews, the commander of 2nd Division, both had the authority to prohibit aerial and artillery bombardments at specific locations, the Canadians followed a general policy in the Netherlands that, when possible, restricted excessive force against cities.[34] In lieu of a specific written order or policy stipulating the use of force in urban centers, the Canadian forces in Northwest Europe followed the directives of Montgomery and 21st Army Group, who took their cues from the British government. London was under pressure from the Dutch government-in-exile to liberate the Netherlands and stave off a full-scale humanitarian crisis, and communication with Dutch authorities reaffirmed the severity of the circumstances and the importance of strategic but cautious military action. Indeed, the health and safety of the civilian population, including local residents, displaced persons, and refugees, was top of mind for the Canadians as they advanced into the Netherlands and carried out their mission to liberate an entire country from German occupation and control.

Despite the general policy restricting heavy armaments and excessive force, the Canadians used artillery and tanks to overcome the defending forces and capture Groningen. To prevent the occupying troops from retreating north toward Delfzijl and northwest Germany, three field regiments of the Royal Canadian Artillery targeted the city's eastern edge with 25-pounders sited at Eelde, a community located 10 kilometers south of Groningen. Strategic artillery strikes also targeted a sugar beet factory, the local electrical plant, water towers, and other strong defensive fortification points located outside the inner city. Old town in the city center escaped largely untouched by artillery, but the Canadians used Sherman tanks and Wasps mounted with flamethrowers to achieve fire superiority and defeat the enemy. This resulted in a relatively precise but destructive attack that caused considerable damage to some 270 specific buildings, dwellings, and businesses—a necessary cost to limit the number of casualties and secure victory.[35]

Grote Markt took the brunt of the fighting and damage to the inner city, but Groningen's infrastructure was largely unscathed. The railway station and adjoining yards remained intact and serviceable, as did all major municipal buildings and hospitals. The large natural gas plant on the edge of the inner city functioned throughout the entire battle, serving its customers during and after the Canadian attack. Artillery fire disrupted electrical power services on Saturday, April 14, but the disruption was temporary and had a negligible impact on city residents. While the battle for Groningen was four days of intense fighting, the relatively cautious approach of the Canadians limited the damages and destruction that occurred.[36]

The ferocity of the street fighting in Groningen led Matthews to commit 5th Brigade, which had been kept in reserve, to a new attack at the city's western edge. Tasked with clearing the northern sector of the city, the Black Watch crossed the canal using a temporary barge and scouted the area for enemy forces. They checked all houses by simply "ringing the doorbell and then standing on the doorstep waiting admission."[37] The first signs of opposition came at dawn, when sniper fire began targeting the first Canadians to cross the canal. Shortly thereafter, 20-millimeter fire prevented the rest of the battalion from advancing into the city, until local Dutch residents positioned a second barge that gave the remaining Canadians the cover they needed to cross.

During the early morning hours of April 15, Les Fusiliers Mont-Royal of 6th Brigade relieved the attacking battalion and continued the advance into Groningen's old town. Soldiers of 6th Brigade passed through the opening created by the Essex Scottish, while 5th Brigade entered the city from the west. Groningen's inner city experienced intense fighting between the attacking Canadian soldiers and the remaining German defenders. "A characteristic of the defence was the siting of [machine gun] posts in the basements of buildings covering neighbouring streets," stated an official battlefield report of Canadian operations in Northwest Europe. "Nevertheless," the report continued, "approximately 900 [prisoners of war] were obtained during the day."[38] Moving into Groningen from the west, with soldiers of the Calgary Highlanders and the Black Watch on the right and left, respectively, 5th Brigade had reported the city clear as far as the 21 easting line by 5:00 p.m. Entering Groningen from the south, with soldiers of Les Fusiliers Mont-Royal and the South Saskatchewan Regiment on the right and left, respectively, 6th Brigade had reached the 14 northing line by the same time.

Despite the severity and intensity of the fighting, Canadian battlefield reports describe large crowds of Dutch civilians in the streets of Groningen during and after the attack on the inner city.[39] Because the Canadians did not shell or bomb the city's main area extensively during the battle, they attacked relatively slowly and cautiously, reducing friendly casualties and limiting any unnecessary harm to the local population. For residents who had lived under German occupation during the war, the sound of rifle and machine-gun fire in the streets of their city represented a hopeful end to the tyranny that had consumed their wartime lives. One signaller of the Calgary Highlanders later recalled the experience of a fellow machine gunner who had set the bipod of his Bren gun atop a small hardwood table inside a city dwelling.[40] Seemingly unaffected by the presence of the Canadian soldier or the danger of the fighting, the homeowner calmly handed the gunner a small cushion and asked that he place it under the legs of the gun to protect her table from damage. The gunner obliged the request and then accepted a cup of coffee, all the while firing through a bay window toward a German vehicle located down the street.

Although the Canadian command issued orders to protect civilian buildings and guard the well-being of city residents, the safety of the soldiers on the ground remained top priority during the attack. Major Sandy Pearson of the Calgary Highlanders received a visit from Groningen's local postmaster during the evening of April 15. In an attempt to save the city's post office from fire and damages, the postmaster informed Pearson about a large contingent of German soldiers concentrated at that location and asked that the Canadian major waive a white flag to persuade the enemy forces to surrender peacefully. Pearson offered to accompany the postmaster to the post office but the postmaster refused. "I told him I'd much sooner burn the post office (with our flamethrowers) than risk any Canadian lives and he left in a bad mood," Pearson later recalled.[41] The next morning, Pearson led the attack on the post office and the German soldiers evacuated the building and surrendered. Surviving records say little about the fate of the post office, but Canadian flamethrowers caused considerable damage and destruction to several city buildings, and it remains unclear if the decision to limit artillery during the battle for Groningen produced the intended result.

The surrender at the post office signaled the end for the occupying forces. Later in the day on April 16, the German commander and his staff in Groningen surrendered unconditionally. "Some fighting continued in the northern outskirts of the city, but the core of the enemy's resistance was gone," noted a Canadian battlefield report. "The divisional task had been completed in record time and 2 Canadian Infantry Division had achieved the distinction of being the first formation to attain its objective in this phase of the campaign [to liberate the city of Groningen]."[42] Elements of the garrison continued the resistance, with forces positioned on the east side of the van Starkenborgh canal at the northeast edge of Groningen. The German defenders had chosen this location for strategic reasons, purposefully opening a drawbridge that could not be closed and lowered from the near side of the canal. Unable to cross, the Canadians opened fire and the two forces engaged in combat from the opposing sides of the canal.

Long and arduous, the firefight at the van Starkenborgh canal was the last stand for the occupying German forces at Groningen. Sherman tanks appeared on site to assist the Canadian ground troops, laying fire from 17-pounders on the German positions and destroying several houses and farms in the process. The Germans had blown up any temporary bridges to stall the Canadian advance, and the tank support fell short of ending the combat. At 4:00 p.m., a local bridge operator appeared on the scene with two family members and offered to help close the bridge for the Canadians. The local residents assisted three soldiers of the Queen's Own Cameron Highlanders, collectively thwarting the German withdrawal on the opposite side of the canal.

Dutch author W. K. J. J. van Ommen Kloeke recorded the events in a book documenting the liberation of Groningen: "Together with three Canadians these gallant men crawled across the lock north of the Bascule bridge and whilst the Canadians by means of a hellish fire barrage diverted the attention of the enemy, they crept along the east back of the van Starkenborgh canal and in this way reached the engine-room of the bridge unscathed. With man and might they were now turning the movement mechanism, whereupon the bridge came down with tantalizing slowness."[43] After lowering the bridge, the Dutch supporters opened the west-end gate. German machine gunners immediately brought the bride-deck under fire and prevented the Canadian advance, at which point the three brave civilians moved underneath the bridge and attempted to open the gate at the other end. The bridge operator took a gunshot to one of his upper arms, and his two relatives jumped into the canal to keep from faring worse. "The tanks did not wait for the end of this drama, but rammed the last gate and thundered across the bridge spitting fire in all directions," van Ommen Kloeke continued. Once across the canal, the Sherman tanks overwhelmed the Germans, and the Canadian ground troops took more than sixty prisoners with relative ease. Any remaining resistance collapsed shortly thereafter and the battle for Groningen came to an end.[44]

In the aftermath of the fighting, Lieutenant General Charles Foulkes, commander of I Canadian Corps, met with German representatives to establish a temporary truce and discuss the opening of Dutch towns to food deliveries. Talks progressed slowly and there was no cease-fire until April 28, even though much of the actual fighting had ended nine days earlier.[45] Beaten and severely reduced in strength, the occupying enemy forces that remained in the Netherlands tended to follow the unofficial cease-fire, save for a fierce clash with 5th Armoured Division at the port of Delfzijl that marked one of the final battles for the Canadians who fought in Northwest Europe.[46] The German forces in Italy surrendered on May 2, followed shortly thereafter, on May 5, by the surrender of all German forces in the Netherlands, northern Germany, and Denmark. Negotiations for the total surrender of Germany came to a close two days later, in the early morning hours of May 7, officially ending the war in Europe.

Civil Affairs in World War II

The responsibility of controlling a civilian population in a theater of war was certainly not unique to World War II. While conquering armies have long discharged the responsibility of civil affairs in different ways and with varying degrees of success, the creation of a special civil-military relations organization in the British Army dates to the Palestine campaign of World War I. Between 1914 and 1918, trained and experienced administrators from the Colonial Service managed civil affairs in Palestine as the Occupied Enemy Territory Administration (OETA).[47] British officials revived the OETA under a different name in the midpoint of World War II, when the conquest of Cyrenaica, Eritrea, and Italian Somaliland necessitated the replacement of the Italian Colonial Administration.

Owing to the diplomatic context of the period, the liberation of Abyssinia prompted the removal of the word "Enemy," and British officers of the Occupied Territory Administration acted as civil advisers to the restored Negus. The organization underwent another name change shortly thereafter, functioning as the British Military Administration until the United States entered the war following the Japanese attack on Pearl Harbor in December 1941. The invasion of Sicily and Italy in 1943 urged a third name change, and the Allied Military Government of Occupied Territory (AMGOT) came into being under the auspices of an organized and coordinated approach to civil affairs in conquered territories.[48]

AMGOT consisted of British, US, and Canadian officers who possessed certain qualifications for, and had received special training in, civil administration. The organization was "unique in the fact that for the first time in history two major powers established on the principle of joint responsibility and equality of effort a single military government for conquered enemy country," explained historian Ralph Gabriel during a lecture on Sicily and Italy, delivered to the University of Charlottesville's School of Military Government in late December 1943.[49] The Allies had signed the armistice with Italy three months earlier, and as a diplomatic gesture, military officials dropped the last two words from the organization's name and settled on Allied Military Government as a final designation.

Senior civil affairs officers managed appropriately qualified staffs who worked with the existing chain of command, from the Supreme Allied Headquarters down to the headquarters of each particular corps. Each staff managed its own group of officers who, working in detachments, served as civil affairs field troops. A detachment consisted of no fewer than two administrative officers and two public safety officers, in addition to other ranks, augmented as required by certain specialist officers with particular knowledge or expertise (e.g., financial, legal, labor, supply, and public health).[50] The civil-military relations division of the Canadian Army had two distinct designations. A service known as "Civil Affairs" maintained civil-military relations in the liberated countries of Northwest Europe, while a branch of the staff called "Military Government" was responsible for civil-military relations involving Canadian personnel and operations in Germany. Both designated units ensured that the civilian population did not interfere with military operations.

Although the same branch of the Canadian military was responsible for civil affairs in the liberated territories of Europe and military government in Germany, the two approaches to civil-military relations imposed and applied different methods. Civil Affairs used friendly and close cooperation with the government representatives, civil authorities, and inhabitants of the liberated territories. Its official functions were twofold:

> First, to assist the military commander's plans in the forward battle area by liaison with the civil authorities and by controlling the activities of the local inhabitants in such a way as to prevent disorganization, disease and unrest hampering the activities of the fighting troops.
>
> Second, at a later stage, to exercise administrative control and supervision, in such areas as may be directed by the competent authority, in order that the civil machinery may be set going as early as possible and in such a way as to benefit the allied war effort and to ensure the preservation of law and order.[51]

The legal basis for most operations was an agreement with, or invitation by, the accredited government of the country concerned. In this regard, the Canadian Army endeavored to reconcile practical military requirements in the field with the welfare of the liberated people, displaced persons, and refugees. Any demands made upon local government authorities or inhabitants were as reasonable and modest as the circumstances permitted. "Only such orders and controls as were necessary have been instituted: they have, on the whole, been tactfully imposed and enforced and they have been lifted as early as possible," stated a July 1945 military brief prepared by Canadian civil affairs officer Major A. K. Reid. "Every encouragement and assistance has been given to the governments to resume their rights and duties and to the civilian populations to resume their normal lives. Only in isolated and extreme cases has Civil Affairs been forced to operate on other than an indirect basis."[52]

In Germany, Military Government officers not only carried out the functions required to maintain civil affairs but also provided the Supreme Commander and appointed military governors with an organization through which the country could be governed and Allied policy enforced. "The velvet glove has been removed," Major Reid explained, "[and] the authority for the establishment of Military Government is force."[53] This does not mean that the Allies operated free of constraint in Germany, however. The customs and usages of war as codified in the Hague Regulations limited the power exercised by the Allied armies in the field, as did any such directives received by the Supreme Commander from government authorities.[54] Military governors issued orders to civil organizations and authorized personnel on the ground, which held the force of law. Failure to obey orders resulted in trial by a Military Government court and punishment for convicted offenders included fine, imprisonment, or death.

Following the establishment of a Civil Affairs directorate and the appointment of Brigadier S. Swinton Lee as deputy chief civil affairs officer in December 1942, officials in the Canadian Army began canvassing and selecting qualified personnel for special training as Civil Affairs officers.[55] Lee and other Allied military officials interviewed suitable candidates at Canadian Military Headquarters in late January 1943, ultimately selecting fourteen Canadians for training at the newly established Civil Affairs Staff Centre in Wimbledon.[56] The requirements for selection stipulated that each successful candidate have experience in at least one specialized area of civil affairs and a "sound knowledge" or working proficiency in one or more European languages other than English.[57] The course curriculum included military and civil studies. All students received instruction in military organization and staff duties, and specialist groups studied the history, governmental structure, and local administrative systems of a particular country. France, Belgium, Holland, Luxembourg, Denmark, Norway, and Germany were among the countries studied by officer candidates at Wimbledon. By the end of 1944, Civil Affairs had trained a sufficient number of officers for duty in the liberated countries of Northwest Europe, and subsequent instruction was confined to Germany. In addition to locality, specialized groups focused on such subjects as finance, international law and usages of war, trade and industry, supply and relief, fire and civil defense, public safety, and labor and engineering.

As the need for trained personnel increased, National Defence Headquarters (NDHQ) received nearly 300 applications from officers who desired to serve in some capacity with Canada's civil affairs organizations.[58] In fact, when the sixth Civil Affairs Staff Centre course opened at Wimbledon in early December 1943, eleven candidates arrived direct from Canada. Later that month, owing to the expectation that the Canadian Army would play an increasing role in forthcoming operations, the War Office suggested the creation of one or multiple Canadian civil affairs units for employment with the Canadian Army.[59] NDHQ approved the formation of one unit, drawing officers from those trained at Wimbledon and graduates of the Canadian civil affairs staff courses conducted at the Royal Military College in Kingston, Ontario, Canada. By the end of September 1944, the Canadian Army filled the quota of other ranks, and the unit readied to join the wider civil affairs pool of the Allied armies in Europe.

Civil Affairs in the Netherlands

Canadian civil affairs personnel received a rude awakening in the Netherlands. In France and Belgium, the Canadians encountered local administrators and organisations capable of maintaining regional civil control with limited outside interference. This enabled Civil Affairs to divest several responsibilities, thus freeing its supporting armies to plan for Germany while keeping any formal civil-military commitments to a minimum in both extent and duration. The Netherlands presented different challenges altogether. "In the Netherlands," explained a second report prepared by Major Reid, "more was expected of it than Civil Affairs was intended to provide and the people never understood that it was merely part of the war machine and existed only to assist in the prosecution of the war and the defeat of the enemy."[60] For this reason alone, the Netherlands produced the most difficult civil affairs problems of the whole campaign. The Canadians encountered a "general feeling of apathy" among the Dutch people and faced unique challenges in working with regional representatives.

In preparing for the liberation of the Netherlands, the Allies planned to limit any period of civil affairs administration. The main reason for this was to prevent military personnel from becoming involved in long-term rehabilitation projects, but equally important was the need to avoid the impression that some form of Allied "occupation," with Civil Affairs acting as a benign military government, had simply replaced the defeated German regime. The exiled Dutch government in London co-opted the plan and formalized the Netherlands Military Administration (NMA), a national civil affairs organisation designed to assume control and act as an interim military government until the liberation was over and normal conditions returned. A preinvasion tripartite agreement among the Netherlands, the United Kingdom, and the United States provided that "in areas affected by military operations it is necessary to contemplate a First, or Military Phase, during which the Supreme Commander, Allied Expeditionary Force, will, to the full extent required by the military situation, de facto possess complete authority to take all necessary measures."[61] Once the first phase was over and the military situation under control, the Dutch government was to reassume full responsibility for regional and national civil administration.

The Allies originally planned to withdraw Civil Affairs from the Netherlands by October 1944. Unfortunately, the liberation proceeded much more slowly than the comparatively rapid advance of the Allied pursuit through France and Belgium, which exposed the relative limitations of the Dutch civil affairs organization. When it became clear that a gradual advance would characterize the liberation, the Allies learned that the NMA lacked sufficient personnel to function independently at the local level. Until the Allies gained considerable territory and made greater resources available to local Dutch authorities, senior officials at Supreme Headquarters Allied Expeditionary Force (SHAEF) determined that the NMA was unsuitable for administering civil affairs independently throughout the Dutch provinces.[62]

The liberation of the Netherlands commenced during the third week of September 1944, when the Allies launched Operation Market Garden. Second British Army initially operated in the Nijmegen sector while two formations of First Canadian Army penetrated Dutch territory south of the River Scheldt. Seven and a half months later, on May 5, 1945, the German military commander in the Netherlands surrendered unconditionally and the liberation ended. Except for the provinces of Limburg and North Brabant, First Canadian Army assumed full control and responsibility for all military operations in the Netherlands.[63] These operations, undertaken in a densely crowded country during the cold and wet winter season of 1944–1945, produced some of the most unique and difficult civil affairs problems of World War II.

Civil-Military Relations in Groningen

Nine million Dutch people lived through the war. Starved of food and fuel in the final winter of 1944–1945, thousands of men, women, and children died under enemy control. Records maintained during the war later showed that the Nazis murdered 104,000 Jews in the Netherlands and about 150,000 Dutch succumbed to death during the occupation.[64] Hundreds of thousands more experienced deportation, forcibly removed from their families and made to work as slave laborers far from home. Those who lived under occupation, if they found the strength, resorted to theft, black market trade, and other dangerous acts of survival to augment their meager rations.

In the Netherlands, the arrival of the Canadians signaled to the Dutch that the long and trying war was finally over. Canadian soldiers freed several villages, towns, and larger urban centers as they punched northward through the Dutch provinces, liberating an entire country from the tyranny of enemy occupation. The Dutch people welcomed their liberators with tears of joy, encircling Canadian convoys with large crowds, throwing flowers, waving flags, and passing out kisses. Young soldiers happily entered Dutch homes on invitation and reciprocated the celebratory kindness by sharing food, chocolate, cigarettes, and even surplus army uniforms. Many Canadians subsided on half rations after the fighting ended, feeding starving children before relief supplies arrived, while others worked on farms and around homes, helping support Dutch families in their transition to postwar life.[65]

After the Germans had surrendered and Groningen was secure, the Canadian Army established a civil-military detachment in the city to administer and support the Dutch reconstruction efforts. Canadian officers maintained close liaison with local authorities while military personnel on the ground assisted Dutch civilians as directed. A report written in late April identified eight priority areas for the Canadian Army: finance, legal, public safety, displaced persons and refugees, civil maintenance, essential services, trades, and industry.[66] Subsections of the report also included the press, flooding, and graves registration on the list of civil-military responsibilities undertaken by the Canadians at Groningen. Coal was the most urgent need, and Canadian soldiers worked on broken rail lines to reopen important transportation routes. Electricity powered by coal supplied homes and businesses throughout the province. Coal also powered grain mills, water supply pumps, food factories, and such other essential services as hospitals.

Not only did transportation infrastructure carry essential supplies and a source of power to Dutch citizens in need, but rail service also returned displaced persons and refugees to the other liberated provinces in the Netherlands. According to Canadian reports, army personnel recorded the presence of no fewer than 15,000 refugees in Groningen from the southern Dutch province of Limburg alone. "Since the allied advance to North Holland displaced persons have commenced to arrive in the Province of Groningen from Germany in fairly large numbers," one army report explained. "They are of varied nationalities including Russians, Poles, Belgians, French and Dutch. Also a small number of displaced persons who have been hiding in Groningen are now coming to the surface."[67] Local Dutch authorities established temporary facilities to receive and accommodate people arriving and traveling through the city, and the Canadians on-site assisted with the processing efforts required to support Groningen's temporary and transient population.

Food rationing was another top priority for the Canadian Army in Groningen. Shortly after securing the city, army officials implemented and administered ration scales to control the proper allocation and distribution of essential supplies. The provincial food commissioner in Groningen restricted children under the age of three to 1,489 calories per day, while persons 3 years and older had a maximum daily intake of 1,285 calories. Canadian military personnel administered each ration scale through distribution, carefully allocating supplies in accordance with the policies and procedures implemented under the authority of the provincial food commissioner.

Food delivery was also central to the Allied relief efforts in the Netherlands. Royal Air Force and Royal Canadian Air Force bombers carried out Operations Faust and Manna in late April and early May, dropping thousands of tons of food and essential provisions to suffering Dutch communities. Pilot Colin Friesen, who happily dropped food to waving and cheering Dutch families, vividly recalled reading the words "Thank-you, Canadians!" painted on several barn rooftops over which he and his crewmembers flew.[68] The Canadian Army maintained a complete record of all food stocks in Groningen, including the supply totals for such products as grain, flour, bread, butter, meat, milk, cheese, barley, sugar, and syrup. The province had surplus potatoes as well as oats and wheat in the weeks following the battle, but deficiencies of salt, yeast, coffee, and chocolate persisted well after the liberation. Military support of this sort benefited Dutch citizens throughout the Netherlands, not the least because Groningen was a valuable source of food for the entire war-torn country.

The Dutch were not the only people to concern the Canadians after the battle for Groningen, however. There were also some 150,000 German soldiers in the Netherlands when the war in Europe ended, and their presence constituted a problem for the liberating forces. The Canadians avoided putting the captured Germans in large camps, saving vital resources and personnel to help support the Dutch elsewhere. Ultimately, the Canadians disarmed the Germans but allowed them to keep their own commanders, who exercised discipline, occasionally through firing squads. Although the Canadians did not interfere with the discipline carried out by the German officers, they did help the Allies hunt down senior Nazis officials to face trial for their wartime actions.[69] As for the liberated people, few Dutch citizens killed their former German overlords out of revenge, opting instead for retribution against fellow citizens who had collaborated with the Nazis. Records show that trials and executions occurred, as did violent public shaming of women who had courted with German occupiers.[70] Although this type of justice made many Canadian soldiers uncomfortable, much like the discipline exerted by German officers, interference was rare.

Friction in Civil Affairs

As could be expected with reconstruction efforts at war's end, several issues characterized the reconstruction efforts in Groningen. In late August 1945, four months after the liberation, Canadian authorities on the army's civil-military relations committee convened a special meeting with city officials to discuss policies governing consensual relations between Canadian soldiers and Dutch women.[71] Late-night parties, illegitimate children, marriage, and venereal disease (VD) constituted the list of priority items on the meeting's agenda. Dutch police brought forward several complaints about the actions and intentions of Canadian soldiers who had allegedly fraternized with young Dutch women in Helpman, Haren, Eelde, and other towns near Groningen. Major J. M. Colthart responded by confirming that the Canadian Army had experienced a rise in the number of VD cases in the Netherlands, citing comparatively lower numbers from Italy and Belgium. "By education we are doing our best to cut down on activity leading to the disease—or if the solider 'must' then the taking of preventive treatment," Colthart said, attempting to reassure the Dutch authorities.[72] "The Canadian Army has been active for five years in this regard and the means and methods of control are put into effect more easily in the Army than in civil life."

Seemingly unsatisfied with Colthart's response, the Dutch authorities pressed the issue further and told the Canadian representatives to take action. Major W. J. Phillips intervened and cast aspersions on local Dutch women to defend the soldiers under his command. "A couple of days ago returning to Assen from Meppel I picked up a 15 year old girl from the Hague [*sic*] who was carrying Army rations and a tent! Need I say more?" Phillips asked, rhetorically. "VD cannot be contracted by shaking hands, and it always takes two! Your papers accuse Canadians of giving VD to the Dutch girl. . . . We know we are aggravating the situation—[but we are] NOT giving VD to Dutch girls."[73] As a solution, Colthart proposed instituting formal public health campaigns and suggested that physicians, chaplains, and social service workers all play a role in controlling VD among Canadians and Dutch alike. "VD in Groningen is extremely high—252 Canadian soldiers named 252 Dutch girls in the Divisional area—102 of who lived in Groningen," Colthart said, reiterating the underlying significance of the growing infection rates.[74] The meeting ended without a mutual agreement governing consensual relations between soldiers and Dutch civilians, but the lines of communication stayed open, and the Canadian forces played an increasing role in the provision of health-care services in the province of Groningen.

Indeed, even liberating armies can wear out their welcome. For a Dutch population ravaged by war, the Canadians represented saviors who brought much-needed relief. For the young Canucks who had survived the fighting and completed their mission as liberators, they found themselves free to breathe and savor the fruits of their labor. The temptations were many and few Canadians were without girlfriends. Dutch campaigns designed to warn young women against developing relations with Canadian soldiers had little success, as did attempts by Dutch men to pressure women to avoid the liberators out of patriotism toward their home country. Joyfulness, youthful independence, and attraction brought many together, often culminating in shotgun marriages. While some new partners built strong and happy families after the war, in both Canada and the Netherlands, other relationships fizzled out and many Canadian soldiers returned home alone. Of the thousands of babies born in late 1946 and early 1947, there remains a group of elderly Dutch still searching for their Canadian fathers.[75]

The looting and pillaging of Dutch homes and property by some soldiers also tarnished the image of the Canadians who liberated the northern Netherlands. In the Groningen-Delfzijl region, several Canadian troops stole valuable goods and money from city dwellings and municipal buildings. One group of soldiers blew open a safe that belonged to a German hospital in Groningen and looted its contents, while no fewer than five cars were stolen in Winschoten.[76] Several local families who fled their homes to escape the fighting returned to ransacked properties, their lives upturned by another unfortunate and cruel reality of war. The civil authorities, for fear of causing unrest among the population, passed this information to the Canadian command. In response, SHAEF-Mission established a court of inquiry to investigate complaints, which represented the Canadian Army's first and only tribunal of World War II.[77] (SHAEF Missions were the formal representation of the Supreme Allied Commander in the liberated countries of Northwest Europe to maintain contact with these countries without interfering in military operations.) Many of the complaints sought compensation for war damages and several others presented evidence of alleged crimes committed by Canadian soldiers. While it would be unfair and disingenuous to characterize all troops as an undisciplined monolith, the illegal actions of some Canadians cast a dark shadow over the liberation.

Although friction between the Dutch people and their Canadian liberators was unavoidable, the successful campaign to liberate the Netherlands resonates as a point of great pride in the history and memory of the Canadian war effort. Young soldiers from across the Dominion of Canada freed an entire country from enemy occupation, fed and saved a starving population, and helped countless communities begin to rebuild. Likewise, the Dutch have remained forever grateful to their liberators. There were tearful farewells when most of the Canadians departed in the months and weeks leading up to Christmas 1945, and the return of veterans to the Netherlands for commemorative ceremonies and events often evokes a strong emotional reaction in wartime survivors and their children and grandchildren.

Conclusion

As one of the largest urban battles of World War II, the unique operational challenges encountered and overcome by the Canadian Army at Groningen illustrate the complexities of warfare for soldiers and military commanders with little training in urban environments. Military records demonstrate the extent to which Canadian officials perceived and approached the wartime situation in Groningen, including actual fighting in a densely populated location, as well as the long and complex process of liberating a large Dutch city from German occupation and control. Communication was paramount. Canadian and British commanders planned and coordinated the attack, relying on intelligence and experience to devise an offensive military strategy and successfully overwhelming a determined force that held several defensive advantages.

In the context of modern warfare, the standard interpretation of civil-military affairs suggests that the liberating force establishes control over civil affairs to ensure that the local civilian population does not interfere with military operations. The experience of the relatively small, outnumbered and underpowered Canadian force committed to Groningen reflects this pattern. Indeed, Groningen was not an outlier. As a small army attacking a big city, the Canadian liberators overcame several challenges. They fought and defeated a strong and determined enemy that was superior in both size and power, but retaking the city and reestablishing civil authority required temporary measures of civilian control. Interference in military operations was antithetical to the liberation of the Netherlands, and the Canadian experience at Groningen is a tangible reminder of this important lesson in the history of civil-military affairs.

Groningen also represents a useful case study for understanding the development and evolution of civil-military relations in Northwest Europe during and after World War II. Military records about the liberation of the Netherlands document extensive efforts on behalf of the Canadians to work in concert with local authorities, establishing rules and procedures governing all interactions between military personnel and Dutch civilians, for both rural and urban centers alike. Canadian officials held regular meetings and developed practical policies to support Dutch rehabilitation efforts, handle relations between service personnel and local residents, and care for displaced persons and refugees, among other important civil-military issues. The Canadian Army stayed in Groningen well after the fighting ended, and the long life of the liberation speaks to the military commitment required of effective and accountable urban operations. Attentiveness and care for the civilian population before, during, and after conflict are prerequisites of responsible and humane military engagement. As evidenced by the Canadian experience at Groningen in 1945, this reality of modern urban warfare is anything but new.

Notes

1. On the Canadian casualties at Groningen, see Stacey, *Victory Campaign*, 556; Cook, *Fight to the Finish*, 401. On the German and Dutch numbers, see van Ommen Kloeke, *De Bevrijding van Groningen*, 152; Ashworth, *City as Battlefield*, 21; Hofman, *Groningen in de Tweede Wereldoorlog*, 108.

2. Stacey, *Victory Campaign*, 556.

3. *Groningen in Vuur en Puin*; Leeninga and Westra, *En T'och Staat de Martini*.

4. Huizinga, *Maple Leaf Up*.

5. Huizinga and van Leusen, *Vier Dagen in April*.

6. Hofman, *Groningen in de Tweede Wereldoorlog*, 103–116; Hillenga, *Groningen 40-45*.

7. Sennema, "Liberation of Groningen."

8. While there are several accounts published in Dutch, Sennema's work stands alongside other firsthand memoirs published in English. See, for instance, Noorbergen, *Shadow of Terror*; Piëst, *True Watcher*; Kaufman and Horn, *Liberation Album*.

9. Stacey, *Victory Campaign*, 554–556.

10. Copp, *The Brigade*, 196–203; Copp, *Cinderella Army*, 270–275.

11. Cook, *Fight to the Finish*, 399–402.

12. Dykstra, "Occupation of Groningen"; Noll, "Restraint in Urban Warfare."

13. Ashworth, *City as Battlefield*, 21.

14. For instance, see Goodspeed, *Battle Royal*; Brown and Greenhous, *Semper Paratus*; Bercuson, *Battalion of Heroes*.

15. Noll, "Restraint in Urban Warfare," 17.

16. Stacey, *Victory Campaign*, 546.

17. War Diary, 2nd Canadian Corps, "Appreciation: In Regard to Employment of 2nd Canadian Corps in the Immediate Future," February 1945 (Library and Archives Canada); War Diary, 1st Canadian Army, 3821st Army Group, "21 A Gp/20748/7/8 (Plans)," March 1945 (Library and Archives Canada).

18. Noll, "Restraint in Urban Warfare," 24.

19. Harry Fox, quoted in Cook, *Fight to the Finish*, 401.

20. HW 1/3631: Government Code and Cypher School: Signals Intelligence Passed to the Prime Minister, Messages and Correspondence, Western Europe: on orders from Berlin, 6th military district is relieved of all defensive and security tasks and charged with reforming the Ersatz Army (National Archives, Kew).

21. Cook, *Fight to the Finish*, 381.

22. Copp, *The Brigade*, 176.

23. Cook, *Fight to the Finish*, 383.

24. CAB 44/315: Committee of Imperial Defence, Historical Branch and Cabinet Office, Historical Section, Canadian Forces, North West Europe, Major A. K. Reid, General List, Historical Officer (Civil Affairs), Canadian Military Headquarters, for Colonel C. P. Stacey, Historical Officer, Canadian Military Headquarters, Report No. 172: *Canadian Participation in Civil Affairs/Military Government Part IV: Belgium and the Netherlands, General Historical Survey*, January 11, 1946, 19–20 (National Archives, Kew). Historian Tim Cook places the daily caloric intake forced upon the Dutch civilians to be as low as 500 calories per day; see Cook, *Fight to the Finish*, 400.

25. Unnamed infantryman, quoted in Cook, *Fight to the Finish*, 400.

26. Cook, *Fight to the Finish*, 400.

27. J. P. Brady, quoted in Cook, *Fight to the Finish*, 400.

28. Copp, *The Brigade*, 196.

29. On Canadian actions near the port of Delfzijl and the German threat in the Ems Estuary, see Byers, "Operation 'Canada.'"

30. Copp, *The Brigade*, 201.

31. Stacey, *Victory Campaign*, 555.

32. Copp, *The Brigade*, 201.

33. Whereas Ralph Dykstra credits the order for restraint to Matthews, David Bercuson credits the order to Simmonds. See Dykstra, "Occupation of Groningen," 71; Bercuson, *Battalion of Heroes*, 232.

34. Noll, "Restraint in Urban Warfare," 59.

35. On the number of buildings destroyed during the battle for Groningen, see Noll, "Restraint in Urban Warfare," 66–67.

36. Dykstra, "Occupation of Groningen," 135.

37. Copp, *The Brigade*, 201.

38. Lieutenant-General P. J. Montague, Chief of Staff, Canadian Military Headquarters, Canadian Operations–North-West Europe: Extracts from War Diaries and Memoranda (Series 25), 17 July 1945, Library and Archives Canada, Record Group RG 24, vol. 10948, file 249C5.(D84), 11.

39. Stacey, *Victory Campaign*, 555.

40. "Groningen," Canadiansoldiers.com, September 2, 2021, https://www.canadiansoldiers.com/history/battlehonours/northwesteurope/groningen.htm. See, also, Holm, *Backwards Glance*.

41. Major Sandy Pearson, quoted in "Groningen." See, also, Williams, *Long Left Flank*.

42. Lieutenant-General P. J. Montague, Library and Archives Canada, RG 24, vol. 10948, file 249C5.(D84), 12.

43. Van Ommen Kloeke, *De Bevrijding van Groningen*, 148. For the English translation of this excerpt, see Translated excerpt from "The Liberation of Groningen," by Dr WKJJ van Ommen Kloeke on action at Bascula Bridges, 16 April 1945, Library and Archives Canada, RG 24, vol. 20297, file 934.013(D10).

44. Stacey, *Victory Campaign*, 555–556.

45. Cook, *Fight to the Finish*, 401.

46. Canadian soldiers fought until the final day of the war in Europe, May 7. The first week of May saw 114 Canadians die, including 12 on the last day of battle. See Cook, *Fight to the Finish*, 402.

47. CAB 44/314: Committee of Imperial Defence, Historical Branch and Cabinet Office, Historical Section, Canadian Forces, North West Europe, Major A. K. Reid, General List, Historical Officer (Civil Affairs), Canadian Military Headquarters, for Colonel C. P. Stacey, Historical Officer, Canadian Military Headquarters, Report No. 140: *Canadian Participation in Civil Affairs/Military Government Part I: Background and Beginning*, July 12, 1945 (Records of the Cabinet Office, National Archives, Kew), 2.

48. Ibid.

49. Ralph Gabriel, quoted by Major A. K. Reid, ibid. See also Gabriel, "American Experience with Military Government."

50. CAB 44/314: Reid, *Canadian Participation in Civil Affairs/Military Government Part I*, 2.

51. Ibid., 3.

52. Ibid., 4.

53. Ibid.

54. On the Hague Regulations, see International Committee of the Red Cross, "Convention (IV) Respecting the Laws and Customs of War on Land and Its Annex."

55. CAB 44/314: Reid, *Canadian Participation in Civil Affairs/Military Government Part I*, 5.
56. On the Civil Affairs Staff Centre at Wimbledon, see Adams, *From Crusade to Hazard*, 27–28.
57. CAB 44/314: Reid, *Canadian Participation in Civil Affairs/Military Government Part I*, 5.
58. Ibid.
59. Ibid., 6–7.
60. CAB 44/315: Reid, *Canadian Participation in Civil Affairs/Military Government Part IV*, 9.
61. Ibid., 12.
62. Ibid., 13–14.
63. Ibid., 14.
64. Cook, *Fight to the Finish*, 407.
65. Ibid., 407–408.
66. Military Government: Activity Reports—Groningen, Canadian Army Headquarters, First Canadian Army, Weekly report by 828 (P) Det. C.A./Mil. Gov. at Groningen, 28 April 1945, Library and Archives Canada, RG 24, vol. 10948, file 249C5.(D66).
67. Ibid.
68. Colin Friesen, quoted in Cook, *Fight to the Finish*, 407.
69. Cook, *Fight to the Finish*, 407.
70. Ibid.
71. Conferences and Debates: Conferences—General, Civil-Military Relations Committee, 21 August 1945, Library and Archives Canada (LAC), RG 24, vol. 10948, file 249C5.(D81).
72. Conferences and Debates: Conferences—General, Major J. M. Colthart, quoted in Civil-Military Relations Committee, 21 August 1945, LAC, RG 24, vol. 10948, file 249C5.(D81).
73. Conferences and Debates: Conferences—General, Major W. J. Phillips, quoted in Civil-Military Relations Committee, 21 August 1945, LAC, RG 24, vol. 10948, file 249C5.(D81).
74. Conferences and Debates: Conferences—General, Major J. M. Colthart, quoted in Civil-Military Relations Committee, 21 August 1945, LAC, RG 24, vol. 10948, file 249C5.(D81).
75. Cook, *Fight to the Finish*, 409.
76. Dykstra, "Occupation of Groningen," 136.
77. Ibid., 137.

8

Influence Operations

Steve Tatham

AS MILITARY AND POLITICAL LEADERS CONTEMPLATE THE CHALLENGING prospect of urban operations, they will consider nonkinetic activities and tools for shaping the attitudes, behaviors, and actions of adversaries and civilian populations of cities. Discussion of "influence" is reminiscent of the "hearts and minds" of the counterinsurgency operations of the past 20 years. By necessity rather than design, counterinsurgency (COIN) has largely been the currency of North American Treaty Organization (NATO) military operations since 2001; unsurprisingly, the literature surrounding COIN, be it formal military doctrine or more academic research, is now substantial. Although urban operations of the future may be of a different character than the COIN operations of the previous two decades, it would be foolish to discard too quickly all the hard-won lessons from Afghanistan, Iraq, and Syria, in particular those surrounding the idea of "influence."

Defining Influence

Like the term *strategic communication* that predates it, *influence* has become a buzzword—a "must have" in the ever shape-shifting lexicon that is military discourse. And like strategic communication, influence is all things to all people; in fact, trying to track down an approved military definition is surprisingly difficult given its now widespread use. In the UK's JDP0-01.1 (UK Terminology Supplement to NATO Terminology) it appears fourteen times—eleven times as a verb, the rest as noun. So, we learn, for example, that hard power is "physical effect to achieve influence"; "information strategy is undertaken to influence approved audiences"; and a "littoral region is susceptible to influence from the sea." Accepting that the JDP is only a supplement, it is reasonable to expect that *influence* may be better defined in the NATO Term database. It is not. This absence of definition for a term that is so extensively used has form but not substance. Take for example, *information warfare* (IW), a term that has now reached out from its military antecedents and has day-to-day public currency. No substantive definition of IW exists in UK Defence Doctrine JDP0.01, in JCN2/18 (Information Advantage), in AJP-3 (Allied Joint Doctrine for the conduct of operations), or in the UK Supplement to NATO Terminology. We need to turn to other sources to understand what influence may be.

The RAND corporation has defined influence operations as "the coordinated, integrated, and synchronized application of national diplomatic, military, economic, and other capabilities in peacetime, crisis, conflict, and postconflict to foster attitudes, behaviors, or decisions by foreign target audiences that further U.S. interest and objectives."[1] As the RAND authors themselves point out, "Somewhat serendipitously, our definition bears a striking resemblance to a recently approved definition of strategic communications. . . . Thus readers can generally conceive of influence operations as synonymous with strategic communications."[2] Why not, therefore, call influence operations strategic communications then? Because, in the view of this author, this definition (coupled with the absence of any meaningful doctrinal definition) is weak and unhelpful, misses the key tenets of what influence really is, and drives readers down a line of thinking that associates influence solely with information and communication. Unfortunately, they are not alone. For example, one of the few textbooks on the subject, *Influence Warfare* (edited by James Forest), devotes most of its chapters to considering how influence can be achieved via media and communications.

A better, simpler, definition is needed, and we have only to turn to experimental social psychologist Robert Cialdini, who defines influence, in his book *Influence: The Psychology of Persuasion,* as "the factors that cause one person to say 'yes' to another person."[3] He then provides six universal principles of influence: reciprocity, commitment, social proof, liking, authority, and scarcity. Although there may be some tangential bleed over of Cialdini's thinking in RAND's definition, what is obvious is the former's focus on the science of psychology and the latter's focus on process and communications. For this author, influence is all about an audience saying "yes" or "no." Such as saying "yes" to moving out of an occupied area, "yes" to considering peace terms, "yes" to laying down arms. Conversely, it is about an audience saying "no" to joining the adversary, "no" to taking up arms, and so on. Or in another words, influence is a tangible, demonstrable, measurable behavior. And in this context, we can perhaps learn a little from previous COIN operations.

Musa Qala, Afghanistan

In December 2007 the hugely emblematic town of Musa Qala in northern Helmand, Afghanistan, was retaken from its Taliban occupiers by British and coalition troops under the command of then brigadier Andrew Mackay. Mackay wanted the population—whom he referred to as "the prize"—to quickly return to the town once it had been liberated; he was determined that Musa Qala would not suffer the same fate as Nowzad, some 15 kilometers to the west, which had largely been destroyed, permanently scattering its inhabitants, during previous operations. By mid-December the operation was over. The town was still standing—largely untouched. As Mackay noted: "Within days the population of Musa Qala returned. The markets reopened."[4]

Musa Qala was not a city, nor a vast urban mega-sprawl; it was not a Delhi, Fallujah, Stalingrad, or Hue. It was a town, underdeveloped but a sprawling network of low-level compounds defined by mud and wattle walls and woven together via tight dirt roads and alleyways. It was a town that was home to a population of nearly 60,000 people; to provide context, the same as Loughborough in the United Kingdom or Kissimmee in Florida in the United States. Former US assistant secretary of state for defense Bing West, a former US Marine and combat veteran, wrote in *The Atlantic* that "urban warfare remains characterized by slow, massive destruction."[5]

Musa Qala would not have been good territory for tanks; infantry would have had to go house to house. Every window and rooftop would provide a vantage point for the enemy; every doorway, a hiding place for improvised explosive devices. And amidst the death, mutilation, and evitable destruction that would follow an assault, old men, women, children, and noncombatants would seek shelter in their homes, largely indistinguishable from the Taliban adversary. The town's destruction—like not just Nowzad before it but so many towns, villages, and cities throughout history—would make a mockery of the coalition's mantra of reconstruction, stability, governance, and peace. No wonder previous commanders had balked at the task; no wonder that Mackay looked for a "different" solution to the problem. "Influence—the subtle art of persuasion—would form the backbone of our efforts," Mackay would later tell the BBC.[6]

Understanding Behavior

When Mackay and I published *Behavioural Conflict* in 2011, it was an attempt to explain what influence was and what needed to be done to operationalize it rather than just talk about it. We wrote at length about the need to understand the population ecosystem, the many groups within it, and their interactions, and critically, the need to understand group and behavioral psychology.[7] Skeptics told us that influence has always been a component of (military) power. Destroying a bridge, launching an artillery barrage, deploying troops—all will influence.[8] But the difficulty is the outcomes from such actions will invariably be uncertain. An airstrike may well have influence on an adversary, but do we have surety in advance on what that influence will be? Will they feel their position is untenable and give themselves up? Or will it just prompt them to become more tenacious in the fight? A former UK chief of air staff once asked, what emotions do Afghan civilians feel when they hear Royal Air Force aircraft above them; he seemed surprised when he was not told "security" and "safety" but instead "fear." We might apply the same principle to the London Blitz. History will have us believe that for the seventy-six days that the Blitz reigned over the capital, Londoners "kept calm and carried on," their determination hardening and the "blitz spirit" growing. Lately, this has been questioned by some historians, who suggest that morale was a misnomer—that the British "keep calm and carry on" concept was in fact a "grim willingness to carry on," or passive morale. This means that they had this supposed fighting spirit because they had to, because they had no other choice.[9] But what it most certainly did not do was endear the population to revolt against Winston Churchill and the British government and ask for peace terms. We might reasonably argue therefore that Hermann Goering's intent—the decimation of British morale and securing of a quick victory—failed to take into consideration the limits of possible behavior by Londoners. Even if they had wanted to give up, they simply could not have done so. It was not within their power. As a tool of destruction, the Blitz was, for a while, highly successful. As a tool of persuasive influence, the Blitz was always doomed to failure because people could not physically behave in the way that Goering hoped.

Indeed, all too often previous influence efforts have failed to understand "reality" on the ground, and often the second- and third-order effects that would likely follow. So, if we wish to wield influence more precisely and to our advantage in future urban conflict, and increasingly this is how influence is being conceptualized in today's doctrine, we need greater clarity of the reasons why people behave (or don't) in particular ways and why groups often coalesce around specific behaviors. Because ultimately, in defense and security, it is always behavior that we should be concerned about. Indeed, the Blitz highlights one of the glaring misconceptions that endures today—the link between attitudes and behaviors. Londoners may well have hated the destruction and the Nazi planes that wrought it. They may well have been angered that the country had been dragged into another world war, some perhaps may even have been angered at the British government entering a war over Poland, a country far to the east. But all of this was largely irrelevant because it did not translate into a useful behavior for the Nazi leadership. David Myers, a psychologist and author of one of the core undergraduate psychology textbooks, tells us: "*The original thesis that attitudes determine actions was countered in the 1960s by the antithesis that attitudes determine virtually nothing*."[10] Or more simply, if we wish to change behavior (which in conflict we will always do), we really need to understand the reasons for and the limits of behavior. And if we need further evidence of this, we need only look at the end of the war. Churchill's personal popularity rating was 84 percent at that time, yet he lost the 1946 general election in a Labour Party landslide.[11]

The counterintuitive nature of Myers's words remains deeply problematic with many senior leaders. Again and again, we see senior officers directing information activity that will "alter perceptions" under the assumption that this will result in an immediate corresponding change in behavior. We know this is not the case. In the summer of 2020, following the death in the United States of George Floyd, the debate over institutional racism and bias in US law enforcement brought thousands of protesters to the streets. In 2018, the New York Police Department (NYPD) had instituted "implicit bias" training among their officers. The training usually consists of "a seminar in the psychological theory that unconscious stereotypes can lead people to make dangerous snap judgments." For instance, unconscious associations of African Americans with crime might make cops quicker to see them as suspects.[12] In the same article Robert E. Worden, director of the John F. Finn Institute for Public Safety in Albany, New York, explained that "we could certainly say that the training can be credited with elevating officers' comprehension of what implicit bias is."[13] But—and it's a big but—the researchers examined data about NYPD officers' actions on the job before and after the training. Specifically, they looked at a breakdown of the ethnic disparities among the people who were arrested and had other kinds of interactions with those officers. And in those numbers, they found no meaningful change. "It's fair to say that we could not detect effects of the training on officers' enforcement behaviors," says Worden.[14] Or in other words there is no guarantee at all that changing attitudes or perceptions will affect behaviors.

Doctrine notes and publications place great emphasis on describing, understanding, and manipulating the conceptual, virtual, and cognitive domains of specific audiences. They are on the right track, but they also inherently imply that influence will be achieved via some clever information campaign by psychological operations (PsyOps) or information operations (IO). This is a significant mistake and one in part perpetuated by the rise of social media. Social media, we are told, was instrumental in influencing the US electorate to vote for Donald Trump as president in 2016; social media swung, apparently, the 2016 Brexit vote. Both are highly debatable. Although the role of social media cannot be discarded, its primacy as a tool of influence is vastly overstated. Social media should be regarded primarily as a media channel, as a way of communicating within an already established network or networks, as a near real-time sensor-to-sensor network, and as a way of encouraging and deepening behaviors and attitudes that are already established. But, as a primary means of influence? No.

I can be this emphatic because after over a century of social science research, we have an extensive scientific evidence base for what binds groups together and how those groups might be influenced in their behavior. And that evidence base tells us that influencing behavior is highly complex with many variables; to presume a few tweets or Facebook posts can alone alter behavior is naive at best and foolish in extreme. Although a review of the entirety of that evidence is beyond the scope of this chapter, it is useful to consider a few basic concepts, if for no other reason than to provide a flavor of the complexity behind this subject. It should also serve to illustrate why a commander's battle staff would be better served by the addition of a psychologist than of creative social media specialists.

Group Behavior

The first concept of interest to us concerns how groups bind together. The US social psychologist Bibb Latané wrote about impact theory in social influence and showed that the strength of social conformity in groups depends on several factors, such as the number of people in a group, the group's strength, and the group's immediacy. As the number of people in a group increases, each individual member has less of an impact on the group as a whole.[15] Psychologist Solomon Asch's conformity experiment demonstrated the power that the group can have over its members. Asch placed a subject in a room with others whom he introduced as fellow subjects, though in fact they were all confederates to the experiment. Each member of the group was asked to say which of three drawn lines was closest in length to another drawn line. When each of the confederates named the same—and self-evidently wrong—line, the subject was between 50 and 80 percent of the time influenced by the group to give the same wrong answer. Such is the power of conformity to lead that subject to give the answer that he or she knew was wrong.[16] YouTube carries a particularly interesting video from the 2009 Sasquatch music festival that shows how groups can build.[17] It shows an exuberant but lone dancer on a hillside listening to a band with many more people sitting passively close by. As the dancer's exuberance intensifies, he is joined by more and more people until it is those sitting passively who are in the minority. The video has been used to illustrate a number of different points about the human condition (e.g., see

Derek Sivers's TED talk on leadership[18]), but in this context it shows the power of attraction and how groups can very quickly form. The conformity and strength of different groups has long been understood as a powerful factor in influencing behavior. Going into urban operations, we probably won't know what groups exist; but we can be quite certain that groups will coalesce around causes, issues, and/or leaders. And those groups will cause us difficulties.

Professor Harry Triandis of the University of Illinois defined four attributes that would need to be understood about a group if it were to be influenced. How do individuals within the group perceive themselves; how do group members relate to each other; what goals do individuals within the group pursue; and what issues drive group behavior?[19] Another US psychologist, Julian Rotter, undertook significant research into what he called "locus of control."[20] Locus of control (LOC) refers to an individual's perception of where control over events that take place in their lives resides, or who or what is responsible for what happens to them. Control can either be internal or external. Groups and individuals with an internal (or high) LOC feel that the circumstances are under their control, while groups and individuals with an external (or low) LOC feel their fate is attributable to luck or determined by people with authority. Psychologists Henir Tajfel and John Turner are responsible for the development of social identity theory,[21] which states that a person's sense of who they are is very strongly related to the groups that they belong to, bringing a sense of pride and self-esteem. As a consequence of this, people can develop a social identity that divides their world into "them" and "us," and in doing so exaggerate both the differences between groups and the similarities within the same groups. It provides us with an understanding of why people chose to join specific groups in the first place. Issues such as security, status, and power all play into the decision-making process. The "them" and "us" philosophy can become quite destructive and a powerful motivator for behaviors. Much research links the extremes of these psychological ideas in the Nazi persecution of the Jews during World War II or the massacre of Tutsi by Hutus in Rwanda. In 1957, the US psychologist Leon Festinger developed his theory of cognitive dissonance. This tells us that people can often hold contradictory beliefs and values, and indeed willingly participate in activities that are contrary to those values but in doing so experience psychological stress. Festinger believed that human beings actually try to find psychological consistency in their real-world experiences and so are prepared to alter their attitudes or behaviors in order to reduce the discomfort that they feel. That change might be eliminating specific behaviors, finding new information to offset the discomfort, or simply setting aside the cognitive element completely—that is, to act without any feeling, sympathy, or empathy. As Gilbert M. Gilbert, the army psychologist at the Nuremburg War Trials, noted, "Evil, I think, is the absence of empathy."[22]

Influence Operations and ISIS

This is but a tiny subset of the psychological research that has gone into groups and behaviors over the past century. What then is its relevance, if any, to military operations? To demonstrate relevance we need only consider the phenomena that was (or is) the Islamic State of the Iraq and Syria (ISIS). ISIS, at its peak, was estimated to comprise around 6,000 fighters in Iraq and around 5,000 in Syria, including as many as 3,000 foreigners. *The Economist*, which published the figures, went on to suggest that as many as 500 fighters hailed from Chechnya and perhaps 500 or so more from France, Britain, and elsewhere in Europe.[23] The exact number will perhaps never be known and is largely irrelevant. What is undeniable is that ISIS comprised many thousands of fighters and supporters drawn from multiple nations and, it is presumed, attracted by the purity of the caliphate and the desire to defend Islam. The influence campaign against ISIS grew in complexity and understanding with time; much of it remains classified, but what has escaped into the public domain would suggest that orthodoxy and communication creativity triumphed over the application of social science research and psychology. Take, for example, the (in)famous PsyOps leaflet dropped over Raqqa in 2015.

PsyOps products and propaganda are traditionally analyzed using a mnemonic, SCAME. Taking each in turn: S refers to source, and it is clear from both the delivery mechanisms (air drop) and the leaflet message that the source is the Western-led coalition. C refers to content, and it shows a meat grinder being used to process what are likely recruits to ISIS, each one having taken a "Now Serving" ticket on arrival and as their number is called stepping forward at the behest of a seasoned ISIS fighter, portrayed as a devilish or demonic figure. M refers to the message—joining ISIS is a terminally bad decision for Western recruits (the hoodies and trainers worn by the figures in the illustration typifying European or Western recruits). Finally, E is the effect of the message. Here we can have less surety—not least as the measurement of effectiveness is a perennial problem for PsyOp practitioners, who struggle to link causality, rather than correlation, to PsyOps products. But we can apply some common-sense analysis that would suggest that the leaflet had little success on the primary intended audience but possibly massive effect on another audience. Let's examine the primary audience first—the aspirant ISIS recruit.

The leaflet was dropped over Raqqa. If the ISIS recruits had reached Raqqa, then they had almost certainly successfully undertaken a hazardous and difficult journey, dodging national security services, counterterrorism officers, and border checks. They will have left behind their families, jobs, homes, and safety. It is not unreasonable to conclude, therefore, that with such significant personal sacrifices behind them the injection of a PsyOps leaflet into their thinking is unlikely to have much effect. We might also make the argument that the leaflet is in Arabic. Quite aside from the inevitable presence of illiterate and semi-illiterate fighters, for whom this would have no effect, we might also wonder what ISIS fighters from Russia, Turkey, Britain, and Sweden might make of it if their command of Arabic was either weak or, more likely, nonexistent. And we could also look at the McDonalds "Now Serving" board and wonder if that experience translates to Uzbekistan, China, Turkmenistan, and the multitude of other nations where the concept of fast food may not be prevalent. But the leaflet's errors and failings run far deeper. Tim Spencer is owner of Cognitif, a cultural intelligence agency specializing in semiotics and strategy. Spencer's comments about the leaflet are insightful:

> The semiotic texture/style of the image is a very western style of cartoon evocative of the long-standing "savage satirist" tradition in the UK and northern Europe (Ralph Steadman, Charlie Hebdo et al). It's a form of polarising political satire that has for generations depicted Muslims as "mad Mullahs," raving lunatics, or backward people. It's similar to the way the west began seeding anti-semitic ideas in very early cartoon narratives of the 20's and 30's. Way before the rise of nazism, Jews were ingrained into the cultural consciousness through mainstream popular cultural materials and narratives, depicted as shifty, untrustworthy hook-nosed shysters. Eventually the "humourous" caricaturisation of Jews became a lazy "colour by numbers" blueprint for villainous characters in all popular cultural material. "Jew" became a pantomime shorthand semiotic code for "rogue." So this style of cartoon comes from a judgemental western tradition of insularity and looking down upon "others" as being lesser people with no developed sense of morality and no guiding values or principles. It's a way of processing xenophobia and emerging from it as the better example. The sophisticate versus the brute . . . at best it may not be a style of caricature/illustration that feels familiar and "local" and is therefore "foreign" . . . so semiotically speaking it's a cultural incursion. The voice of the west, teaching the less enlightened. At worst—if it is familiar as a European, politicised "savage wit" style of communication associated with liberal-thinking west, then it makes the arrogant assumption that not choosing ISIS is a binary choice between supporting ISIS or siding with "The West" and its entire liberal political mindset. It's interesting to me that the people in the queue look very European with western clothing and boy band hairstyles, and the ISIS people have more "Arabic/Islamic" traits whilst also being depicted as grotesque monsters. To my mind the cartoon screams "be like us."[24]

PsyOps

It does not seem unreasonable to conclude that as an instrument of influence on the primary target audience—ISIS recruits—the leaflet is singularly unsuitable. But there is another audience for whom the leaflet may well have been hugely successful: the international media. The perception of military inactivity by the media is often interpreted as a failure—that somehow people do not know what they are doing or don't have the right equipment or the necessary political support. In Western democracies, this can cause significant problems; might it therefore be argued that the deployment of the PsyOps leaflet—so perfectly attending to Western tastes—was a deliberate act to influence the Western media that "stuff" was happening? Former US PsyOpers offer knowing smiles, but never confirmation, when asked just such a question—for of course a key tenant of PsyOps is that it is never used against your own government or population.

So if we are to use influence principles from psychology rather than the creativity of leaflet designers, we need to ask many more questions. In March 2015 I presented to the UK Foreign and Commonwealth Office (FCO) and provided a list of eight questions:

- Are there different groups within ISIS?
- If so, who are they? Do they fall into original nationalities, or are they grouped by their degree of extremism or something else?
- Is there a hierarchical pecking order?
- How do the different groups get on with each other?
- Have the new recruits to Daesh completely abandoned their families in favor of the new Caliphate or do they still feel a familiar pull?
- Is guilt a powerful motivator?
- What do foreign fighters find difficult—food, heat, violence, language?
- Are Daesh fighters motivated by money . . . worldly goods?

These questions provide an indicator of some of the types of information that will be needed by commanders if they are to apply genuine social science–based influence into adversarial groups. But it would be a mistake to presume that there are not other actors who could be as—or possibly more—important within the wider population. The problem is how to understand, and how is this research to be undertaken? Cleary some of it will be evidenced in routine J2 reporting, and human intelligence assets can be tasked to ask relatively innocuous questions in day-to-day interactions. For example, ISIS retained rigid control over its occupied populations, but even in heavily secured areas access can never be totally denied. For 2 years, Omar Mohamed ran the *Mosul Eye* blog, reporting from inside ISIS-controlled territory.[25] From inside Mosul,

> he attended Friday sermons with feigned enthusiasm. He collected and posted propaganda leaflets, including one on July 27, 2014, that claimed the Islamic State leader was a descendant of the Prophet Mohammed's daughter. Back home, writing on his blog in his other, secret identity, he decried the leaflet as a blatant attempt "to distort history" to justify the fanatics' actions. He drank glass after glass of tea at the hospital, talking to people who worked there. Much of the information he collected went up online.[26]

There are other likely assets. Doctors see every sector of society; they treat all patients regardless of background, ideology, or ethnicity. Taxi drivers make their living transporting people around. Teachers interact with not just children, but parents and communities. All routinely, perhaps unknowingly, use diagnostic questioning techniques in their everyday jobs. To be clear, we are not asking them the positions of weapons, of ammunition dumps, or supply lines. What we are asking is for social commentary and what they see and hear around them—the way that their fellow residents interact with each other; the tensions they observe; the relationships and power structures that define their day-to-day lives. This is the rather mundane and benign information that can be so helpful in building a picture of the group structures in complex environments. In part this was recognized by the US Department of Defense when in 2007 it introduced the Human Terrain System (HTS)—deployed anthropologists and psychologists—to Iraq and Afghanistan. At its cessation in 2014, it had cost the US military $750 million, making it the "largest investment in a single social science project in U.S. government history."[27] That it was not continued was, according to Professor Montgomery McFate, who started the program, more a testament to the organizational, academic, and personnel challenges it faced, than to any specific weakness in the principle.[28] The United Kingdom had a smaller but not dissimilar program called Cultural Advisors (CulAds). At their heart, the HTS and CulAds were seeking to develop an understanding of the population ecosystem that existed nowhere else—so-called Tier 1 Target Audience Analysis (AJP3.10.1[29] in the NATO and UK doctrine publication for PsyOps). In section 1-3 it describes target audience analysis (TAA) as "the critical enabler," defining it as "the systemic study of people to enhance understanding and identify vulnerability and susceptibility to behavioral and attitudinal influence." Three tiers are defined, from basic desk-based research (Tier 3) to a multi-source methodology undertaken in country to identify specific latent behaviors. Tier 1 TAA data is the key to successful psychologically driven influence.

So, presuming we have this information, how can it be used to our military advantage? Let's revisit Rotter's locus of control. LOC is the degree to which people believe they have control over the outcome of events in their lives, as opposed to external forces beyond their control. It is useful to know whether a group has an internal or external LOC because there are important differences between how the two groups are motivated, how they respond to external messages, and how they respond to external influences. Groups with high, or internal, LOC feel responsible for their own fate, and messages need to focus on logic and cost-benefit analysis (on a group level) and demonstrate what is best for the group. Time frames can be short or longer term. It is important to emphasize outcomes and reinforcement of future rewards if the desired action is taken. Groups with low, or external LOC respond to authoritarian and directive interventions. Time frames should be short—focus on the present and what happens now. It may also be useful to show what might happen if the desired outcome is not taken or show possible undesirable outcomes. How do we measure it in a target audience? Rotter developed a thirteen-item questionnaire. Clearly its deployment against an adversary is impossible. But a more basic distillation of the questions is perfectly possible—particularly if it is deployed among not just the fighting adversary but among prisoners, escapees, deserters, or just groups exhibiting very similar characteristics to the adversary. A simple question such as "What do you expect your future holds?" or "Would you be able to rebuild your life if you returned to your people/town/region?" will elicit information that will greatly aid the formation of the influence campaign. These are also questions that are significantly different from conventional tactical military questioning and, to a questioned recipient, quite benign and nonthreatening.

We can also measure the strength of conformity of groups and the internal dynamics of their structure and power. By developing specific question sets, it should be possible to measure the following: (1) how individuals within the group perceive themselves (a sample question might be, please describe the men in your community); (2) how people relate to others (a sample question: If someone lost their wallet and wanted people in the community to help look for it, what would happen?); (3) goals individuals pursue (a sample question: Other than essentials, what do people typically spend extra income on?); and (4) what concerns drive behavior (a sample question: Is it important for people in your community to work hard?). Thus we begin to build an emic-[30] rather than etic-[31] based field research methodology based on qualitative and, potentially, quantitative research.

Social Media

Conspicuous by its absence in this discussion is social media and in particular so-called "scraping." Social media scraping[32] is defined as "collecting online data from social media and other Web sites in the form of unstructured text."[33] Many have suggested that social media is the key to future influence. This perception has been intensified by deep concerns over the use of social media data for elections and, in particular, by the now infamous chief executive officer of Cambridge Analytica, Alexander Nix, who said that "by having hundreds and hundreds of thousands of Americans undertake this survey,[[34]] we are able to form a model to predict the personality of every single adult in the United States of America."[35] Nix's assertion seemed highly dubious at the time to most people who understood the complexity of human behavior and influence, but driven by a rabid global press, incensed at Donald Trump's election, psychographic modeling took on a sinister and Orwellian reputation for manipulation and influence.

Some 18 months after Cambridge Analytica's collapse, more sensible, critical analysis has been undertaken. Facebook's Andrew Bosworth referred to Nix and Cambridge Analytica as "snake oil salespeople. The tools they used didn't work, and the scale they used them at wasn't meaningful. Every claim they have made about themselves is garbage. Data of the kind they had isn't that valuable to begin with and worse it degrades quickly, so much so as to be effectively useless in 12–18 months."[36] Brad Parscale, Trump's election manager, told *CBS News* that psychographics "don't work."[37] Jamie Bartlett, author of *People Vs Tech*, notes that "there are no studies about whether this works for voting behaviour. . . . I don't think millions of minds can be manipulated by . . . psychographic messaging."[38] Numerous other studies suggest negligible impact.[39]

But there are more basic reasons why social media must not be considered a panacea for military influence. First, access to social media is neither universal nor guaranteed. Facebook, for example, is currently inaccessible in China, North Korea, and Iran and has been banned, for various periods of time, in Vietnam, Pakistan, Bangladesh, Syria, Egypt, and Tajikistan. Second, as the "father" of US psychology education and research, philosopher William James, once observed, we have as many personalities as the number of situations we are in. How representative of the real character is a digital personality? The truth is, we don't really know. But there are many who believe there is little meaningful correlation. "I don't think that you could have any type of accurate or even semi-accurate personality analysis based on what people are writing in their Twitter streams. Probably the same case goes for Facebook statuses as well," said John Grohol, an online mental health expert and founder of PsychCentral.com. "The persona online may be much more fabulous, much more exciting than the everyday life that they're leading," said Julie Albright, a digital sociologist at the University of Southern California, "because they see everybody else doing it."[40] Or put another way, how do we know that the audience we are dealing with is the audience we think it is? Next, context is extremely important in shaping attitudes and behaviors. And context can often change. A research experiment asked Egyptians to describe, in one word, Egypt. Asked outside a mosque, a majority described Egypt as Islamic; asked in a high street, the majority described Egypt as Arabic; and when asked on the beach, a majority described Egypt as Mediterranean. So when we look at people's postings on social media, we cannot ignore context—even if we don't know what it is. Finally, the very reach of social media can be not only overstated but also faked. In an experiment run by the NATO Centre of Excellence for Strategic Communication in Latvia, 3,500 comments, 25,000 likes, 20,000 views, and 5,000 followers were bought for $333 by eleven Russian and five European social media companies.[41] As already stated, social media is never going to be a primary means of military influence.

Lessons from Case Studies

Other chapters in this book contain a number of useful case studies from which we might draw some helpful lessons on operationalizing influence. In his study of Crossmaglen in Northern Ireland in Chapter 6, Patrick Finnegan notes that "the population of Crossmaglen was nationalist from the outset of the conflict, although not automatically republican." This is very reminiscent of Helmand in Afghanistan, where International Security Assistance Force (ISAF) forces took a long time to learn that although the population disliked foreign soldiers with an intensity, this did not mean they were automatically Taliban. The devil, as the old expression tells us, is always in the details, and taking a too simplistic view of a population is a precursor to influence failure. Finnegan goes on to say that the British Army quickly antagonized the local population by using their sports field as a helicopter landing zone. This again is so reminiscent of influence failures where the relative importance of a building or a specific area is not understood, and as a consequence the army is immediately placed on the back foot (put on the defensive) in their influence campaign. The placement of sangars behind heavily armored fences serves only to insulate the soldiers from the population yet further; their understanding of the population and everyday life is based largely on their view out of heavily armored positions and their occasional patrols into "bandit country." The "them and us" mentality is not conducive to a proper understanding of either the population or its behavior. Finnegan quotes one former brigade commander as saying that they were engaged in a counterinsurgency in "what should have been a sleepy corner of the British Isles." Finnegan explores the social cohesion of Provisional Irish Republican Army (PIRA) Active Service Units and, in particular, the age demographics and the importance of the family and its status. This social connectivity is very important in understanding influence. The ties that bind a group together—what we today might refer to as the normative affiliation and the locus of control—are powerful motivators for behavior. As Finnegan's interview with a PIRA volunteer notes, the historic family relations bred trust and confidence and made it very difficult for outsiders to penetrate. Finnegan himself rightly refers to this as the "normative shield."

British troops may or may not have deployed with the expectation that "not a shot would be fired in anger,"[42] but their lack of understanding of what faced them is not in debate. In 2012, the *Journal of the Royal United Services Institute* carried an article by former ministerial adviser Matt Cavanagh. He wrote: "No one inside the British [government] system knew much about the insurgency, the opium trade or the local politics and tribal dynamics—or just as importantly about how these different elements fitted into each other. Planners and policy makers [in 2006] did not know much about the human or even physical geography of Helmand."[43] This lack of understanding was by no means confined to the United Kingdom. In *The Operators*, Michael Hastings wrote: "In one meeting [Andrew] Exum drills down on the briefers. Who controls the water? Who are the local power brokers? Tell me how they *are related to the insurgency. The Intel Officers shrug. The questions "scare the hell out of them."*[44] Unfortunately these are the questions we need to be asking.

Matthew Wiseman's study of the liberation of Groningen in Chapter 7 is interesting in that why the operation took place at all is contested. He notes that some scholars felt that liberating the city was a military necessity but he also cites commentators who believe its liberation was purely political. This dichotomy reminds us of Musa Qala in Afghanistan, which militarily was not necessarily a key target but which politically was deeply problematic. It had become emblematic of the failings of ISAF, established as it was by the Taliban as a parallel seat of governance for Helmand. Wiseman notes that the opposition in Groningen was a mixture of different personnel, from a contingent of Dutch SS troops who "would die before giving up," to German air force and navy personnel, who we can assume might not have had the same behavioral drivers as their SS colleagues. And the forthcoming battle would be played out against a backdrop of a desperate population surviving on starvation rations, whose safety would become an issue of paramount importance to the advance Canadian forces and critical in their selection of artillery targets. In short, it was a complex environment, with multiple actors and with a variety of behaviors likely to be exhibited.

Wiseman describes the reaction of the population at liberation: "The Dutch people welcomed their liberators with tears of joy, encircling Canadian convoys with large crowds, throwing flowers, waving flags, and passing out kisses." However, within months, frictions began to arise. Relationships between young Canadian soldiers and Dutch girls upset locals and also saw a dramatic rise in instances of venereal diseases, which put pressure on the relationship between Dutch and Canadian authorities. Petty crimes also served to antagonize locals, and Wiseman recalls that "the illegal actions of some Canadians cast a dark shadow over the liberation." However, none of this could detract from the massive sense of relief that the population felt at the end of the Nazi occupation and their gratitude to their Canadian liberators. As Wiseman notes, that gratitude has endured even until today: the Dutch have remained forever grateful to their liberators. There were tearful farewells when most of the Canadians departed in the months and weeks leading up to Christmas 1945, and the return of veterans to the Netherlands for commemorative ceremonies and events often evokes a strong emotional reaction in wartime survivors and their children and grandchildren. We might conclude, therefore, that influence, if properly understood and deployed, can stand the test of time.

Urban warfare often conjures up images of the bitter hand-to-hand fighting of Stalingrad, Hue, or Mosul—but not Musa Qala. Forty-eight hours after "liberation," the marketplace was again trading; people were back in their homes, and the influence approach had allowed the British military to touch the face of the population with a feather, not a hammer.

Conclusion

So what does a future British Army—or indeed other army—influence capability look like? The United Kingdom has made small and tentative steps toward operationalizing influence, but in comparison to some other NATO nations and to Russia and China in particular it is as yet very immature. It is simply not enough to create structures such as 77X and 6XX without staffing them with people who understand what they are doing. (X is the NATO abbreviation for brigade; XX is the NATO abbreviation for division. 77X and 6XX were stood up by the British Army post-Afghanistan/Iraq for information activity and outreach, to provide nontraditional military capability for contemporary threats.) The continued absence of any professionalization in the information domain remains a significant weakness in order of battle, and so too the absence of a robust understand capability. Time and time again, 77X have fielded people at exercises who have demonstrated almost negligible knowledge of operational influence. This weakness extends to the very senior-most leadership, who consistently have no background in information and yet are placed in positions of responsibility in determining the UK's capability. Muddling through is not a great strategy, and some of our NATO colleagues are rightly perplexed at why the United Kingdom seems so out of step with evolving thinking of what influence is and how it should be executed. This is exacerbated by senior leaders' unhealthy fixation on social media influence, largely a result of some of the contractors it has employed.

Former commander of coalition troops in Afghanistan General Stanley McChrystal told the US Council on Foreign Relations that "the US and its NATO allies are only 'a little better than half way' to achieving their military goals, partly due to a frighteningly simplistic understanding of the country."[45] Yet a nonmilitary reader might well be surprised at this admission; even if they don't understand the details it is widely known that the armed forces have huge intelligence collection organizations. The number of "ints" is substantial: OSINT (open source intelligence), HUMINT (human intelligence), GEOINT (geo-spatial intelligence), SIGINT (signals intelligence), IMINT (imagery intelligence), MARINT (maritime intelligence)—and the list goes on. But for all of these capabilities, the wars in Iraq and Afghanistan more than amply demonstrated that one critical source of intelligence is missing—what this author calls POPINT, or population intelligence. The simple fact is that most military intelligence is aimed at the adversary and the adversaries' equipment, not at understanding a population. In *Behavioural Conflict* we ended with our own view of the future. More than 10 years after publication, its seems remarkably accurate even today:

> The future is not primarily about tailoring messages—it will be about nudging, shoving and shaping behaviours through a variety of means, of which conventional messages will be only one. We will require a detailed and quantifiable understanding of the behavioural, psychological and social environment, and the forces and pressure that create that. This cannot be performed by hackneyed marketing techniques: this is the domain of advanced social science. Finding the messages that exert strong normative pressures is crucial.[46]

If the United Kingdom is serious about influence, it needs to immediately build its TAA capability and professionalize a cadre of personnel from across the three services and their reserves that endures. We accept amateurism in no other part of the UK armed services; why it remains embedded in our information activities is a mystery.

Notes

1. Larson et al., *Foundations of Effective Influence Operations*, 2, xii.
2. Ibid, xii–xiii.
3. Cialdini, *Influence.*
4. Mackay, "Art of Influence."
5. West, "Urban Warfare, Then and Now."
6. Mackay, "Art of Influence."
7. To reinforce that point, the last chapter of that book was written by a behavioral psychologist, who explained that it was (social) science and not marketing or communication creativity that would win future influence campaigns.
8. Indeed Mackay's operations in Musa Qala used many more conventional military tactics, not least deception, in influencing the Taliban to leave the town than revolutionary social science ones; his greatest success, however, was in moving the often rigid thinking of military planners away from their vice-like attachment to kinetics and a military orthodoxy of thinking. As he himself said, "We would embed Influence into all of our thinking, planning and execution. Influence became the first element of any operational briefing and all planning." In this regard, Mackay was, rightly, ruthless in his leadership—if his staff could not grasp the new influence lexicon of language, and support it, they were out.
9. Shannon Bent, "The Blitz Spirit," Historic UK, n.d., https://www.historic-uk.com/HistoryUK/HistoryofBritain/Blitz-Spirit/.
10. Myers, *Psychology* (emphasis added).

11. Paul Addison, "Why Churchill Lost in 1945," April 29, 2005, https://www.bbc.co.uk/history/worldwars/wwtwo/election_01.shtml.

12. Martin Kaste, "NYPD Study: Implicit Bias Training Changes Minds, Not Necessarily Behavior," National Public Radio, September 10, 2020, https://www.npr.org/2020/09/10/909380525/nypd-study-implicit-bias-training-changes-minds-not-necessarily-behavior.

13. Ibid.

14. Ibid.

15. Latané, "Dynamic Social Impact."

16. "Asch Conformity Experiment," Eqivideos, video, December 22, 2007, https://www.youtube.com/watch?v=TYIh4MkcfJA.

17. "Guy Dancing on the Hill at Sasquatch!" Checkitoutmusic, video, June 9, 2009, https://www.youtube.com/watch?v=nU7dxkIz1Vs.

18. Derek Sivers, "How to Start a Movement," TED, video, February 2010, https://www.ted.com/talks/derek_sivers_how_to_start_a_movement?language=en#t-16531.

19. Triandis, *Individualism and Collectivism*.

20. Rotter, "Generalized Expectancies"; Rotter, "Internal Versus External Control."

21. Turner and Reynolds, "Story of Social Identity."

22. Gilbert, *Nuremburg Diary*.

23. *The Economist*, "Two Arab Countries Fall Apart."

24. Tim Spencer, email correspondence with author, May 11, 2018, and January 15, 2020.

25. Hinnant and Michael, "Undercover Blogger Lived a Double Life."

26. Ibid.

27. Sims, "Academics in Foxholes."

28. In early 2020, McFate told the author that "HTS was shut down when the US pulled out of Iraq, along with the PRTs and the Counterinsurgency Advise and Assist Teams. There was no need for the program any more, especially since Afghanistan was transitioned from COIN to counterterrorism." Tellingly, she also hinted at a problem that has bedeviled military information operations, psychological operations, and influence operations for years—conspiracy. "Most academic anthropologists don't understand how the military works and what exactly 'counterinsurgency' means. The objective in COIN is to *reduce* the level of violence and increase the functioning of government. One important means to do that is to increase the military's level of understanding of the local social-political-economic system. But academic anthropologists always seemed to think it was just a sneaky way to kill more civilians. For the most part, I think their so-called ethical concerns are overblown." Montgomery McFate, email correspondence with author, 24 January 2020.

29. Allied Joint Doctrine for Psychological Operations, Allied Joint Publication 3.10.1, Edition B, Version 1, September 2014.

30. A research approach that is characterized by "insider," "inductive," or "bottom-up" takes as its starting point the perspectives and words of research participants.

31. A research approach that is characterized by "outsider," "deductive," or "top-down" starts with theories, hypotheses, perspectives, and concepts from outside of the setting being studied.

32. It is also known as *site scraping*, *web harvesting*, and *web data extraction*.

33. Batrinca and Treleaven, "Social Media Analytics," 89.

34. This survey was based on the five-factor model. The traits include "extroversion," "emotional stability," "agreeableness," "conscientiousness," and "intellect/imagination."

35. Kai Ryssdal, "I Asked a Security Expert to Reveal How Cambridge Analytica Might Target Me Based on My Personality," *Marketplace,* March 26, 2018, https://www.marketplace.org/2018/03/26/i-asked-security-expert-reveal-how-cambridge-analytica-might-target-me-based-my/.

36. Lecher, "Facebook Executive: We Got Trump Elected."

37. Stahl, "Facebook 'Embeds.'"

38. Bartlett, "Psychographics."

39. Bailey and Hopkins, "Unresponsive and Unpersuaded"; Hersh and Schaffner, "Targeted Campaign Appeals"; Nickerson and Rogers, "Political Campaigns and Big Data."

40. Mark Milian, "Online Personas Rarely Match Real-Life Behaviours Observers Say," May 10, 2010, https://phys.org/news/2010-05-online-personas-rarely-real-life-behavior.html.

41. Wollacott, "Social Media Platforms Easy to Manipulate."

42. This is a popular misquote ascribed in 2009 to the UK defence secretary John Reid, who actually told parliament that he "would be perfectly happy to leave [Afghanistan] without firing a shot." The misquote is deliberately included because, regardless of the nuance, it and subsequent quotes show how ill-informed senior leaders were about the environment they were committing troops to.

43. Matt Cavanagh, "Ministerial Decision-Making in the Run-up to the Helmand Deployment," *RUSI Journal* 157, no. 2 (2012): 48–54.

44. Hastings, *The Operators* (emphasis added).

45. Stanley McCrystal, Speech to Council on Foreign Relations, July 2011.

46. Mackay and Tatham, *Behavioural Conflict*, 173.

9

Siege and Intrusive Operations

John W. Spencer

FROM MARCH 23 TO MAY 12, 2008, US ARMY AND IRAQI SECURITY FORCES engaged in an intense urban battle in and around Baghdad's Shiite neighborhood of Sadr City. The battle forced units that had been mainly focused on counterinsurgency operations to convert overnight into combined-arms teams of armor and mechanized infantry fighting high-intensity offensive combat. The battle saw a small main-effort brigade task force of roughly 3,000 soldiers defeat a numerically superior enemy numbering up to 4,000 fighters in a major fight held in some of the densest urban terrain in Iraq containing over 2 million residents. The adaptions and unique approaches these units made during this short but pivotal battle provide important lessons for future combat in urban environments where small units will be required to accomplish complex tasks in dense urban terrain possibly containing tens of millions of residents.

Sadr City is a dense Shiite area in eastern Baghdad whose boundaries with adjacent Sunni-dominated neighborhoods represent sectarian fault lines. At the time of the battle, Sadr City contained an estimated population of between 1.6 and 2.4 million residents.[1] Purpose-built in 1958 to house a large population of migrants moving to Baghdad to look for work, it maintained near slum-like conditions from the time of its construction. It stands out on a map of Baghdad as a solid, compact square of streets arranged in a grid pattern. The neighborhood is made up of narrow streets and alleyways, spider webs of power lines, and mostly buildings of three stories or less that range from simple cinder blocks to reinforced cement construction.

Containing around a quarter of Baghdad's estimated population of 7 million in less than 5 square miles, Sadr City is one of the most densely populated areas where the United States has ever fought. Its population surpasses those of the sites of other urban battles, such as Manila in 1945, whose pre–World War II population was an estimated 1.1 million, or Seoul in 1950, with 1 million residents. As a battlefield (both in 2008 and an earlier battle in 2004), Sadr City exceeded the population and structural density of any other part of Baghdad.

Amid Iraq's worsening sectarian violence between 2006 and 2008, in Baghdad the two major combatants were the Sunni al-Qaeda in Iraq (AQI)[2] and Shiite Jaysh al-Mahdi (JAM).[3] There were many other terrorist groups, criminal gangs, and war profiteers involved, but most either aligned with, or at a minimum chose not to compete with, AQI or JAM—largely out of fear for their own survival. AQI was a militant Sunni terrorist network led by Abu Muṣab al-Zarqawi. JAM was the Shiite paramilitary wing of fiery cleric Muqtada al-Sadr's national political movement. Al-Sadr had a major influence on the large Shiite populations and associated JAM militias concentrated in Sadr City, along with those in the southern cities of Najaf, Nasiriya, and Basra. Northeastern Baghdad contained a large JAM presence within Sadr City, but also a large AQI presence in neighboring Sunni areas—Adhamiya to the northwest, Rusafa to the southwest, and Tisa Nissan to the southeast—and the boundaries between Sunni and Shiite neighborhoods were often only demarcated by single roads, making it a fertile sectarian battleground. It did not take long, however, for Americans stationed in Baghdad to learn and recognize the boundaries from local residents, governance officials, and bloody incidents of sectarian violence.

In late 2006, Thawra District—Sadr City and two smaller, adjacent areas—saw fierce cycles of attacks and retaliations between Sunni and Shiite groups. Both sides attacked civilians, Iraqi security forces (ISF), and US troops. AQI frequently employed vehicle-borne improvised explosive devices (VBIEDs) and other mass-casualty attacks that had strategic implications on perceptions of the war in Iraq. The rising violence was frequently aired on international television, and many US politicians were strongly voicing concerns and questions on the effectiveness and ultimately reason for US involvement in an Iraq that seemed on the verge of civil war. The bodies of victims of sectarian murders were found dumped in the streets daily, averaging 200–250 per month.[4]

In early 2007, the United States implemented a surge of five additional brigade combat teams (BCTs)—about 30,000 troops—to Baghdad and Anbar Province in the country's west to stop the sectarian violence that had pushed Iraq into chaos. Under what was called the Baghdad Security Plan, one of the surge brigades, 2nd Brigade Combat Team, 82nd Airborne Division, led by Colonel Billy Don Farris, was assigned to Thawra District, although no US forces would be based in Sadr City itself. The unit implemented an aggressive plan to reduce overall levels of violence that consisted of moving into small outposts within the neighborhoods; using concrete barriers to construct checkpoints, safe neighborhoods, and safe markets that reduced the enemy's ability to transport resources and to conduct mass-casualty attacks with VBIEDs and other weapons; targeting key insurgent leaders and groups; and improving ISF capabilities.

The role and effectiveness of concrete in reducing violence across Baghdad cannot be overstated. Concrete barriers had been used throughout Iraq for years to reduce the threat of improvised explosive devices (IEDs) emplaced along major roads. But during the surge, concrete barriers were widely used to limit the enemy's ability to maneuver freely across Baghdad. The 2nd Brigade Combat Team, 82nd Airborne Division, led by Colonel Farris, emplaced over 30 miles of concrete barriers.[5] One of the first uses of concrete was to protect the Adhamiya Sunni neighborhood by building a 3-mile wall around it—which would earn a nickname: the "Great Wall of Adhamiya."[6]

The barriers used to form the walls were named after American states to denote their progressive size.[7] The smallest, Jersey barriers (3 feet tall; 2 tons), were used to block roads and slow traffic approaching checkpoints. Medium to large barriers—Colorado (6 feet tall; 3.5 tons), Texas (6 feet, 8 inches tall; 6 tons), and Alaska (12 feet tall; 7 tons)—were used to construct checkpoints and protective walls around markets, mosques, and other areas where crowds were being targeted by bombs and shootings. The Texas barrier, due to its width and ease of transport, among other reasons, was predominantly used to create the safe neighborhoods. But it was the massive T-walls that were used to create coalition and ISF bases and to maximize protection and prevent infiltration. Similar in size to Alaska barriers, the massive 12-foot-tall, 6-ton T-wall, with its interconnecting edges, created an effective barrier.

The walls turned Baghdad's densely populated, sprawling urban areas into more manageable "gated communities." Many of these communities were built around the already established Sunni and Shia concentrations where sectarian violence was occuring, although there were also mixed neighborhoods. Once a wall was completed, Iraqi Army, police, or paid local neighborhood militias (called the Sons of Iraq) could occupy checkpoints that allowed them to identify outsiders, search vehicles, and restrict the insurgents' ability to bring large explosives and other weapons from one area to another.

Insurgents and guerrilla fighters rely on their ability to move undetected among the population to survive and operate. The Chinese communist revolutionary Mao Zedong counseled fighters to move among the people as a fish swims in the sea.[8] In Baghdad, concrete walls in the sea made it a lot easier to kill or catch fish by creating choke points in the terrain that restricted enemy freedom of movement. Using obstacles to restrict insurgent movement in urban military operations is not new.[9] British forces used blockhouses, fences, or other barriers to interdict insurgents in the Second Boer War (1899–1902) and during the 1950s in Malaya. French forces used wire and other obstacles to cordon off entire urban areas in the Casbah during the 1957 Battle of Algiers. Concrete was the barrier of choice in the Iraq War.

By the end of 2007, the Baghdad Security Plan had significantly defeated, or at least neutralized, AQI and had decreased sectarian violence in Baghdad considerably. But JAM continued to use Sadr City as a base of operations to carry out strikes against ISF and coalition forces. While the US and ISF units in Thawra District had begun to isolate Sadr City with their safe neighborhoods and other lines of operations, JAM retained firm control over the population and the terrain.[10]

Special operations and conventional forces conducted some raids against JAM leadership in Sadr City, but these were extremely dangerous missions because of how fast the militias could respond to any outside visitors. While these responses varied, a common tactic used by JAM fighters (similar to insurgent tactics in the 1993 Battle of Mogadishu and 1994 Battle of Grozny) was to allow the raiding force to enter Sadr City and then barricade the streets behind them (using tires and other debris). This funneled the raiding force into IEDs and prepared ambushes. JAM would also use cranes to put car hulks on roofs, which JAM fighters would use as improvised fighting positions.

In October 2007, US special operations forces in Sadr City close to being overrun had to call in an air strike. In so doing they killed a number of Iraqi civilians and sparked public outrage that was fueled by wide media reporting of the event.[11] In response, Iraqi prime minister Nouri al-Maliki placed Sadr City off limits to US conventional units except under the most extraordinary circumstances. It would remain under such restrictions for the duration of the US military's presence in Iraq.

With these political restrictions in place, one of the only things US forces could do to affect Sadr City was work to influence the flows—of fighters and weapons—going in and out by blocking some roads and building small joint security stations (outposts manned by both US and Iraqi forces) and checkpoints along the major routes.

Successes in reducing the leadership, capabilities, and influence of AQI after the surge allowed the Iraqi government, mainly under the direction of Prime Minister al-Maliki, to focus on other threats to Iraq's stability, specifically the Shiite militias. The Iraqi government made plans as early as January 2008 to conduct a major offensive against the Shiite militia stronghold in Basra, Iraq. Two days before the offensive, planned to begin on March 25, JAM militias launched attacks in Baghdad with the hope of halting or diverting resources away from the planned offensive.

The Battle Begins

The violence that erupted on March 23 surprised the US forces in Thawra District; the strategic and operational context that drove the violence's logic didn't filter down to their tactical level. The soldiers patrolling the street in Baghdad had no idea that the reason for the sudden attacks against them and the Iraqi security forces in their neighborhoods was due to an Iraqi government planned operation in a different part of Iraq. Their only warning was when their garbage wasn't picked up by a contracted trash collector at the same time it typically was each week. When they called him, he was adamant that he was not coming. Then all hell broke loose—not just in Thawra District, but all around Baghdad, and especially in Sadr City.

JAM's uprising included launching 107-millimeter rockets against targets in and around Baghdad, including the International Zone (commonly referred to as the Green Zone) that served as the central location of Iraqi government offices. They also attacked many joint security stations, ISF checkpoints, and police stations around Sadr City. The ISF, police, and Sons of Iraq manning these locations mostly collapsed or joined forces with the insurgents. By the time the Basra offensive commenced as planned on March 25, JAM forces had cleared the Iraqi Army and police from Sadr City, overrun half of the checkpoints that surrounded Sadr City, and increased their rocket attacks against the Green Zone. The Americans defended and reinforced where they could and began to receive reports on the size and magnitude of the attacks.

Prime Minister al-Maliki could not allow JAM's rockets and uprisings to have a strategic effect on national governance by either derailing the Basra offensive or making the Iraqi government seem ineffectual in providing security. On March 25, he authorized ISF and coalition forces to undertake operations against JAM. The mission was simple: stop the rockets and restore government control in Sadr City. Importantly, he did not change the standing order preventing coalition ground forces from going inside Sadr City itself. US units would have to fight an enemy who maintained a real physical safe haven.

JAM's estimated strength within Sadr City ranged between 2,000 and 4,000 fighters armed with AK-47s; Soviet PKM machine guns; .50-caliber sniper rifles; rocket-propelled grenade (RPG) launchers; 60-, 82-, and 120-millimeter mortars; 107- and 122-millimeter rockets; and an unknown number of SA-7 surface-to-air missiles.[12] The group's fighters also had a seemingly unending supply of IEDs, especially explosively formed penetrators (EFPs) made of machined copper plates and homemade explosives that form a shaped charge capable of penetrating several inches of steel. JAM fighters ranged from local men with little training and equipped with small arms to foreign-trained "Special Groups" and snipers with advanced individual military skills.

The mission of responding to JAM fell predominantly to the 3rd Brigade Combat Team, 4th Infantry Division, led by Colonel John Hort. The brigade had deployed to Iraq 3 months earlier and was responsible for the area around Sadr City. The two main units that would conduct the majority of the fighting were the 1st Battalion, 68th Armored Regiment (1-68 CAB), a combined arms battalion commanded by Lieutenant Colonel Mike Pappal, and 1st Squadron, 2nd Stryker Cavalry Regiment (1-2 SCR), commanded by Lieutenant Colonel Dan Barnett.

The combat during the uprising was fierce. The 1-68 CAB and 1-2 SCR responded immediately by attempting to resecure the lost or abandoned checkpoints along Sadr City's boundaries. Instead of insurgents fleeing when threatened with battle, they found JAM militants ready to fight, laying in hasty defenses, and attempting to mass against coalition forces. US platoons had to call in attack helicopters for support after finding themselves trapped by large groups of enemy fighters bunkered in multistory buildings. Soldiers in one unit had 75–100 JAM fighters try to overrun them as they attempted to secure buildings around a soccer field that was being used as a rocket launching site.[13] Luckily, army attack helicopter teams were able to support the platoon and push the enemy back.

The intense nature of the fight caused 1-68 CAB to immediately transition to its fleet of M1 Abrams tanks and M2 Bradley infantry fighting vehicles (IFVs). They had parked these vehicles at Camp Taji, a large US base about a 45-minute drive away. Prior to the start of the battle, the battalion had traveled in up-armored high mobility multipurpose wheeled vehicles (HMMWVs), which were more appropriate to population-centric counterinsurgency operations. A normal patrol consisted of four HMMWVs and sixteen to twenty soldiers. It took three to four days for 1-68 CAB to move and fully convert from a HMMWV-based counterinsurgency formation to an armored task force of Abrams and Bradleys. Patrols were organized at the company level to best use their equipment for the situation and urban environment. Bradleys, tanks, and armored HMMWVs were configured for best effect. For example, patrols consisted of one Bradley and two tanks or two Bradleys and one tank, with dismounted infantry in hunter-killer teams designed more for major urban fighting, rather than counterinsurgency.

Prior to the battle, 1-68 CAB had made numerous changes to the normal task organization of a battalion within a heavy brigade combat team (HBCT), which facilitated a quick transition to small-unit mixed armor and infantry teams. A standard CAB from an HBCT had two armor and two infantry companies as its major fighting forces. Normally these companies trained and deployed together, but they were also designed to be capable of further task-organizing to any variation the mission may require. When 1-68 CAB deployed into Baghdad, the battalion's leadership quickly realized their armor companies did not have sufficient troops or vehicles to produce enough patrols to manage the considerable battlespace each was responsible for. Each company was therefore reorganized into three platoons—either one armor and two mechanized infantry platoons or two armor and one mechanized infantry platoon. Some companies were further stretched to four platoons. This gave each company the personnel needed to form large-enough patrols and to have the total number of patrols needed to maintain a sustainable work/rest balance. But what it also did was create diverse teams of infantry and armor soldiers. Thus when 1-68 CAB converted back to armor and mechanized infantry teams, the command and personnel relationships were already well established.

As 1-68 CAB and 1-2 SCR continued to fight, the battalions' higher headquarters during the battle, 3rd BCT, 4th Infantry Division, began to plan a movement-to-contact operation around and into parts of Sadr City to recapture all of the lost checkpoints and police stations. This course of action would have been a very destructive, high-risk operation. But the plan was shelved when ISF reinforcements arrived and quickly pushed into many of those locations so that at least the original checkpoints on the major routes into Sadr City were reestablished. The mission was successful at pushing JAM back into Sadr City and even allowing for the return of some civilian activity that had ceased during the initial days of the battle. Portions of Baghdad's second largest market, the Jamila Market on al-Quds Street on the border between Sadr City and the Jamila neighborhood, slowly began reopening. But enemy forces maintained their sanctuary and showed their ability to attack at times of their choosing.

The primary mission for 3rd BCT was to stop JAM rockets from hitting the Green Zone. From March 23 to March 31, JAM launched over eighty-five rockets at the Green Zone.[14] A major element of the BCT's ability to go after JAM and the rockets was an unprecedented range of intelligence, surveillance, and reconnaissance (ISR) and strike capabilities, including the following: joint surveillance and target attack radar system aircraft; counter-battery radar; Global Hawk, Green Dart, Shadow, and Predator unmanned aerial systems, many armed with Hellfire missiles; guided multiple launch rocket system (GMLRS); air weapons teams of Army attack helicopters; constant air force close-air support; and Aerostat balloons and Rapid Aerostat Initial Deployment (RAID) camera feeds. Many of these were theater-level assets pushed all the way down to the brigade level. Previously, these would be considered division-level and above capabilities. But, in the fight for Sadr City, they were all under the BCT's control. It took 3rd BCT's headquarters at least three or four days to reconfigure itself to manage the large number of information feeds and assets now under the BCT's control. Just identifying personnel to watch the different television feeds from the drones was a challenge.

Even with the most advanced observation capabilities available, stopping the rockets still required ground maneuver to seize rocket points of origin (POO). The plan to seize these locations, Operation Striker Denial, ran from March 26 to April 14.[15] The US and ISF units conducted combined cordon-and-search operations on the neighborhoods immediately south of Sadr City that were within the maximum range of the enemy 107-millimeter rockets, taking advantage of securing key terrain that provided observation and fields of fire over the relatively level terrain. They occupied one of the few five-story buildings in the area immediately southwest of Sadr City, the Jamila neighborhood, establishing a patrol base with snipers on the roof. The building gave the snipers the ability to overwatch multiple POO sites and engage any positively identified combatants. Once the POOs were secured, they became a challenge to hold because of the enemy's sanctuary. JAM fighters could easily slip in and out of Sadr City, attack coalition forces, and then retreat back into their sanctuary. The routes to the POOs also had IEDs emplaced along them daily; US and ISF units would clear the IEDs one day only to find them there again the next.

Even though coalition forces were prohibited from going inside Sadr City, they maintained an impressive capability to see inside the city and strike when necessary. Persistent observation of all open areas in the city was aided by everything from balloons equipped with modern cameras and positioned just outside the city to loitering drones. The aforementioned strike capabilities gave coalition forces the ability to strike any JAM forces that came into the open from their concrete sanctuaries.

The changing nature of the fight revealed that the Stryker vehicles used by 1-2 SCR were highly vulnerable to the EFPs and RPGs employed by JAM; 1-2 SCR lost six Strykers in six days.[16] Not only were the vehicles not survivable, but their width (especially when fitted with RPG cages due to their vulnerability) limited them to driving on the main roads, making their potential locations predictable and susceptible to ambush. For this and other reasons, 3rd BCT, with assets provided from its parent division, made changes to both 1-68 CAB's and 1-2 SCR's task organization and battlespace. Major parts of 1-2 SCR's assigned battlespace were given to 1-68 CAB, whose Abrams tanks and Bradley IFVs provided soldiers more protection and firepower to fight JAM. The 1-68 CAB received an additional tank company and two additional Stryker companies; 1-2 SCR additions included an armor platoon that began leading patrols so the tanks could bear the brunt of attacks before they could destroy the Stryker vehicles. After reorganization, the 1-68 CAB and 1-2 SCR combined task force included nine companies. The total US forces numbered around 3,000, with an additional few thousand ISF, depending on the phase of the operation.

Both of the assigned missions—to stop rockets hitting the Green Zone and to restore government control of Sadr City—were constrained by JAM's ability to slip in and out of Sadr City to launch rockets from sites closer to their intended targets, lay IEDs, and attack coalition forces. One of the first coalition responses was to use concrete barriers to seal the individual routes leading out of Sadr City to the southwest, increase patrols of these areas, conduct ambushes on known firing points, and secure key terrain by, among other things, reinforcing ISF checkpoints.

These initial efforts were troop-intensive and not fully effective in stopping the flows of fighters, weapons, and rocket launches. The 3rd BCT decided to construct a continuous wall made up of 12-foot-tall, 6-ton T-wall sections along the entire southern border of Sadr City on al-Quds street, named Route Gold on US military maps. Previous walls had been built along other portions of Sadr City's boundaries that set the conditions for the planned operation. By this time, the areas bounded by Routes Grizzlies, Gold, Predator, and Pluto were mostly secured, but the effort was troop-intensive and JAM could still infiltrate in and launch rockets as well as plant IEDs and conduct other types of attacks.

On April 15, Operation Gold Wall was started; 1-68 CAB and 1-2 SCR conducted daily concrete-laying missions to construct the continuous wall. The desired purpose of the wall was different than that of previous walls built. Instead of protecting populations by keeping insurgents out of areas, the Gold Wall would be built to affect JAM while keeping them in Sadr City. Like medieval siege engines, the units formed each night in massive convoys consisting of a tank in the lead, flatbed trucks with concrete barriers, a civilian or military crane, Bradleys, and other vehicles.

The construction of the wall was heavily contested from the beginning. JAM's first tactic was to establish a defensive zone by emplacing IEDs along the roads leading up to where construction of the wall would start that day. Because of these IEDs, the concrete-laying patrols had to first do a deliberate breach into the battle area. Streets were lined with trash, each pile a possible IED hiding spot. Combat engineers started by doing route clearance, but the normal procedures required them to stop at each suspected IED location, conduct a detailed search of the area, and if an IED was present, neutralize the bomb (preferably done by attached explosive ordnance disposal teams) and try to render it safe for collection so that it could be exploited for evidence. The process could take hours for a single device, but JAM had lined the roads with IEDs. Some patrols would encounter twenty IEDs on a single street. US forces adapted, with convoys developing new and faster methods. Tanks began to fire 120-millimeter canister rounds (in essence, shotgun-type rounds that, once fired, open into hundreds of tiny pieces) down streets from their main guns. The canister rounds blew the trash off the streets and, in many cases, exploded IEDs lying in wait for the patrol. If an IED was found, they would also shoot at it with 25-millimeter cannons and 7.62-millimeter machine guns to detonate it.

Once a patrol reached the site where the walls would be emplaced, the tanks, Bradleys, and infantry established overwatch positions for the crane and team of soldiers that would guide the massive concrete walls into place. Initially, JAM heavily targeted these patrols from within the city. This revealed their locations, though, and they were easily engaged. Some JAM militia fired from cover provided by buildings along Route Gold. Snipers would also shoot at the crane cable or the lone soldier that was forced to climb a ladder to unhook each concrete wall. Special operations forces snipers were extremely useful in a countersniper role. Nevertheless, there were situations where confirmed snipers and fighters that could not be engaged by US snipers had to be targeted with air-delivered, precision-guided bombs or GMLRS rounds that could penetrate the buildings' layers of concrete. One instance that required a concrete-penetrating option was when a JAM sniper occupied and fortified one of the few five-story buildings north of where the Gold Wall was being constructed and at a key intersection where he was able to engage the wall-building team. Direct fire placed on the building did not affect the sniper's fortified position. The 1-68 CAB's solution was to drop the building using the GMLRS. But for the duration of the Battle of Sadr City, coalition forces used fewer than three mortar or artillery fire missions because of the risk of collateral damage, the prospect of injuring civilians, and the potential political ramifications both locally for the government of Iraq and internationally for political support to the coalition forces.

The wall construction went slowly. At first, the wall was only constructed at night to take advantage of the American night-vision advantage. But when these forces would return to their bases after the nighttime construction operations, JAM emplaced more IEDs on the routes to the wall construction area. A decision was quickly made to transition to 24-hour operations, rotating 1-68 CAB and 1-2 SCR every 13 hours to give each patrol both daylight and nighttime hours. The number of concrete barriers that could be emplaced in a single shift varied greatly, depending on the enemy situation, number of tractor-trailer vehicles to transport concrete from staging areas, and other factors. A bad night might not see a single barrier emplaced; during the most productive 13-hour shift, 105 were laid.

The logistical planning and execution of each barrier emplacement is an impressive case study in and of itself. The number of personnel, vehicles, and refined tactics to safely emplace an average of over seventy barriers in a single shift is a lesson that must be captured for any similar future operations. A concrete holding area had to be constructed near Route Gold with a continuous flow of barriers and activity. Production at Iraqi concrete batching sites had to be increased. Each operation included a route clearance team to clear the route up to the wall; a security team for the holding area; a holding area team that included a crane, flatbed tractor trailers, and even two armored forklifts that in some cases transported walls all the way down to the wall site using already cleared routes; and a building site team (the main effort responsible for IED clearing around the site and construction of the wall) that included a crane, emplacement crew, and combined-arms team.

As JAM realized that IEDs were not going to stop the construction of the wall, the group's fighters came out in force to fight the besiegers with direct fire. By so doing, they gave up one of the biggest advantages of being an urban fighter: remaining hidden by the city. Once JAM militia engaged the wall construction team, they became visible, easily identified, and targeted by the coalition's far-superior weapons and aerial ISR and strike capabilities.

While the decision regarding where to place the wall was based primarily on the maximum range of rockets targeting the Green Zone, one of the surprising effects the wall had on the enemy was cutting them off from their financial support. The Gold Wall on al-Quds Street not only separated Sadr City from nearby neighborhoods, which were more affluent and vulnerable to JAM extortion, but also cut JAM off from the Jamila Market, which was a major source of money. JAM extorted residents, merchants, and customers around the market area. The group also sold its own goods and weapons in the market. Coalition forces were not fully aware of the consequent effects on JAM finances before they started constructing the wall. They did have the full understanding of the urban flows (economic, services, social) to both the local residents and of those the enemy was relying on.

The militia had little choice but to "fight the wall." Being sealed into the city would eventually force them to enter through heavily armed Iraqi Army checkpoints, which would allow government forces to individually target them. It also would prevent their access to the Jamila Market and the population south of the wall. But most importantly, the wall would prevent the group's ability to move fighters and weapons, conduct rocket attacks, and maintain a flow of resources, all of which were vital to JAM's ability to function and retain influence.

As JAM members fought the wall, they continuously diminished their resources. In the beginning, JAM contested the wall with large groups of fighters. As the days and weeks progressed and the wall continued to grow, the numbers of fighters that would attack the siege party progressively got smaller. Coalition ground forces were successful at killing enemy fighters that contested the wall. At the same time, 3rd BCT also became increasingly effective at destroying rocket teams with the suite of persistent ISR and immediate precision air strikes (from attack helicopters, fixed-wing aircraft, and drones) that they controlled. As the final barriers of the wall were emplaced, JAM appeared to have exhausted its supplies and fighters.

The positive effect of the Gold Wall is hard to argue. From the start of the operations on April 15 to late May, when the wall was connected to other walls that had been previously built or were simultaneously built around Sadr City, the number of attacks and other significant events recorded in 1-68 CAB's area of responsibility went from a high of 138 a week to just 8.

In just over a month, 3rd BCT had emplaced 4,000 12-foot-tall, 6-ton T-wall sections to construct a wall nearly 3 miles long. During the battle, the BCT fired 120 Hellfire missiles, six GMLRS rounds, eight air force guided bombs, and over 800 120-millimeter tank and 12,000 25-millimeter

Bradley cannon rounds.[17] Precise JAM casualty numbers are hard to determine (especially as coalition forces were not able to go inside Sadr City to conduct battle damage assessments during the operation). Some reports estimated the number of JAM fighters killed to be only 700, but it is likely that US and ISF soldiers killed, wounded, or captured thousands of JAM fighters.[18]

On May 12, al-Sadr announced a cease-fire.[19] The final barrier of the Gold Wall was emplaced on May 15. Al-Sadr's decision was clearly made from a position of weakness. JAM fighters within Sadr City had been neutralized. They were surrounded and effectively contained inside a concrete fortress. While the wall's completion didn't mean JAM was defeated, it did set the conditions for reestablishing government control of Sadr City. The wall achieved the strategic objective of limiting JAM's ability to exit Sadr City to fire rockets at the Green Zone and reduced the fighters, weapons, and critical resources flowing in and out of the city. The reduced flow of resources quickly took many of the strengths away from JAM, including their ability to freely move in and out of the city as well as the resources they needed to actually fight. Iraq Army forces were ordered into Sadr City but faced minimal resistance from JAM's exhausted forces. Both Iraqi and coalition forces then exploited their success with major reconstruction projects that focused on local economics, government services, and infrastructure. They also transitioned back to counterinsurgency operations to maintain control of the gains they had achieved. The Gold Wall remained in place, the Iraqi Army maintained their checkpoints at the entry points into Sadr City, the JAM militia discontinued their fighting, and overall a new normal was established.

Lessons from the Battle of Sadr City

Context in war is everything. The political and military factors involved in the Battle of Sadr City—from the tactical to the strategic level—were in many ways unique. But there are also a number of urban warfare lessons that were learned (and in many cases relearned) that need to be captured for consideration in future urban combat operations.

The history of urban warfare includes a wide variety of operations along the full spectrum of conflict (offense, defense, and stability). Since 9/11, the US military has predominantly been engaged in low-intensity counterinsurgency and stability operations, but the Battle of Sadr City was a high-intensity offensive mission against an enemy-held area. This rapid transition between peacekeeping, nation-building, and other low-intensity combat coupled with intense episodes of violence in dense urban terrain is a very likely reality for future operations. All militaries will be required to be prepared to adapt overnight to such variety.

Urban terrain gives a distinct advantage to a defender. Military or militarized forces that defend from cities seek to diminish the technological and numerical superiorities of an attacking force, especially ISR and aerial-enabled strike capabilities. Dense urban terrain limits the ability of military forces to maneuver. Hiding among buildings that effectively become concrete bunkers, defenders can funnel military formations into ambushes and terrain that breaks them apart, allowing them to be defeated piecemeal. A city's maze of fortified structures also allows defenders to maneuver from building to building, often with prepared tunnels or holes in walls, to avoid air-delivered munitions or fixing forces.

Populated urban terrain, where it is extremely difficult to distinguish enemy forces from the civilian population, allows an insurgent to employ guerrilla tactics. These include anti-armor near ambushes, sniper attacks, and bombings. If engaged, enemy forces can slip back into the population. The entire environment establishes a military dilemma. The advantages that occupying a city gives to a defender will drive them to withdraw into cities. Likewise, the disadvantages for an attacker encourages most modern military forces to adopt an approach used historically—avoiding or bypassing cities whenever possible. But warfare includes chance, and militaries often do not get a choice. When a war's political objectives include destroying an enemy in a city, reclaiming a besieged city, or both, the military must accomplish the mission.

There are many mission-specific variables that will determine a military's approach to accomplishing an urban operation. Among these are the amount of time and forces available to accomplish the mission, the viability of removing the city's population, and political will in the face of the destructive nature of an urban battle. History shows that the most common approach a military force will take against any enemy-held urban area is to isolate or surround the city, remove as much of the population as possible, and then conduct costly, block-by-block, house-by-house operations until the enemy is cleared from the city. Because of the advantages afforded the defender, the attacker will typically have to use massive amounts of concrete-penetrating artillery and bombs that destroy most of the structures in the area of operations. This explains a US Army major's remark during the Vietnam War's Tet Offensive that "it became necessary to destroy the town to save it."[20]

These methods were used in most urban operations to retake an enemy-captured city in recent history: Stalingrad in 1943, Hue in 1968, Grozny in 1994, and Fallujah in 2004. This is also the approach the world has witnessed in recent operations in cities occupied by the Islamic State, from Mosul to Marawi. But the 2008 Battle of Sadr City revealed a different approach that stressed the integration of nonkinetic means to influence the urban flows of a city to achieve their desired effect. Political conditions on the ground, such as those of Sadr City, in the future will likely limit the use of force in dense urban areas, but militaries will still be asked to achieve the political objective.

Many of the lessons of urban warfare discovered in the Battle of Sadr City are enduring; others are unique to the context of the fight. Either way, it is important to ensure the lessons learned, or relearned, are considered by military thinkers when they face the next complex urban mission.

Strategically, the Battle of Sadr City reminds us that war is the continuation of policy by other means and subject to political control, objectives, and rationale.[21] Politics and war are always intertwined. Military forces must account for national political objectives and constraints on the use of force when deciding on military strategy and tactics. In the case of Sadr City, significant political constraints were placed on the military. The relative size of the population of Sadr City residents (estimated at 1.6 to 2.4 million and already in opposition or potentially combative to the Iraqi government) and of the suspected armed resistance in the city was a major constraint on actual military options as the size of US and Iraqi forces were significantly outnumbered.

Because coalition forces were not allowed to go into the city the enemy was using as a base of operations and safe haven, they were forced to create different options. They could not adopt the historical tactics of house-to-house clearing, particularly given that the option to evacuate the city's 2 million residents was not feasible because of both the fragile sectarian security situation and political environment in Iraq. While 3rd BCT did execute information operations, telling civilians to leave or stay clear of the battle areas (especially the vicinity of the wall-building operations), the reality of the situation was that there was nowhere for the residents of Sadr City to go. Managing 2 million internally displaced persons was not something the government of Iraq was capable of, politically or logistically. Unable to clear the enemy safe haven, the military sought to isolate it.

Because of the local, national, and international political dynamics, coalition forces also further self-restrained their methods by only using wide-area effects weapons such as mortars and artillery in very limited circumstances out of concern for civilian casualties and the destruction of infrastructure. The risk to the wall construction could have been reduced with such fire, but it was not used. However, the need for concrete-penetrating munitions, which drove the use of GMLRS, for example, on enemy snipers deep within a five-story building, was consistent with previous urban battles.

Maintaining the political will to continue urban operations will often prove challenging but will always be crucial. In the Battle of Sadr City, coalition forces had to worry about the will of their political leaders and casualty-weary populations; six US soldiers died during this battle.[22] They also had to worry about Iraqi political will, as al-Maliki balanced fragile sectarian relationships and populations. Moreover, all perceptions were influenced by a globally connected world and 24/7 media reporting. Journalists were embedded with coalition forces during much of the battle. The total number of civilian casualties is unclear, but had destruction and civilian casualties reached a critical threshold, the military operation would have been impacted by further political constraints.

Operationally, it is important to remember that the Battle of Sadr City was fought within a greater campaign. It was influenced by all the other operations going on around the country (and vice versa). There were reports of JAM leadership asking for reinforcements from other strongholds, but those fighters were too busy fighting coalition forces and ISF in their areas. The Iraqi-planned offensive against Basra was the spark of JAM's uprising in Baghdad. A military commander's ability to understand, visualize, and describe these connections will always be a challenge, and an important part of a commander's staff's focus in combat.

The Battle of Sadr City also strongly contests the overly simplistic notion that cities are sponges that soak up troops.[23] Determining force requirements for urban environments cannot only focus on the size of the city and its population. The Battle of Sadr City saw a coalition force of around 3,000 defeat an enemy of between 2,000 and 5,000 within an urban area inhabited by 2 million residents. The political objective, military mission, forces and tools available, enemy capabilities, security environment, terrain, and time needed to accomplish the mission all are more important factors than just the size of the city. A small military force can accomplish many missions in extremely large urban areas given the right circumstances, clear objectives, and accurate knowledge of key physical and human terrain.

The small US force in the Battle of Sadr City would not have been able to accomplish its mission without the unprecedented theater-level ISR and strike capabilities pushed down to 3rd BCT. The organizational flexibility necessary to make such changes, the skills required to manage those assets, and the prudent risk acceptance of commanders at all levels are all important lessons to institutionalize and consider when thinking about force size and design.

Tactically, there are many lessons from the Battle of Sadr City. Adaptions made by the units involved—across all of the army's warfighting functions (mission command, movement and maneuver, intelligence, fires, sustainment, and protection)—could be discussed at length.[24] Selecting available capabilities, accounting for terrain considerations, and the use of concrete obstacles stand out as necessary highlights.

Military capabilities are about having the best tools for the job. Military capabilities are designed so that when a formation faces a situation or environment it isn't prepared for, small-unit leaders can adapt their methods and their modular equipment. When a force faces an unfamiliar environment, it must be able to adapt. But good leaders and units adapt; great ones adapt quickly. The units of the Battle of Sadr City quickly adapted from counterinsurgency formations and methods to high-intensity urban combat.

One advantage 1-68 CAB had was that they had increased their adaptability inherently when they had redesigned their formations of mechanized and armor teams to fight the manpower-intensive demands of counterinsurgency operations. This formed command relationships, increased awareness of combined arms capabilities, and overall teamwork bonds that became invaluable when the unit converted formations into combined arms teams for offensive operations. The benefits of this rapid cross-organization and mission-specific combining of combat arms down to the company and platoon levels would benefit any formation in training and combat operations.

The reconfiguration of the 3rd BCT's headquarters to manage the large number of information feeds and assets they were given—from more drones and attack helicopters, to national surveillance assets—has significant implications for a smaller army that may be limited in size of command centers. The ability to establish command center roles, processes, personnel duties, and flexibility is critical. This occurs in training. Command centers must know their maximum capabilities as well as how they can redesign themselves based on mission requirements.

These adaptations required employing equipment and weapons for different purposes than they were designed for, such as using tank canister rounds to clear streets of IEDs. Adaptations also can call for completely different equipment. The fact that 1-68 CAB had its tanks and armored vehicles within reach—and had trained for high-intensity operations before its deployment to Iraq—allowed the unit to quickly transition from a counterinsurgent force into a combined arms team capable of high-intensity urban combat. Tanks and infantry fighting vehicles proved invaluable, while Stryker vehicles were found wanting in this environment.

It is not that the Stryker is not a useful vehicle. It offers a combination of speed, mobility, and transport options unlike that of other ground combat vehicles. But in "iron triangle" considerations (payload, performance, and protection) for vehicles in this environment, it did not provide the necessary protection against EFPs and was too wide for most of the urban terrain.[25] Adding tanks that could lead the Stryker patrols was a successful adaptation to the Stryker organization. Other vehicle mixes and tactical changes were also required, such as emplacing Bradley IFVs in overwatch positions on high ground and using highly maneuverable armored HMMWVs to secure the rear of formations. Until the military deploys a combat vehicle more suited for urban terrain with mobile protected firepower capabilities, adaptable leaders and equipment sets will be required.

These important decisions and actions to rapidly adapt to the conditions of urban combat require leaders to assume risk. These risks must be deliberately thought out and be prudentially measured against the perceived benefits. When 1-68 CAB tanks fired 120-millimeter canister rounds down streets to explode potential IEDs, the leaders assumed risk for the potential collateral damage but viewed the reduced risk to soldier and civilian casualties as more beneficial. The empowerment of leaders to make these types of risk decisions will be required and should be trained as often as possible.

Concrete barriers deserve their own place in urban combat history as one of the most effective weapons of modern warfare—especially in the Battle of Sadr City.[26] Operation Gold Wall took many of the advantages of fighting in a city away from the enemy. The wall took away the enemy's ability to hide among the population, took away their ability to defeat the attacker in detail with guerrilla tactics, separated them from multiple vital support lines, and forced them to fight. If JAM did not come out to resist the emplacement of the wall, they faced isolation and being cut off from resupply, firing sites, and their critical financial base. When they did come out to resist, they became visible and were easily targeted by US and ISR ground forces. Instead of fish in the sea, they became fish in a barrel. Building a wall around the enemy's base of operation created the conditions that made it both impossible for enemy fighters to operate effectively and possible to restore security to the broader population of Baghdad.

The use of concrete before and during the battle must be considered for future urban offensive, defensive, and stability operations. These lessons will quickly be lost if not written into doctrine. Creating a formation with effective siege-engine capabilities, able to emplace concrete walls safely and quickly, is both a science and an art. The technologies and tactics all played a role in the speed of wall emplacement and should be used as a starting point. But these methods could also be improved with planning, practice, and possibly new innovations.

Operation Gold Wall employed medieval siege warfare tactics with a twist. Instead of a city's population withdrawing behind castle walls to wait out the besieging army, coalition forces brought a modern version of a siege engine up to the edge of JAM's safe haven and built a wall around the enemy. This was very similar to the ancient tactic of circumvallation, an example of which was seen when Julius Caesar built a 12-foot-high, 11-mile wall around Alesia to defeat 60,000 Gauls in 52 BC.[27]

Indeed, there are also lessons from ancient siege operations that could be applied to future modern urban combat operations. For instance, during the Middle Ages, a large wicker or wooden screen called a *mantlet*[28] was placed in front of a besieging force's engineers digging trenches up to a castle wall (an activity called *sapping*,[29] the origin of the modern term *sapper*) to provide concealment and some cover from archers on top of the walls. The coalition forces' concrete emplacement teams, especially the lone soldier that had to unhook the crane cable from the T-wall section, were extremely vulnerable. A version of the mantlet or modern sniper screens could be placed in front of the barrier operation to provide concealment from attack.

Even without a historical review, advancements can be made in countermobility, siege, and isolation tactics, especially those involving concrete barrier emplacement operations. Despite the years and miles of concrete operations in Iraq, no notable new technologies were deployed to assist in concrete emplacement. Concrete wall design did not change. Cranes were required to get T-walls off trucks, and soldiers had to scale ladders to unhook them from the cranes. None of the cranes, crane cabling, or procedures for unhooking concrete blocks changed during the operation. A closer examination is warranted, with the aim of enhancing the construction, transportation, and emplacement of barriers and walls. Research, experimentation, and the incorporation of new automated devices and protection for soldiers unhooking concrete blocks are simple but critical aspects to consider in future obstacle emplacement in combat operations.

The logistical requirements of the building of the Gold Wall cannot be overlooked. It required a 24-hour flow of personnel, vehicles, and supplies. A smaller army or a unit with fewer resources would clearly require external assistance. But as demonstrated by the 3rd BCT, many of the wall-building requirements can be contracted to local civilian support. The key is early planning of any such operations, being sure to incorporate the lessons and tactics from operations such as the building of the Gold Wall.

The psychological impacts of the Gold Wall also warrant discussion. The wall clearly agitated the enemy and made JAM fighters come out and fight. The concept of imposing one's will on the enemy goes back at least to Prussian philosopher Carl von Clausewitz. The actual act of imposing their will on the enemy by building the wall had a positive psychological effect on the coalition soldiers. It was a proactive action, fundamentally different than the reactive responses that had characterized day-to-day operations prior to the battle and especially the emotionally draining experience of waiting to get hit by IEDs without an enemy to strike back at.

The Battle of Sadr City also serves as a vignette to analyze urban terrain, both for the physical advantages and disadvantages it provides to both sides of a battle and in relation to the complex, social-cultural "system of systems" that are cities. Existing key terrain has to be properly identified and then utilized. Because Sadr City was purpose built, its buildings designed and constructed relatively at the same time, its uniform layout of largely three-story structures made the few taller buildings key terrain that could be fortified and used as overwatch positions for key locations such as major intersections, ambush sites, or rocket-launching sites. Key terrain can also be created to shape a battlefield. The Gold Wall fell into this category, but attacking forces might also build a new piece of high ground with a new building, tower, or other structure. Finally, key terrain can also be removed, such as when, during the Battle of Sadr City, attacking forces used a GMLRS to reduce a sniper bunker in a five-story building. Beyond physical terrain analysis, understanding and balancing the relationship of a city's systems to enemy and friendly forces will continue to be a vital requirement for the military. The financial and popular support provided by a very specific part of the city was a critical requirement for JAM's survival, so much so the group gave up all the advantages of city warfare in attempting to maintain control of it.

The US military prefers to avoid cities. But being a global superpower with complex and wide-ranging national interests means the military often does not get to choose where, who, or when it will fight. The Battle of Sadr City provides one alternative to traditional approaches to defeating an enemy force that has chosen to fight from within a city. It breaks the mold of avoiding cities because of the advantages it gives a defender, the troop-intensive requirements of attacking a city, and the destructive results of doing so. The use of combined arms maneuver (tanks and armored infantry fighting vehicles), massive ISR and precision-strike capabilities, and elements of ancient siege warfare (walls) allowed a relatively small US force to contain, exhaust, and set the conditions to defeat a large enemy force within a populated dense urban area without destroying the city.

Notes

Special thanks to Dr. Dave Johnson and US Army Col. (ret) Mike Pappal, Lt. Col. (ret) Rob MacMillan, Maj. John Chambers, and Command Sgt. Maj. (ret) John Kurak, who contributed to this chapter with their experience and research.

1. Johnson, Markel, and Shannon, *2008 Battle of Sadr City*, 17.
2. "Al-Qaeda in Iraq: Militant Group," *Encyclopedia Britannica*, February 2, 2022, https://www.britannica.com/topic/al-Qaeda-in-Iraq.
3. "Mahdi Army: Iraqi Militia Group," *Encyclopedia Britannica*, n.d., https://www.britannica.com/topic/Mahdi-Army.
4. Johnson, Markel, and Shannon, *2008 Battle of Sadr City*, 9.
5. Ibid., 14.
6. Mike Nizza, "Baghdad's 'Great Wall of Adhamiya,'" *Lede* (blog), *New York Times*, April 20, 2007, https://thelede.blogs.nytimes.com/2007/04/20/baghdads-great-wall-of-adhamiya/.
7. Spencer, "Most Effective Weapon."
8. Russell, "Mao Zedong's *On Guerrilla Warfare*."
9. Spencer, "Most Effective Weapon."
10. Johnson, Markel, and Shannon, *2008 Battle of Sadr City*, xiv.
11. Hurst, "US: Raid of Baghdad's Sadr City Kills 49."
12. Johnson, Markel, and Shannon, *2008 Battle of Sadr City*, 100.
13. Ibid., 60.
14. Ibid., 57.
15. Ibid., xvi.
16. Ibid.

17. Ibid., 57.
18. Ibid., xvii.
19. Ibid., 3.
20. *New York Times*, "Major Describes Move," 14.
21. Echevarria, "War and Politics."
22. Johnson, Markel, and Shannon, *2008 Battle of Sadr City*, xvii.
23. Kilcullen, "Urban Combat."
24. Department of the Army, *ADP 3-0 Operations*.
25. Kelly et al., *U.S. Combat and Tactical Wheeled Vehicle Fleets*, xviii.
26. Spencer, "Most Effective Weapon."
27. Hudson, "Battle of Alesia."
28. Giles, "Sun Tzu on the Art of War."
29. "Medieval Siege Warfare," Ancient Fortresses educational site, n.d., http://www.ancientfortresses.org/medieval-siege-warfare.htm.

10

Drone Warfare

Paul Lushenko and John Hardy

IN 2002, THEN US PRESIDENT GEORGE W. BUSH AUTHORIZED THE FIRST-known use of an unmanned aerial vehicle (UAV), now commonly referred to as a "drone," to kill an al-Qaeda leader. The strike, launched from a General Atomics MQ-1 Predator, killed Ali Qaed Senyan al-Harthi in Yemen, who had a role in al-Qaeda's bombing of the *USS Cole* several years earlier. Bush's use of a drone for targeted killing set a dangerous precedent in global security. Somewhere between nineteen and thirty-five states now possess armed drones.[1] More problematic, a litany of nonstate actors have acquired or manufactured armed drones, chief among them the Islamic State, and have turned them especially against US forces operating in the Middle East and across Central and South Asia.[2] Indeed, US Marine Corps General Kenneth McKenzie Jr., head of the US Central Command, recently stated that drones are the number one threat to US forces operating in the region.[3] This development has also encouraged the UN Special Rapporteur on Extrajudicial, Summary, and Arbitrary Executions to declare the onset of a "second drone age," which also suggests that urban conflict may be the next frontier for the use of armed drones.[4]

Since 2002, the literature on drone warfare has evolved over three phases that reflect scholars' shifting focus. First, scholars attempted to understand drone proliferation.[5] The debate crystallized around demand-side considerations, such as the putative appetite among US allies and partners for armed drones, as well as the implications of great power competition between the United States and China for enhanced foreign military sales.[6] However, scholars paid less attention to the supply side of proliferation. Supply-side proliferation consists of, inter alia, lobbying of the US Congress by the largest US manufacturer of armed drones, General Atomics, for greater authorization of UAVs sold overseas. As the most prolific user of armed drones, the United States's export of them provides a useful barometer for scholars on the prospects of a broader global diffusion, as the country's trading practice has been for the international sale of conventional arms more generally.[7]

Next, scholars sought to measure the effectiveness of strikes. Suffice it to say, and not for lack of trying, the debate is largely unsettled. While some claim that drone warfare can redress the military capability of targeted networks, others contend that strikes do not address grievances that engender political violence. Much worse, strikes can intensify grievances that prolong violence within and across political communities, primarily states.[8] Finally, the disagreement over the effectiveness of strikes encouraged many scholars to shift their focus to the normative dimensions of drone warfare, or the ethical, moral, and legal permissibility of strikes.[9]

Some scrutinize the existing scholarship as a veritable "drone-a-rama."[10] Indeed, Hugh Gusterson recently mused that "it seems to me that the public has largely lost interest in drones."[11] This narrative implies that studying drone warfare may constitute the death knell for junior scholars, including for those of us who are also experts on drone warfare. A review of the literature suggests that scholars have yet to address the broader implications of drone warfare across multiple levels of analysis. Until recently, scholars did not study the impact of drone warfare for the legitimacy of global order.[12] Similarly, scholars are puzzled by variation in the public's perception of legitimate strikes, which is the subject of ongoing research. To the extent scholars have treated legitimacy as a dependent variable, their analyses are interpretative, nonfalsifiable, and monocausal, all of which threatens the generalizability of results.[13] Despite the importance of presidential decisionmaking in America's use of drone warfare, scholars also reify the practice to an opaque bureaucratic process referred to as the "Interagency."[14] Doing so discounts the agency of leaders in authorizing strikes and reinforces the need for scholars to better account for how individual preferences, beliefs, and decisionmaking styles shape policy and strategy.[15]

At the same time, scholars also conflate drone warfare as a tactic and strategy; use drone warfare interchangeably with decapitation; eschew a discussion of drone warfare in urban conflict; and fail to realize that drone warfare, as merely one form of robotics-enabled warfare, is not simply the preserve of technologically sophisticated Western armies. "Small armies," as we might describe many within the North Atlantic Treaty Organization alliance, as well as a panoply of paramilitary forces and transregional terrorists, increasingly use drones based on the appeal of "riskless" war.[16] Drones promise to help small armies achieve their aims while protecting against the risks, including a reduction in friendly casualties, an appealing premise when there is little political appetite for casualties. US Air Force

General David Deptula once argued, for instance, "the real advantage of unmanned aerial systems is that they allow you to project power without projecting vulnerability."[17] Such vulnerability is especially acute in urban conflict marked by the use of force in canalizing terrain—alleyways, bazaars, and tenements—teeming with noncombatants. These features put a usually larger advancing force at a positional disadvantage, which small armies can exacerbate by employing drones that help establish interlocking fires from hardened battle positions.

The assumption of riskless war, particularly in urban terrain that creates unique risks for maneuver forces, is not without cause. A recent survey experiment by Janina Dill and Livia Schubiger found that most Americans adjudicate their support toward the use of force abroad through the lens of US military casualties.[18] An unfortunate consequence of this moral psychology of war, however, is that the intended purpose of drone warfare has largely been assumed despite the reality that state and nonstate actors alike use strikes for a variety of reasons. Indeed, Sarah Kreps and Sarah Maxey show that the strategic context of violence, which can range from counterterrorism to humanitarian intervention to interstate wars, mediates how strikes are used.[19]

In this chapter, we address the still outstanding question of how scholars and practitioners should best understand the purpose of drone warfare in urban conflict, as well as the associated advantages and disadvantages. These outcomes are especially germane for small armies attempting to capitalize on drones to achieve parity with larger forces maneuvering in complex urban terrain. Though drone warfare is routinely reduced to decapitation, or leadership targeting, commanders use strikes for three purposes: pressure, leverage, or desynchronization. Drone warfare can harass a clandestine network, remove critical enabling personnel and assets from a network, or neutralize symbolic leaders that are integral to a network's longevity. Commanders can pursue more than one purpose at a time, and they often integrate all three simultaneously. This model helps us better understand not just how commanders use drone warfare in urban conflict but why. It is predicated on linking the intended purpose of strikes to the targeting effects, and outcomes to the operational considerations that condition action.[20]

Defining Drone Warfare

Existing definitions of drone warfare largely address one aspect or another of states' use of strikes. They do not integrate considerations of *what* armed drones do, *how* they do it, and *where*. Scholars are also guilty of assuming *why* strikes are conducted in the first place. The trade-off in focusing on one dimension of drone warfare at the expense of others is the lack of an intuitive definition that helps especially casual or nonexpert observers understand the intended purpose of remote warfare and in different contexts. Below, we trace the ways that scholars have defined drone warfare before offering an original definition that anchors the remainder of our analysis.

Scholars often assume that armed drones constitute a revolution in military affairs (RMA) that irrevocably alters both the character and nature of war.[21] Armed drones do not simply shape how war is waged but also its underlining social and political dimensions.[22] Yet others, especially military practitioners, caution "that war is an intensely human endeavor, no matter how the trappings of technology might seem to put us at some distance from the horror."[23] Conflating the practice and nature of war in terms of drone warfare is largely the result of states' use of armed drones for targeted killing. Although international law does not address the practice, scholars concur that targeted killing is the "premeditated killing by military and intelligence officials of named and identified individuals without the benefit of any judicial process."[24]

The prevailing RMA narrative has encouraged scholars to equate drone warfare to the platform itself. Armed drones, or what the Missile Technology Control Regime classifies as Category I weapons, are epitomized by the General Atomics MQ-9 Reaper. These platforms fly above 497 miles per hour, reach beyond 186 miles, and deliver payloads over 1,110 pounds.[25] Reducing drone warfare to "the agentic capacity of drones," which one scholar criticizes as "drone essentialism," fails to appreciate that drones constitute one component of a digital system that enables militaries to shorten the sensor-to-shooter timeline.[26] By integrating sensors enhanced with artificial intelligence and linked to a global intelligence collection and analysis architecture, armed drones acquire and strike targets based on commanders' approval.[27] To the extent that scholars attempt to define drone warfare more broadly, they interpret the practice in one of three ways.

Some typologize drone warfare based on *what* missions armed drones can perform. The most insightful analysis recognizes that armed drones can "kill," "watch," or "aid."[28] That is, armed drones can enable humanitarian assistance and disaster relief, perform surveillance and reconnaissance, deliver preparatory fires prior to ground missions, and provide close air support to soldiers. Because most states lack the resources and technical wherewithal to buy, maintain, and integrate armed drones such as the MQ-9 Reaper into their arsenals, they also employ tactical or nonarmed drones that are predominantly operated by radio signals rather than satellite downlinks. According to Stefan Borg, these cheaper, smaller, and unarmed drones are useful for conducting short-duration intelligence-gathering and surveillance missions. Irredentist disputes, such as the imbroglios between Armenia and Azerbaijan in Nagorno-Karabakh, as well as Russian-backed separatists and Ukrainian forces in eastern Ukraine, are representative of this operational profile. (For more on eastern Ukraine, see Chapter 4.)[29]

Precisely *how* armed drones engage a target is also a matter of special inquiry that links to a broader study for the moral permissibility of strikes. Neil Renic argues that technological advancements in war epitomized by armed drones threaten radically asymmetric violence. The imbalance of liability to be harmed among combatants, Renic contends, imposes a "crossover point between unproblematic and problematic killing."[30] Gusterson engages the moral status of drone warfare by bifurcating the practice into "pure" and "mixed" types.[31] Pure drone warfare relates to the use of strikes separate from ground operations conducted by soldiers including patrols and raids. Mixed drone warfare characterizes the use of strikes in support of patrols and raids. This latter mode implies that soldiers' assumption of greater physical risk on the battlefield helps shape the public's perception of morally legitimate drone warfare. Emerging research suggests, however, that states' use of drones as a tactic or strategy, coupled with how strikes are unilaterally or multilaterally constrained to protect against unintended consequences, can also shape the public's intuitions for morally legitimate strikes. Behind this finding lies a constellation of moral norms—including soldiers' personal courage on the battlefield, the protection of soldiers, and noncombatant immunity—that the public recalls based on how drones are used to adjudicate legitimate strikes.

Nevertheless, Gusterson's construction is useful to help inform *where* drone strikes take place, which can be in "hot" or "cold" theaters of operations such as Afghanistan and Yemen, respectively. While this discussion largely circumvents a consideration of drone warfare in urban terrain, it helps clarify how member states of international society understand the legality of strikes. The laws of armed conflict consist of two main components: *jus ad bellum* (just recourse to war) and *jus in bello* (just use of force). In the context of drone warfare, scholars often conflate these dimensions and fail to submit each to an empirical study for the public's support to and perceived legitimacy of strikes.[32] Doing so is important, however. States can flout *jus ad bellum* norms, such as just cause and proper authority, while also attempting to fulfill *jus in bello* norms, including distinction and proportionality. The mismatch between these two levels of norms is often a key characteristic of drone warfare in urban terrain. This was the case with US president Donald Trump's drone strike against Iranian major general Qasem Soleimani in 2020.[33] The strike breached Iraq's sovereignty but killed Soleimani near the Baghdad International Airport while resulting in no civilian casualties.

Though the legal merits of drone warfare are important, especially for global governance, this discussion tells us little about how strikes are actually conducted, especially in urban terrain. Similar to raids and cyberattacks, drone warfare presupposes a rigorous targeting process that incorporates what Daniel Brunstetter refers to as the "human element," including enabling and combat forces.[34] It takes over 200 personnel, for instance, to operate an MQ-9 Reaper for every 24-hour period. Approximately 60 of these are deployed within a theater of operations.[35] Although drone warfare is often portrayed as a less manpower intensive form of conflict, it simply shifts personnel away from the frontlines, which offers important force protection dividends for especially small armies attempting to impose costs on larger forces operating within urban terrain. Also, while some confine drone warfare to use against military targets exclusively, or what Dill refers to as the "logic of sufficiency," it can represent a form of coercive diplomacy and limited war.[36]

This discussion suggests that scholars often emphasize merely one aspect of drone warfare. As a result, drone warfare is often depicted as strictly lethal, unethical, immoral, and illegal. Tapping into the underlining purpose of strikes suggests that drone warfare is a more complicated concept than most experts realize. A broader depiction of drone warfare requires us to synthesize its various dimensions. A more satisfactory definition of drone warfare is *the use of armed drones in concert with expeditionary forces to achieve military and/or political objectives, limited or maximal, across the continuum of competition and in conflict.* Adopting this definition is beneficial for several reasons. It is broad enough to account for the lethal and nonlethal aspects of drone warfare, the diverse missions armed drones can conduct, and what types of conflict they can be involved in. The latter can range from irregular warfare to urban warfare to interstate warfare in both "hot" (declared) and "cold" (undeclared) theaters of operations. This definition leads to a more complex and nuanced understanding of the utility of drone warfare. As such, the next section presents a model for the operational use of drones in urban conflict.

Drone Warfare in Urban Conflict

The model presented below provides a framework for analyzing drone warfare in urban conflict against a networked opponent.[37] The purpose of the model is to better enable assessment of the effectiveness of strikes in urban terrain. Though drones can perform other missions, their use for targeted killing is the most common operation. Focusing on the effectiveness of these strikes allows us to better relate our model to the ongoing debate for drones in urban conflict.

Indeed, our model provides a comprehensive framework for assessing drone warfare in a manner that is informed by doctrine and practices that relate to intelligence collection, analysis, and sharing and by targeting cycles recommended by the US Army.[38] As a holistic model, it is not reducible to a targeting cycle, such as "find, fix, finish, exploit, analyze, and disseminate." Although this has been described as an aggressive leadership targeting cycle, it encompasses the primary methods through which drone warfare is conducted and ignores the operational and strategic contexts of strikes.[39] In this model, drone warfare is situated at a level of analysis above the targeting cycle. Drone warfare drives this cycle by linking intelligence collection and analysis to the targeting methods applied, the intended effects and outcomes, and the conditions that shape action.

The model is focused on utilizing drone warfare against networked opponents in urban conflict (see Table 10.1). For the purposes of the model, drone warfare is employed against clandestine networks.[40] These are illicit groups that conduct illegal activities and operate with a decentralized network structure to conceal the identity of their members and organization from authorities.[41] While many may think of clandestine networks in terms of terrorist and insurgent organizations only, they do not have to be. By virtue of their equally cagey quality, criminal organizations such as drug cartels can also constitute clandestine networks. Clandestine networks rely on social networks to manage their security-to-efficiency trade-off, which is essential to enable operatives to avoid detection and regroup when targeted.[42] Internal connectivity increases communication and functionality, but also increases the risk of detection.[43]

Nevertheless, clandestine networks require the rapid and accurate exchange of information to accomplish tasks.[44] Because the presence of a large number of dispersed members reduces the overall efficiency of the organization due to unnecessary redundancy and profligacy,[45] clandestine networks often rely on established, trust-based social connections.[46] These connections form a social-relational hierarchy within the network that is based on "relationships and acquaintances, reputations, and fame" rather than organizational hierarchy.[47] The small-world phenomena in social networks account for links between and among small clusters of individuals.[48] While social connections within cells are thought to be stronger than those between cells, clandestine networks may be forged by strong ties that remain mostly dormant for the purposes of operational security.[49] The countervailing requirements for security and communication, then, can cause inactive social ties that appear weak even when they are strong.[50]

This tension undermines the assumption that clandestine networks have moved "away from controlling leaders and toward multiple, loosely linked, dispersed agents and cells."[51] This presumption belies a materialist orientation and discounts the social context in which clandestine networks are constituted, reconstituted, and endure.[52] Clandestine networks rely heavily on social connections that unite members and help them coordinate within and across cells given guidance from organizational leaders.[53] Two widely discussed examples are the Islamic State and al-Qaeda. Their various organizational structures rely heavily on social hierarchy, with key personalities guiding and inspiring action at lower levels, as has been thoroughly documented elsewhere.[54] This suggests that while clandestine networks are increasingly decentralized in a material sense, they remain hierarchical in social-relational terms.

Ostensibly, decentralized social networks in urban terrain can be targeted by drone strikes in at least three ways. Commanders can use drone warfare to (1) pressure the network in order to harass it; (2) leverage network elements, including "middlemen"[55] or critical enabling capabilities that facilitate communication and action; and (3) desynchronize the network by targeting pivotal decisionmakers and/or figureheads to alienate operatives and leaders from one another.

Pressure

Applying pressure to a clandestine network involves harassing the organization and its members to complicate operations, force errors, and delay communications. Commanders use strikes to disrupt the clandestine network and its operations. Doctrinally, disruption is the interruption of the enemy's operational capacity, tempo, formation, or initiative.[56] In practical terms, this means that strikes are used against various components of a clandestine network to confuse the network's critical support activities and operations.

Table 10.1 Drone Warfare in Urban Conflict

	Actions	Effects	Outcomes
Pressure	High volume and frequency of strikes against targets at any network point	Harass the network Shape the urban environment Enable other operations	Disrupt
Leverage	Selective strikes against facilitators, highly skilled or highly connected individuals, and critical capabilities	Deprofessionalize the network Increase the costs of operations Degrade the network's functions	Disrupt Destroy
Desynchronization	Strikes against the most visible and/or influential leaders and figureheads	Induce discord across the network Alienate network members from leaders Deskill and delegitimize the network's leaders	Disrupt Destroy Defeat

Source: Paul Lushenko, "Reconsidering the Theory and Practice of High Value Targeting," *Counter Terrorist Trends and Analysis* 7, no. 7 (2015).

Applying pressure to a clandestine network is similar but not analogous to swarming. This is a "seemingly amorphous" but deliberate, sustained, and coordinated series of strikes intended to disorient a network.[57] In some corners of the US military, swarming has also been described as "whack-a-mole."[58] However, applying pressure accords commanders a degree of flexibility to concentrate, surge, or redirect combat power to disrupt a clandestine network in a piecemeal or systematic fashion.

Because the pressure method similarly prioritizes targets, or casts the widest net across the depth and breadth of a clandestine network, it also increases the probability of killing consequential members, including financiers and couriers. This approach has the added benefit of generating second-order effects such as intelligence that facilitates further operations. It also shapes the urban environment for both military and nonmilitary follow-on operations, some of which are designed to ameliorate grievances to legitimize a political authority.[59] Finally, pressure fatigues clandestine networks by increasing the transaction costs of communication, planning, and operations. This can encourage members to take greater risks and expose themselves by conducting operations to demonstrate resolve and garner support. The results can be devastating, as the effects of US and coalition forces' strikes on the Islamic State in the Khorasan Province from 2015 to 2018 will show.

Leverage

Rather than directing action, leverage points are assets that enable a clandestine network's activities. In the US military, these are often referred to as critical capabilities.[60] They may include individuals with favorable reputations and therefore a high degree of social connections. They also often facilitate communications; possess niche skills, such as bomb-making, forgery, or media production; are ideological authorities; or facilitate lethal aid including componentry, money, and supplies. Leverage points may also consist of critical capabilities that enable and protect the network, including ammunition and explosives, communications devices, and sanctuaries. For some defense analysts, leverage points represent a network's center of gravity, which is defined as "the hub of all power and movement, on which everything depends."[61] Network analysis indicates that leverage points are also often associated with task organization.[62] The Islamic State and al-Qaeda, for instance, organize themselves across a range of operational and support specialties that require "facilitators, financiers, computer specialists, or bomb-makers."[63]

Commanders strike leverage points to destroy a clandestine network's functionality or to damage it "so badly that it cannot perform any function or be restored to a usable condition without being entirely rebuilt."[64] This deprofessionalizes the network and imposes additional recruitment and training costs that further diminish operational capacity.[65] The impact of such skills shortages on short-term offensive capacity is evidenced by US and coalition forces' use of drone strikes against the Taliban in the Panjwai District of Kandahar Province in 2012, which we explain below.

Desynchronization

Desynchronization, often referred to as decapitation, is the most championed but scrutinized goal of drone warfare in urban conflict. It involves killing visible and symbolic network leaders in order to disintegrate the group, isolate members from each other, and protect against civilian casualties. The successive removal of leaders impedes guidance and coherent messaging, which threatens to weaken and delegitimize the network. Although replacement leaders are often found, they typically lack the same charisma, experience, relationships, and skills as their predecessors. Meanwhile, the loss of leadership disrupts a network's capacity to plan, prepare, and execute attacks given the emphasis on defensive measures and poor guidance.[66] Networks that rely on central guidance to autonomous and decentralized cells also suffer setbacks from the targeting of highly visible or influential figures. Interference with a network's key decisionmaking body can induce power struggles, confuse organizational direction, and encourage infighting. This diverts time and resources from offensive operations to enable a preferred yet possibly contentious line of succession.[67] Such internecine violence can engender competing factions and fracture a network.

The intent of desynchronizing is to both disrupt and destroy a clandestine network, considering that internal confusion impedes and degrades critical capabilities. In certain cases, desynchronization may also enable the defeat of a network. Defeat "occurs when an enemy force has temporarily or permanently lost the physical means or the will to fight" and may be induced by a lack of coherent organizational purpose following the killing of an influential leader.[68] Price's empirical study of leadership decapitation in terrorist groups found that religious organizations are vulnerable to leadership targeting, perhaps because of leaders' integral role in setting goals, aligning resources to achieve these desired aims, and managing risks to the organization during operations.[69] It is important to note, however, that leadership targeting should not be considered a means to victory by itself. Rather, it ought to be used as a tactic subordinate to an overall strategy. Yet, the stature and centrality of especially ordained leaders—such as the Islamic State's outspoken leader, Abu Bakr al-Baghdadi—means that their removal is capable of at least disrupting clandestine networks. For instance, coalition forces removed the second-in-command of al-Qaeda in Iraq in 2008—effectively a regent of the organization given the reclusiveness of the overall leader, Abu-Hamzah al-Muhajir—that temporarily defeated the network in northern Iraq.[70]

Integrating Pressure, Leverage, and Desynchronization

Although presented discretely, commanders in urban conflict can pursue all three purposes of drone warfare and often do so simultaneously within the same area of operations against the same network and related groups. The cumulative effects of drone warfare on a clandestine network in urban terrain is usually prefigured on a high operations tempo designed to exploit the intelligence dividends of previous operations, particularly information on the intent, capability, and disposition of networks gleaned from captured enemy material. This can include communication devices, correspondence, and documents such as ledgers, maps, and war plans.

Drone warfare also produces chatter across a network that can enable future operations. Chatter is a spike in communication across network members after a drone strike. Heightened communication helps commanders and their staffs identify which members are sending and receiving information, the nature of the information, and the network's recourse. Chatter is captured through multiple—human, imagery, measurement and signature, and signals—intelligence disciplines. The exploitation of captured enemy material and chatter is critical to informing measures of effectiveness that justify commanders' decisions to recalibrate the purpose of drone warfare. The ability to shift to another purpose, pursue an additional one simultaneously, or redirect resources toward an altogether different target helps commanders create opportunities and exploit a network's vulnerabilities. This operational flexibility enables commanders to apply drone warfare to target clandestine networks during urban conflict in a variety of ways in pursuit of unique outcomes.

The Utility of Drone Warfare in Urban Conflict

The model of drone warfare presented above enables a purposive case selection strategy to demonstrate the utility of drone strikes in urban conflict. This case selection strategy is useful because it emphasizes variation in the purpose of strikes, which is the key unit of analysis; enhances control over commanders' use of drone warfare in urban terrain; and maximizes the observable implications. It also reflects an established research tradition in political science of analyzing cases that are similar in some respects but different in others. This has been called the "most similar systems design."[71] It has also been referred to, confusingly, as the "method of difference."[72] Adopting this case selection design helps to control for exogenous variables when studying the reasons why commanders use drone warfare to impose costs on clandestine networks in urban conflict.

Before introducing the cases, it is important to address several methodological issues that threaten to confound our analysis. First, because drone warfare amounts to what the historian Michael Howard notes as "modern history," data may be difficult to obtain.[73] The sensitivity of decisionmaking and the strikes evaluated may bias the results, as well as any evidence that is available for discovery.[74] This suggests a potentially more problematic concern. If the typology relies on limited data, how is it possible to trace the different purposes of drone warfare in an urban environment? Assuming we can, what justifies exploring one instance of drone warfare at the expense of another example?

These challenges can be resolved by triangulating data. The case studies rely on primary and secondary sources, including news reports, to substantiate the intended purpose of drone warfare in urban conflict. They also rely on recorded firsthand accounts from combat operations in each case study. This evidence has been publicly documented in other investigations of America's approach to countering insurgencies and transregional terrorists. The cases addressed below are crucial and pathway cases. They show the intended purpose of strikes that are most explanatory of commanders' use of drone warfare against clandestine networks in urban conflict. Each structured and focused case study seeks to answer four key questions in a narrative format. First, what was the threat, including its intent and capability? Second, what strategy did US and coalition forces adopt to counter the threat? Third, how did commanders incorporate drone warfare? Finally, what were the effects and overall security outcomes?

Pressure and Desynchronize: Islamic State in the Khorasan Province (2015–2018)

The Islamic State in the Khorasan Province, or IS-KP, is the Islamic State's regional affiliate across Central and South Asia and is headquartered in Nangarhar Province in eastern Afghanistan.[75] The intent of the group is to secure Afghanistan to legitimize the Islamic State's transregional caliphate across the Khorasan Province, which consists of parts of Central Asia, China, Iran, the Indian Subcontinent, and Southeast Asia. The group has a largely antagonistic relationship with other extremist groups, operating in former al-Qaeda territory and frequently clashing with the Taliban in recent years.[76] IS-KP operates across the Afghanistan-Pakistan border and is regularly targeted in both states by security forces. The overwhelming majority of offensive operations against IS-KP were undertaken in Afghanistan, primarily in Nangarhar Province.[77] Despite heavy losses suffered in these operations, IS-KP continued to pose a significant threat to civilians and security forces and was responsible for numerous lethal attacks in Afghanistan, where it retains the majority of its operational capability.[78]

The group frequently conducts coordinated attacks against security forces and civilian targets, often using remotely detonated explosives and small arms. Although IS-KP has primarily killed civilians in Afghanistan, it has also attacked Afghan security forces, US and coalition military personnel, law enforcement, the Taliban, and local warlords.[79] Throughout 2015–2018, IS-KP attacks primarily occurred in Afghanistan. Attacks largely targeted state security forces, mostly police officers and law enforcement infrastructure, followed by military forces and infrastructure, despite killing more civilians than state personnel. Where attacks targeted civilians, they were often along sectarian lines, with Shiite communities and infrastructure sustaining the greatest proportion of violence.[80] One of the group's recent attacks, which took place at the Hamid Karzai International Airport on August 26, 2021, and resulted in the death of at least 183 people, including 13 US Marines, reflects a broader operational capability that suggests IS-KP is the most virulent extremist organization in Afghanistan. More importantly, IS-KP's expanded operational reach beyond Afghanistan, into India, Myanmar, and Pakistan, indicates the group also constitutes a regional and global security challenge that cannot be ignored.

When IS-KP emerged in 2015, most US defense officials discounted the group as merely a "nuisance."[81] Not until 2017, after which the group demonstrated the ability to inspire, enable, and direct attacks abroad in Europe and the United States, did then US secretary of defense James Mattis direct the coalition to "annihilate" IS-KP.[82] In an effort to destroy the group, US and coalition forces adopted a largely lethal counterterrorism strategy consisting of drone strikes, raids, and clearance operations in concert with Afghan National Security Forces, especially the Afghan Special Operations Forces. Targeting across four tiers of leadership included high-level leaders (emirs), provincial level commanders, critical midlevel leaders, and local (subdistrict) level leaders.[83] The operations, aligned against the "Greensword" campaign plan, unfolded across three distinct phases. Integral to each was the removal of the group's emir via a drone strike. This pattern of drone warfare was not only designed to desynchronize the group, even given Mattis's clear guidance, but also to pressure IS-KP to afford Afghanistan's military forces the time and space to consolidate security gains in key urban terrain, especially Jalalabad and Kabul.

On January 10, 2015, the coalition killed IS-KP's inaugural emir, Hafiz Saeed Khan, with a drone strike near Jalalabad. To maintain the pressure, the coalition then employed the GBU-43 in April 2017, the largest nonnuclear bomb ever deployed in combat, against IS-KP's headquarters south of Jalalabad.[84] The bomb destroyed the nucleus of the group's operational planning; it reportedly killed 100 members and pulverized nearly $10 million in cash reserves. That same month, the coalition also used a drone strike to kill the group's second emir, Abdul Hasib, which further disrupted the network's operations.[85] The strike sowed distrust among IS-KP's leaders and members, which the coalition stoked through its information operations. The heightened suspicion that followed resulted in greater operational security across the group, which stymied communications, delayed payment of salaries and arms to members, and stalled offensive operations in Jalalabad and Kabul.

During July 2018, the coalition killed the group's third emir, Saad Erhabi, in the vicinity of Jalalabad. This strike coincided with the coalition's removal of a key foreign fighter recruiter, Qari Hikmatullah, with another drone strike. At the time, Qari was located in Moghol Village in the Jowzjan Province of western Afghanistan. His removal was especially consequential for the coalition's counterterrorism strategy. Qari persuaded many Uzbeks to defect from the Islamic Movement of Uzbekistan to join ranks with IS-KP. As such, his death created a vacuum that the Taliban exploited to counterattack IS-KP in Moghol Village. The clash claimed the lives of 135 IS-KP members, injured 100 more, and resulted in the capture of nearly 400 remaining loyalists.

Throughout the period 2015–2018, IS-KP sustained heavy losses at all levels. Drone strikes accounted for 21 percent of operations against IS-KP and were responsible for 20 percent of total losses, including 21 percent of militant deaths.[86] The spring of 2017 was the most significant period for IS-KP leadership losses, with a high volume of targeting leading to substantial losses at all levels of leadership throughout the year. Leadership losses peaked in April 2017, coinciding with the GBU-43 strike and reaching 7 percent of total IS-KP losses in Afghanistan.[87] Combat data show that leadership losses typically correlate with a significant drop in the volume of attacks in months coinciding with or following the death or capture of leaders.[88] Leadership losses did not result in a similar reduction of lethality per IS-KP attack over this period, however. This indicates that critical skills and experience were diffusing to lower levels of command, even where the erosion of internal leadership reduced IS-KP's operational tempo.[89] These findings are consistent with broader research on the effectiveness of drone strikes in reducing the operational tempo of terrorist and insurgent networks.[90]

The coalition's use of drone warfare to pressure and desynchronize IS-KP momentarily disrupted the group and destroyed its external operations capability. Because US and coalition forces privileged the use of lethal force in pursuing a revitalized counterterrorism strategy, however, they were unable to redress grievances that allowed IS-KP to emerge in Afghanistan in the first place. Sadly, it is now clear that the strategy did little to influence the Afghan National Security Forces to consolidate gains by facilitating governance in areas once occupied by IS-KP, including in key villages and cities across eastern and western Afghanistan. The group, similar to the Taliban, exploited the Afghan government's neo-patrimonial practices responsible for institutionalizing corruption through a bureaucracy that represented a tool of avarice for officials rather than a way to provide security and prosperity for Afghans.[91] The government's reticence to occupy terrain after IS-KP's dislocation allowed the group to complete their task and expand. Following its recent attack at the Kabul airport, the group's spokesman stated, "let the crusaders and their agents know that the soldiers of the caliphate will continue to fight them until God decrees a matter that has been done."[92]

Pressure and Leverage: The Taliban in Panjwai District (2012–2013)

After defeating the Taliban in 2001, US and coalition forces oscillated between two approaches to countering the subsequent insurgency in Afghanistan. One approach, which can be termed "hard" counterinsurgency, focused on killing or capturing insurgents to prevent them from achieving their goals and primarily understood the population as a means to an end. The worth of citizens was measured against their ability to provide intelligence that enabled the prosecution of elusive insurgents. The other approach, which can be termed "soft" counterinsurgency, was designed to protect the population from insurgents through the judicious application of force with the overall intent of improving the public's perceptions of the government's legitimacy.

Whereas US doctrine, namely Field Manual 3-24, *Counterinsurgency*, recommended that commanders reconcile these two approaches, the record suggests that rarely happened. The army's strategic culture, which Andrew Krepinevich argues is based on a reliance "on mid-intensity, or conventional war, and a reliance on high volumes of firepower to minimize casualties," prevented a blended approach to countering the insurgency in Afghanistan. Despite or because of the urban context of much of the fighting, especially in eastern and northern Afghanistan, commanders privileged lethal force, including the use of armed drones, at the expense of other initiatives designed to improve governance and development.[93]

The countervailing "hard" and "soft" strategies framed the deployment of Combined Task Force (CTF) 4-9 Infantry to the Panjwai District of Kandahar Province in southern Afghanistan from 2012 to 2013. Kandahar, but especially Panjwai, is the ideological bedrock of the Taliban. It provided a strong recruitment base given enduring support from the Noorzai and Ishaqzai tribes that are prolific in the area.[94] The area, therefore, afforded the Taliban a platform to "Talibanize" Afghanistan. The Taliban's approach consisted of three tenants. First, it was based on an ideology of absolution amid an ostensibly morally corrupt society. Second, a vanguard of Taliban officials led organized formations that were responsible for expunging impurities across society. Third, absolution was sustained by pursuing the consolidation of an Afghan state under Taliban control.[95]

While deployed to Panjwai, CTF 4-9 adopted a "hard" counterinsurgency approach in concert with US Special Operations Forces. The latter conducted raids and strikes to pressure and desynchronize the Taliban, whereas the former used strikes to both pressure and leverage the network. Armed drones used by CTF 4-9 were strikingly effective, at least in the near-term. They resulted in the death of sixty-five insurgent commanders, fighters, and facilitators, and wounded nineteen more. By killing explosive handlers faster than Taliban leaders could replace them, CTF 4-9 eroded institutional knowledge regarding the whereabouts of improvised explosive devices (IEDs). This resulted in Taliban members killed or maimed by their own mines and helped reduce enemy activity in the district by 77 percent. CTF 4-9 also used strikes while protecting against civilian casualties. One Afghan explained that the strikes were remarkably precise. "Americans do not kill everyone," he argued, "they first ensure that the person in question is 100 percent bad."[96]

At the time, the security gains seemed to create the perception of enhanced safety that encouraged some Afghans to rise up against the Taliban in a movement that became known as the "Peshingan Awakening," named for the village in which it originated. The *New York Times* considered the event "the most significant popular turning against the Islamic insurgents in recent years," especially because it evolved "in the spiritual heartland of the Taliban movement."[97] In hindsight, even the use by CTF 4-9 of tactically effective drone warfare to pressure and leverage the Taliban in Panjwai distorts the movement's complicated genesis that reflects the limits of "hard" counterinsurgency conducted through drone warfare in urban conflict.

A combination of two factors is more responsible for engendering collective action against the Taliban by downtrodden villagers in Peshingan. First, anger transformed villagers' private preference for "everyday resistance," defined by James Scott as "the prosaic but constant struggle between the peasantry and those who seek to extract labor, food, taxes, rents, and interest from them," into support for active opposition against the Taliban.[98] In February 2013, the Taliban crippled Peshingan's *mirab* (water purveyor) after he remonstrated members for brutalizing locals. The removal of Peshingan's *mirab* threatened the seasonal poppy cultivation, a mainstay of the narco-economy. The potential loss of a critical market good, poppy, mobilized an economic grievance among the villagers, even given their long-standing association with the Taliban. Thus enthralled, villagers adopted a form of collective action, violent protests, against the Taliban that echoed the "water wars" in medieval Japan and Bolivia in the late twentieth century.[99]

Second, the hasty protests were harnessed by influential community leaders presumably intent on shepherding a broader movement across Panjwai. Sensing the sudden upheaval constituted a tipping point or "revolutionary threshold," in which villagers' preference for everyday resistance became more psychologically costly than directly confronting the Taliban, a venerated village elder, Abdul Wadood, helped coordinate the movement's otherwise slipshod activities.[100] The arrival of two additional figures compounded the uprising's momentum. Sultan Mohammad, the charismatic yet opportunistic chief of police, dispatched a quick reaction force to consolidate gains against the Taliban. The operation resulted in at least one extrajudicial killing. While deplorable by Western standards, internecine violence has real purchase in Afghanistan as a form of signaling, precisely because it symbolizes grave power disparities. "Running roughshod over the village, Mohammad and his entourage detained several suspected Taliban members. At the point of their AK-47s, members of the ANP [Afghan National Police] marched these detainees . . . to Panjwai's main east-west running road. After quarantining the insurgents, a member of the ANP outfit summarily executed one of the detainees."[101] Following this extrajudicial killing, the district governor, Haji Faizal Mohammad, arrived with members of the international press corps to frame the Taliban's injustices as a means to engender additional pockets of resistance across Panjwai. Whether he knew it or not, the governor attempted to frame a shared identity among villagers to "maintain sustained challenges against powerful opponents."[102]

The Peshingan Awakening collapsed several months later. Why, with such an auspicious beginning, did this occur? Sidney Tarrow notes that the central "problem for movement organizers is to create organizational models that are sufficiently robust to structure sustained relations with opponents, but are flexible enough to permit the informal connections that link people and networks to one another to aggregate and coordinate contention."[103] The movement never bridged this challenge for three interrelated reasons that help explain the dissipation of a palpable "crowd feeling" and "pleasure of agency" among protesting villagers.[104]

First, the villagers, including Wadood, overly relied on CTF 4-9's drone strikes in anticipation that they would help institutionalize the uprising. To be sure, Sultan Mohammad declared independence from CTF 4-9. "If we need help from our counterparts," he claimed, "we are asking for medical services, air support they are helping us, but we aren't asking for help in our operations."[105] The evidence contradicts Mohammad's statement. By pressuring and leveraging the Taliban through drone warfare in urban terrain, while also minimizing civilian casualties, CTF 4-9 provided overwatch that strengthened villagers' resolve. Afghan police and military forces were incapable of conducting independent operations to buoy the uprising on account of training and capability shortfalls, as well as ethnic and sectarian cleavages across their ranks.

Second, whereas CTF 4-9 continued to overwatch Afghan forces and the uprising, security gains were not matched by governance and development improvements. Panjwai's leaders were unwilling to mobilize support from their powerful patrons. Over time, this dispirited villagers who previously risked their safety by revolting. The two Mohammads—that is, the police chief and governor—did little to encourage elites in Kabul to invest in Panjwai. One villager complained that elites "do not work for the people because they are not from our village. They all live in the city."[106] This demonstrates another weakness of community-based movements noted by Tarrow: "autonomy at the base sometimes excludes strong connective ties between center and periphery, making it difficult for leaders to implement coherent strategies."[107] The lack of a coherent strategy allowed the insurgency to consolidate and reorganize. The Taliban reinstituted sharia law, which further evidenced the state's fecklessness and renewed villagers' belief in the insurgency as a governing alternative. Finally, community leaders, especially Sultan Mohammad, capitalized on the awakening to extend patrimonial networks that redressed formation of a unified identity among protesters by isolating them from each other.[108] Under the security umbrella of CTF 4-9, Mohammad cultivated a racketeering operation that condoned criminal activity by other "greedy spoilers," including political and security officials, elites, and insurgents.[109] His "privatization of authority" was so blatant that the governor, his boss, noted, "I am not for it."[110]

The use of drone warfare by CTF 4-9 to pressure and leverage the Taliban resulted in important security gains. By privileging a "hard" counterinsurgency approach through drone warfare, CTF 4-9 afforded time and space for Afghans to reject the Taliban, resulting in the Peshingan Awakening. There was a crucial trade-off, however. CTF 4-9 ignored development, demonstrated through an unwillingness to invest in civic works projects. Had CTF 4-9 been more interested in development and governance improvements, it might have been able to expand the uprising's appeal by providing incentives that encouraged some cautious fence-sitters to lend support. There is a larger lesson, still. Even given the material support provided by CTF 4-9, the revolutionary impulse among local Afghans was fleeting. Villagers never mobilized social structures to ensure the movement's longevity. Instead, they relied on caretakers with competing interests. The focus of CTF 4-9 on security through drone warfare, the reticence of elites to provide support, and the rapacious behavior of the chief of police—these factors suffocated the movement. Because the Taliban have the watch, or so the aphorism goes, they bided their time in Pakistan before rekindling a dormant shadow government that shifted Panjwai from its momentary status as an incumbent-controlled zone back to an insurgent stronghold.

Conclusion

These two case studies exemplify the multipurposed and overlapping rationales that may be used by commanders to employ drone strikes operationally in urban terrain. Our analysis of the cases demonstrates that commanders applied strikes in a considered and deliberate way to pursue specific outcomes. This finding is consistent with the targeting framework that we introduced and applied to the cases. It indicates that commanders can shift and integrate different purposes of drone warfare in urban conflict. However, these purposes are not sufficient in themselves to assure or consolidate tactical successes. Despite unremitting strikes to pressure and desynchronize IS-KP, the network continues to demonstrate remarkable resolve. Immediately following the US withdrawal, the group attacked a Shiite mosque in northern Afghanistan, resulting in the deaths of nearly fifty Afghans and injuries to hundreds more.[111] Similarly, the Taliban's rapid seizure of Kabul amid the coalition's withdrawal in August 2021 indicates strikes intended to pressure and leverage the organization did little to address its critical capabilities. Extrapolating across these cases suggests three conclusions for the successful application of drone warfare in urban conflicts.

First, the utility and purpose of drone warfare cannot be reduced to simple metrics, such as leadership targeting, alone. Operational utility is linked to the purpose of strikes, which can include pressuring, leveraging, or desynchronizing a clandestine network in an urban conflict.[112] US and coalition forces attempted to pressure and desynchronize IS-KP from 2015 to 2018 with mixed results. As shown, pressure applied to IS-KP, combined with the successful removal of emirs in quick succession to one another following their appointments to overall leadership of the network, preceded a reduction in the total number of attacks. However, with little focus on leveraging the network to erode critical skills, the lethality of attacks increased over time. The coalition's focus on lethal force also did not redress grievances held by the local population in urban environments and did not undermine the support that IS-KP garnered in the cities it operated from.

Coalition forces attempted to pressure and leverage the Taliban in Panjwai from 2012 to 2013 with mixed results as well. Drone strikes were more effective in the short-term in Panjwai, especially when leveraging the Taliban's local network. Lethal strikes degraded the Taliban's critical skills and reduced their operational effectiveness. CTF 4-9 killed explosives handlers faster than the Taliban could replace them, eroding knowledge about the location of IEDs and expertise in building and emplacing mines. The net result was a decrease in enemy activity and an increase in Taliban deaths and injuries from mishandling explosives. These gains were not matched by a similar effort to desynchronize the network, however, by targeting Taliban leaders, many of which absconded to Pakistan. Overall, an emphasis on drone warfare did not lead to the desired strategic outcomes in Panjwai. Both of these cases highlight that the operational effectiveness of drone strikes is necessary but not sufficient against clandestine networks in urban conflict.

Second, maximizing the effectiveness of drone strikes in urban conflict also requires not only knowledge of, but investment in, the social terrain. Arguably, US and coalition forces possessed such knowledge. An elaborate array of human and signals intelligence collection capabilities enabled the use of drones to attack IS-KP and the Taliban. The Islamic State reeled from strikes that removed emir after emir from 2015 to 2018. The Taliban traded space for time in Pakistan to gain a reprieve from the targeted killings. Unfortunately, US and coalition forces did not match these promising outcomes with broader social and political initiatives to enhance the government's legitimacy. In Panjwai, this was a matter of choice. CTF 4-9 favored lethal force during the development. Consequently, the Peshingan Awakening petered out, and drone warfare could not resolve broader deficiencies in the movement that, if addressed, may have sustained the uprising. Afghan National Security Forces also cowered from assuming the liability to consolidate gains against IS-KP in the key cities of Jalalabad and Kabul. They failed to address local grievances by safeguarding the provision of social welfare programs in the same urban terrain IS-KP sought to control. The net effect was that the Taliban and IS-KP capitalized on continued frustrations to peel away support from the government that enabled both groups to survive.

This suggests that though drone warfare has real operational utility, it is not an adequate substitute for a comprehensive strategy for conducting operations in urban conflict. Promoting drone warfare as a strategy amounts to "strategic monism," which Huntington once related to a government's singular focus on one tactic to defeat a clandestine network.[113] The use of strikes, as shown in the cases we explored above, often results in short-term security gains that do not translate into long-term strategic success. Ironically, US officials continue to view drone warfare as a strategy that promises hearty returns, long-term military and political outcomes, for little investment, especially in terms of soldiers' lives.[114] This is evidenced by the adoption by President Joe Biden's administration of an "over-the-horizon" counterterrorism strategy in Afghanistan that is based on drone strikes. If the initial application of this strategy is any indication, it is likely to result in mixed outcomes as well. On August 29, 2021, the United States conducted its first over-the-horizon drone strike in Kabul. Rather than nabbing a suspected IS-KP terrorist, the operation killed ten civilians.[115]

This example reiterates that enduring security gains against clandestine networks in urban terrain are only possible when drone warfare is integrated with other elements of national power. It is clear that drone warfare had an impact on urban warfare in Afghanistan. However, it is now clear that drone warfare, absent a broader institution-building approach to support the Afghan government in opposing the Taliban and IS-KP, was insufficient. As such, drone warfare remains a useful component of urban operations to the extent that it supports an overarching strategy to end a specific conflict. This is a potentially more important consideration for small armies with comparatively fewer resources to withstand tactical mishaps that can pile up to constitute strategic blunders, as the war in Afghanistan arguably represents.

Notes

The views expressed in this chapter are those of the authors and do not necessarily reflect the official policy or position of the US Department of the Army, Department of Defense, or Government.

1. Vilmer, "Not So Remote Drone Warfare"; Borg, "Below the Radar"; Horowitz, Kreps, and Fuhrmann, "Separating Fact from Fiction," 11.
2. Carter, "Coming Soon to a Theater (of War) Near You."
3. Schogol, "Drones Pose the Biggest Threat."
4. "Targeted Killings Through Armed Drones and the Case of Iranian General Qassam Soleimani," *Report of the Special Rapporteur on Extrajudicial, Summary or Arbitrary Executions*, June 29, 2020, https://www.ohchr.org/en/issues/executions/pages/srexecuti onsindex.aspx.
5. Kreps, *Drones*.
6. Chavez and Swed, "The Proliferation of Drones"; Horowitz, Schwartz, and Fuhrmann, "Who's Prone to Drone?"; and Gilli and Gilli, "Diffusion of Drone Warfare?"
7. Erickson, "Saint or Sinner?"
8. Price, "Targeting Top Terrorists"; and Jordan, *Leadership Decapitation*.
9. Renic, *Asymmetric Killing*; Meisels and Waldron, *Debating Targeted Killing*; Galliott, *Military Robots*; and Kaag and Kreps, *Drone Warfare*.
10. Kaplan, "Drone-o-Rama."
11. Hugh Gusterson, email correspondence with the author, September 15, 2021.
12. Lushenko, Bose, and Maley, *Drones and Global Order*; Senn and Troy, *Transformation of Targeted Killing*; and Senn and Troy, "Transformation of Targeted Killing and International Order."
13. Renic, *Asymmetric Killing*; and Galliott, *Military Robots*.
14. Lushenko, "U.S. Presidents' Use of Drone Warfare."
15. Hafner-Burton et al., "Behavioral Revolution and International Relations."
16. Kahn, "Paradox of Riskless Warfare," 2–7.
17. Quote in Chamayou, *Theory of the Drone*, 12.
18. Dill and Schubiger, "Attitudes Towards the Use of Force."
19. Kreps and Maxey, "Context Matters."
20. Hardy and Lushenko, "High Value of Targeting."
21. Lushenko, "Review of the Special Issue."
22. McMaster, "Continuity and Change."
23. Dempsey, "Introduction," 18.
24. Meisels and Waldron, *Debating Targeted Killing*; and Solis, "Targeted Killing."
25. Mistry, *Containing Missile Proliferation*.
26. Demmers and Gould, "Remote Warfare Paradox." See also Gusterson, "Drone Warfare in Waziristan."
27. Kreuzer, *Drones and the Future of Air Warfare*.

28. Welsh, "The Morality of 'Drone Warfare,'" 25.

29. Borg, "Below the Radar." See also Spencer and Ghoorhoo, "Battle of Shusha City."

30. Renic, *Asymmetric Killing*, 5

31. Gusterson, *Drone: Remote Control Warfare*, 14–15.

32. Kreps and Wallace, "International Law, Military Effectiveness."

33. Jahanbani, *Beyond Soleimani.*

34. Brunstetter, "Drones: The Future of Warfare."

35. Elish, "Remote Split."

36. Dill, *Legitimate Targets?*

37. This section pulls heavily from Hardy and Lushenko, "High Value of Targeting." We thank Taylor & Francis for allowing us permission to adopt material from our earlier article.

38. Flynn, Juergens, and Cantrell, "Employing ISR."

39. Ibid.

40. Raab and Milward, "Dark Networks as Problems."

41. Molnar, *Undergrounds in Insurgent*, 23–24; Miksche, *Secret Forces.*

42. Morselli, Giguere, and Petit, "The Efficiency/Security Trade-Off"; Shapiro, "Terrorist's Challenge."

43. McCormick and Owen, "Security and Coordination," 190.

44. Baker and Faulkner, "Social Organization of Conspiracy," 844.

45. Memon and Larsen, "Practical Algorithms for Destabilizing Terrorist Networks," 393.

46. Krebs, "Mapping Networks of Terrorist Cells," 49–51.

47. McChrystal, "Becoming the Enemy."

48. Watts, "Networks, Dynamics, and the Small-World Phenomenon."

49. Barabasi, *Linked*; and Krebs, "Mapping Networks of Terrorist Cells," 49.

50. Granovetter, "Strength of Weak Ties."

51. Stohl and Stohl, "Secret Agents," 11.

52. Provan and Kenis, "Modes of Network Governance"; and Gutfraind, "Understanding Terrorist Organizations."

53. Lindelauf, Borm, and Hames, *Understanding Terrorist Network Topologies*; McCormick and Owen, "Security and Coordination"; and Stohl and Stohl, "Secret Agents."

54. Lushenko, Van Auken, and Stebbins, "ISIS-K."

55. Neumann, Evans, and Pantucci, "Locating Al Qaeda's Center of Gravity."

56. Headquarters, Department of the US Army, *Combat Tactics*, B-16-7.

57. Arquilla and Ronfeldt, "Advent of Network (Revisited)."

58. Flynn, Juergens, and Cantrell, "Employing ISR," 59.

59. Gellman, "William McRaven."

60. Smith, Jeter, and Westgaard, "Three Approaches to Center of Gravity Analysis."

61. Meadows, *Thinking in Systems.*

62. Davis and Jenkins, "System Approach," 8.

63. Frankel, "ABCs of HVT," 26–27.

64. Headquarters, Department of the US Army, *Combat Tactics*, B-16-7.

65. Wilner, "Targeted Killing in Afghanistan," 312.

66. Ibid.

67. Wilner, *Best Defense*, 5.

68. Headquarters, Department of the US Army, *Combat Tactics*, B-15.

69. Price, "Targeting Top Terrorists."

70. Lushenko, "Reconsidering the Theory and Practice of High Value Targeting," 27.
71. Collier, "Symposium: Case Selection, Case Studies, and Causal Inference."
72. Przeworski and Teune, *Logic of Comparative Social.*
73. Howard, "The Lessons of History," 491.
74. Cirone and Spirling, "Turning History into Data."
75. See Lushenko, Van Auken, and Stebbins, "ISIS-K."
76. Ibrahimi and Akbarzadeh. "Intra-Jihadist Conflict and Cooperation."
77. During the period 2015–2018, 98.8 percent of IS-KP losses (killed) were in Afghanistan, and 1.2 percent were in Pakistan. For further details, see Jadoon and Mines, *Broken, but Not Defeated*, 20, 60.
78. Jadoon and Mines, *Broken, but Not Defeated*, 3; and Jadoon, *Allied and Lethal*, 18–30.
79. Catrina Doxsee, Jared Thompson, and Grace Hwang, "Examining Extremism: Islamic State Khorasan Province (ISKP)," blog post, Center for Strategic and International Studies, September 8, 2021, https://www.csis.org/blogs/examining-extremism/examining-extremism-islamic-state-khorasan-province-iskp.
80. Jadoon, *Allied and Lethal*, 18–30.
81. Brimelow, "ISIS Wants to Be as Dangerous as the Taliban."
82. Baron and Weisgerber, "New Tactics, Quicker Decisions."
83. Jadoon and Mines, *Broken, but Not Defeated*, 8–9.
84. Katzman and Thomas, *Afghanistan*, 20–21.
85. Katzman and Thomas, *Afghanistan*, 21.
86. Jadoon and Mines, *Broken, but Not Defeated*, 20, 41–43.
87. Ibid., 37–39.
88. Ibid., 40–49.
89. Ibid., 44–49.
90. Mir and Moore, "Drones, Surveillance, and Violence," 846–862.
91. Maley, "State Strength and Rule of Law."
92. O'Connor, "ISIS Takes Credit."
93. Krepinevich, *Army and Vietnam*, 5.
94. Lushenko and Hardy, "Panjwai."
95. Maley, "Talibanisation and Pakistan."
96. Major Paul Lushenko's Field Notes from 2012–2013. Interview with an Afghan citizen at the Panjwai District Center, March 30, 2013.
97. Gall, "Afghan Villagers Take on Taliban."
98. Scott, *Weapons of the Weak*, 29.
99. Berry, *Culture of Civil War in Kyoto*, 7. See also Simmons, "Market Reforms and Water Wars"; and Peters, *Seeds of Terror.*
100. Kuran, "Now out of Never."
101. Lushenko, "Coining an Ethical Dilemma," 273.
102. Tarrow, *Power in Movement*, 2.
103. Ibid., 124.
104. Pearlman, "Emotions and Microfoundations of the Arab Uprisings."
105. Ardolino, "Panjwai District Chief of Police."
106. Major Paul Lushenko's Panjwai Field Notes from 2012–2013. Interview with an Afghan citizen at the Panjwai District Center, April 18, 2013.
107. Tarrow, *Power in Movement*, 134.
108. Ibid., 138.
109. Stedman, "Spoiler Problems in the Peace Process."

110. Bill Ardolino, "The Taliban Are 'Still Dreaming' of a Return to Power: An Interview with Afghan Army Officers in Panjwai," *Chainsoff's Blog*, October 3, 2013, http://chainsoff.wordpress.com.

111. Al Jazeera, "Afghanistan: Dozens Killed."

112. Hardy and Lushenko, "High Value of Targeting," 423.

113. Huntington, *Soldier and the State*, 400, 418–427.

114. Cook, "Drone Warfare and Military Ethics," 54–56; Martel, *Grand Strategy in Theory and Practice*; and Strachan, "Introductory Essay," 12.

115. Kreps and Lushenko, "What Happens Now?"

11

Surrogate Warfare

Tyrone L. Groh

POLITICS AND PASSION SHAPE THE NATURE OF BATTLE, AND THOSE TWO factors add enormous uncertainty to any form of warfare.[1] An urban setting makes it harder.

In urban warfare, efforts to manage the use of violence in a way that defeats an enemy and protects the inhabitants and infrastructure of a city can evaporate in a moment, should the enemy choose to hold the inhabitants and infrastructure at risk. A preeminent lesson of warfare, however, is that the enemy gets a vote. A second lesson, courtesy of Helmuth von Moltke, is that "no plan survives first contact with the enemy."[2] The fact that some groups have made urban warfare look easy in recent history does require some consideration.

In 2014, a significantly outnumbered Islamic State (IS) force took the city of Mosul from two divisions arrayed inside the city in less than a week. In 2021, Taliban forces took several cities in rapid succession with ease, despite being outnumbered and outgunned by Afghanistan's forces. Both cases suggest that efforts to better equip and better train the forces of professional armies remain relevant but may not be the principal means of reducing the risk of fighting wars in cities. Shaping the battlespace and gaining information superiority in a city become paramount in urban warfare.[3] A more contemporary development is the imperative to reduce the risk to a state's professional army. Exploring a way to balance those factors is the focus of this chapter.

Surrogate Warfare

Surrogate warfare is about the "externalization of the strategic, operational, or tactical burden of warfare, partially or wholly, to a delegate or substitute." The term *surrogate* refers to a broad spectrum of delegation.[4] For example, a state's special operations forces (SOF) provide a way to delegate aspects of a military operation that its main conventional force could not perform. The capabilities of SOF allow for a smaller number of fighters to take a larger portion of the risk. Remotely piloted aircraft (RPAs) represent a more technical version of a surrogate. Both of those examples are surrogates that afford the sponsor ample control and do not negatively affect the cohesion of the policy. Other types of surrogates such as proxies, another state's armed forces, or private military corporations offer far less control and can create stronger negative effects on the policy's cohesion. Regardless of the situation, availability becomes one of the most important factors when choosing a surrogate.

Urban warfare, characterized by close quarters and complicated social terrain, levels the playing field for a vastly superior force. In a more modern context (think post–World War II), a group chooses to fight in an urban environment due to a significant asymmetry in capability or capacity. The professional army will have superior equipment and training. Cities offer advantages to forces engaged in a defensive conflict that include ready-made fortifications, camouflage, and "high concentrations of targets and resources."[5] The group choosing to fight in a city will also use the city's social terrain as a means to level the playing field. A surrogate may help unlevel the playing field. Nick Reynolds argues that "other allies and partners will provide much of the mass" needed for urban warfare in the future.[6] Although this notion will help manage the risks to a state's professional army, it fails to address the other capabilities.

In 1994, James Quinn and Frederick Hilmer addressed the idea that firms use strategic outsourcing to allow an organization to focus on the "core competencies" that give it a "definable preeminence" and outsource the other necessary, but highly specialized capabilities that do not provide a strategic advantage in the firm's likely arenas of competition, to include those "traditionally considered integral."[7] According to Reynolds, surrogates will provide the mass considered integral for urban warfare in the future. I suggest this is short-sighted.

Surrogates have far more to offer than mass when it comes to addressing shortfalls caused by the political limitations caused by an urban setting. I agree with Reynolds that a surrogate can help address problems with a state's capacity to engage in urban warfare. Bridging shortfalls in capacity focuses on having enough stuff. With an asymmetric advantage against a weaker enemy, concerns about committing a sufficient number of forces to engage in urban warfare emerge due to governments and citizens being risk averse when it comes to employing their own forces in such an environment.[8]

Bridging shortfalls in capability focuses on having the right kind of stuff. In violent conflict, that often means having capable offensive weapons that can hold an adversary at risk, while incurring less risk to one's own forces. The neutralizing effect of an urban setting on such advantages, however, shifts the focus on capability to having the right kind of information. Knowing one's adversary and the context in which the conflict will take place remains age-old wisdom from as far back as Sun Tzu.[9] In urban warfare, the need for information increases exponentially due to the complexity of the social terrain.

When we consider improving our ability to achieve political objectives in an urban setting, are we thinking about how surrogates add the capacity needed to outsource a large portion of the risk to our own forces, or are we looking for surrogates with specialized capabilities that our current forces lack? Andreas Krieg and Jean-Marc Rickli argue that "surrogate warfare is the continuation of politics by other means in an era when the state as a sociopolitical complex has to redefine its role as a communal security provider amid a globalized context, against intangible threats, in competition with nonstate actors, under the constant surveillance of a global public sphere, and under the domestic pressure of fiscal austerity and growing welfare expenditure."[10]

Outsourcing urban warfare can create an advantage for a professional army, but those advantages come with costs. The balance between cost and benefit rests heavily on the way urban warfare is outsourced. Obviously, every situation is somewhat unique and requires highly effective intelligence operations to help decisionmakers, strategists, and operational planners understand the operational environment.[11] As described in Chapter 6, urban warfare introduces complicated physical and political barriers, including the melding of noncombatants into the battlespace. More specifically, Patrick Finnegan shows how the dense, opaque nature of the social terrain in an urban area creates localized effects in geographically separated areas.

Introducing another actor into urban warfare can help, hinder, or add tinder. Help comes from adding capacity and capability. Hinder comes from the potential for adverse selection—the surrogate's realized inability to live up to its promised capacity and capability—and how that contributes to the unknowns that affect operations and initiative. Tinder comes from the risks associated with moral hazard—a surrogate's ability to hide its actions and intentions from its sponsor—and how it affects the operation's outcomes, such as complicating negotiations or altering the social terrain.[12]

Choosing to Use Surrogates

Regardless of regime type or the structure of a nonstate actor's organization, the rational use of violence has barriers. Obvious considerations, such as comparing one's relative capability to that of an adversary or the level of an actor's political will to engage in violent conflict, immediately come to the fore. Assuming that war is a rational choice, then direct intervention demands that the threat to vital and/or desirable interests breaches a specific threshold.[13] That threshold varies for each state depending on factors such as time, domestic or internal influences, and international influences. When an actor perceives the need to intervene militarily but lacks the sufficient level of threat needed to commit completely, then an additional factor that may influence the decision might be the availability, capacity, and capability of a surrogate.[14]

When it comes to urban warfare, I assume that the core tactics of fighting possessed by today's professional armies remain sufficient. Given that, I argue that surrogates have a role to play in overcoming the reality that urban warfare demands—that the forces engaged in the conflict must "cede the initiative to the city." One of the main principles of counterinsurgency explains that a population is not something you can fight around. You have to fight through it and with it. Unlike physical terrain, such as the notion suggested in *The Jungle Is Neutral*, urban areas are anything but neutral. Frederick Chapman suggests that the jungle "provides any amount of fresh water, and unlimited cover for friend as well as foe—an armed neutrality, if you like, but neutrality nevertheless."[15] Although urban areas have physical features like a jungle, the defining characteristic lies in an urban area's social terrain—an opaque yet extremely tangible essence that possesses strong agency and fungible resources. Surrogates can help and hinder in this endeavor, but it is also possible that a surrogate may just provide the tinder needed to make an urban fight unwinnable.

The choice to use a surrogate requires careful consideration regarding its agency. Simply put, the greater the surrogate's agency, the greater the odds that the surrogate will pursue its own agenda to the detriment of the sponsor's agenda. As the surrogate's capability to influence an urban campaign increases, so does its ability to alter the outcome. In a perfect world, a surrogate's objectives match up perfectly with its sponsors, negating the negative effects associated with the surrogate's agency.

How Surrogates Help

Getting someone else to fight for you has obvious benefits.[16] In terms of urban warfare, using a surrogate offers three consistent benefits that help bridge shortfalls in terms of capacity and capability. First, a surrogate increases the capacity (the number of forces) of a professional army, while minimizing the actor's political commitment to the conflict. For example, if the surrogate sustains unacceptable losses or commits an atrocity that damages the public's view of the conflict, a professional army can cut ties with the surrogate or place blame on the surrogate to avoid having to appear responsible for either of those outcomes. Second, a surrogate can provide mission-critical skills that add to a professional army's capability. More specifically, a surrogate can possess a much higher level of knowledge regarding the social and physical terrain of the city, as well as a better ability to collect useful and actionable intelligence. Third, employing a surrogate can engender support from the city's inhabitants. If the surrogate maintains a social connection with a significant part of the city's population, or if the surrogate only has connections to a small group in a strategically important part of the city, then a professional army can benefit.

The notion that a surrogate can provide capacity when you don't have it available is only a small part of how surrogates benefit urban warfare. The notion that they change risk calculations is important, but theorists of surrogate warfare get it somewhat wrong when it comes to urban warfare. For example, the idea that "the surrogate brings violence to bear where necessary without causing negative repercussions for the patron that employs him—as long as the link between the patron and surrogate remains under the cloak of plausible deniability and discretion" suggests that the use of a surrogate becomes enticing because it relaxes the pressure from a democratic state's population to use force judiciously and appropriately.[17] The real value of a surrogate in urban warfare stems not from its ability to fight well in close quarters nor from its expendability, but from its ability to navigate the social terrain of a city. A surrogate's ability, from this perspective, can be measured by the quality of intelligence it can provide regarding the social terrain of the city and the legitimacy it wields in the selected urban setting.

How Surrogates Hinder

Generally speaking, using surrogates in urban warfare hinders operations because it introduces the potential of adverse selection, an additional source of the uncertainty in an already complex and uncertain endeavor. Even with a highly trained and well-equipped professional army, Carl von Clausewitz's dictum that "war is a game of chance" remains a fixed aspect of war. Like it or not, the enemy gets a vote in the outcome of war. Using a surrogate means the surrogate also gets a vote. In terms of adverse selection, a professional army must accept the fact that they cannot know the quality of the surrogate's ability in terms of its suitability for both the physical and political environment, its fighting ability, or its ability to get valuable and actionable intelligence. Perhaps more dangerous, the surrogate may conceal information or thwart intelligence operations. Outsourcing shortfalls in capacity and/or capability introduces additional risk into an already risky endeavor.

In an urban setting, the problem of adverse selection becomes especially problematic. The density of a city and the close proximity of noncombatants make monitoring a surrogate especially challenging, especially because the need for a surrogate has already come from political barriers to the urban-based conflict. With density, even the smallest incursions or actions can have widespread effects. For example, in the long-running insurgency in Crossmaglen (presented in Chapter 6), a geographically isolated incident had effects in areas untouched by the violence. In that sense, social terrain reflects something more akin to a spider's web than a collection of physical spaces. A man killed in Belfast awakened a resistance movement in Crossmaglen, a village 50 miles away.

Surrogates as Tinder

When choosing a surrogate, professional armies must consider second- and third-order effects, especially those that are unintended. Moral hazard is a serious issue when it comes to employing surrogates.[18] Knowing the true intentions of a surrogate, or the depth of its commitment to assist in the fight, remains difficult, if not impossible. A surrogate's intentions for agreeing to support a professional army in an urban war can have profound effects on the operation's outcome. Negotiations, concessions, or even outright victory can all be confounded by a surrogate's pursuit of its own objectives. Added to that, the negative effects of a surrogate's agency depend on the amount of control the professional army maintains over the surrogate. If the sponsor can exert little control in an effort to maintain plausible deniability or to manage escalation, then the surrogate has more freedom to act as it sees fit, creating a moral hazard for the sponsor. A cost attributable to adverse selection can also cause the use of proxy to burn out of control. If the proxy does not have access to the information needed to wield the social terrain of the city for advantage, actions can be taken that undermine the sponsor's objectives.[19] This adds an additional cost of the operation: resources the professional army must spend to keep the surrogate in line and to ensure the surrogate lives up to its promised capabilities.

Iraqi Security Forces as a Surrogate for Operation Inherent Resolve

The two cases presented in this chapter serve as a stark reminder of the enigmatic character of urban warfare. Fewer than a thousand soldiers of the Islamic state, with minimal training and resources, needed only 3 days to take Mosul from two divisions of the Iraqi security forces (ISF) entrenched inside the city. A little over 2 years later, four divisions of the ISF, with infinitely more resources and support, needed 9 months to recapture Mosul from 3,000 to 12,000 IS fighters well-established inside the city. In both situations, the ISF served as the principal surrogate. Interestingly, the inhabitants of Mosul felt disenfranchised by the forces in control of the city. The social terrain and the forces' ability to engage in urban warfare in Mosul, however, were vastly different.

The Loss of Mosul, 2014

The city is not neutral. In just 3 days of fighting, the Islamic State's forces, numbering well under a thousand, took the city of Mosul from two ISF divisions quartered within the city and numbering between 25,000 and 30,000 men.[20] Two divisions lost Mosul while engaging in a defensive operation! The ISF served as the principal surrogate for the coalition of states supporting the fledgling central government of Iraq. The advantages of a superior defensive position and the challenges of close-quarter combat for an attacker had little effect. The two most critical elements of the surrogate, its commitment to the cause and its ability to gain and maintain the superiority of intelligence, were all but ignored.

Background. Iraq, in its modern context, stands as an amalgamation of "three historically distinct Ottoman provinces: Basra, Baghdad, and Mosul. . . . Arab Sunnis historically played leading roles under the rule of the Ottomans with whom they shared the official religion (Sunni Islam). . . . Arab Shias of Iraq, on the other hand, occupy most of the southern areas and have a strong presence in Baghdad."[21] Following Saddam Hussein's failed attempt to take and hold Kuwait in 1990–1991, Shia uprisings revealed a starkly divided country along three separate lines: Sunni, Shia, and Kurd.[22] When Hussein's regime fell in 2003, northern Iraq fractured into a sectarian divide between the Sunni and Shia sects of Islam and a nationalistic divide between Arabs and Kurds.[23] The new Iraqi constitution ratified in 2005 attempted to democratize Iraq's three constituencies but ultimately led to Iraq's Sunni population feeling alienated and powerless in the new construct. The US government's demand for de-Baathification heightened Sunni insecurity, leading marginalized Sunnis to turn to more radical elements.[24]

In 2004, al-Qaeda in Iraq (AQI) made its first appearance in the city of Mosul when it occupied over thirty police stations. Although this effort ultimately failed, AQI and its successor IS gained enough of a foothold in the city to grow its influence over the next 10 years.[25] After US combat forces completed their withdrawal in Iraq in 2011, Prime Minister Nouri al-Maliki felt compelled to enact extensive measures to remove any remaining

Baathists in an effort to shore up his administration's control of the country—an effort that only heightened Sunni concerns. To prevent a Sunni resurgence, Prime Minister al-Maliki began "coup-proofing" the Iraqi government by "purging rivals, ethnic stacking, creating a parallel security force, and increasing intra-government surveillance. These measures were highly effective in preventing a possible military putsch but did greatly reduce the Iraqi troops' capacity and willingness to fight."[26]

Baghdad's hold on Mosul grew ever more tenuous due to its neglect. Al-Maliki's government reportedly received warning of an IS attack on Mosul in February 2014, yet Baghdad made no move to respond. In May, al-Maliki's government received warnings that there were six IS training camps outside the city of Mosul and that an attack was planned for June 6. Instead of investigating or sending out a force to attack the camps, the commander of ISF's military operations in Mosul went on holiday on June 3.[27]

As reported, 200 IS fighters arrived at the gates of Mosul to confront 25,000 soldiers from the ISF. When an additional 400 IS fighters joined the fray 2 days later, IS captured Iraq's Sunni capital, a city of 2 million people with huge supplies of weapons, ammunition, and wealth. Overnight, IS became the wealthiest terrorist organization on the planet.[28] This calamitous advance sparked the beginning of Operation Inherent Resolve, a global coalition of seventy-four states.[29] Two months later, Prime Minister al-Maliki resigned from office.[30]

Why use the ISF as a surrogate? When IS emerged as a real threat to northern Iraq, the political will for most coalition countries had been exhausted. Political constraints regarding the number of "boots on the ground for the United States ultimately led to the politically mandated drawdown of Operation Iraqi Freedom in 2010, driving a switch to an advisory role and then the redeployment of US forces in 2011. The US-led coalition, including the NATO Training Mission in Iraq, continued with advisory and training assistance until 2013, but the lion's share of Iraq's security fell to the al-Maliki government and the ISF starting in 2014. It is relevant to note, however, that between 2005 and 2014, coalition forces spent over $20 billion training and equipping the ISF.[31]

By June of 2014, the ISF was the only viable force to serve as a surrogate for the coalition and for the government in Baghdad. The Kurdish Peshmerga, evenly split into its two factions—the Kurdish Democratic Party and the Patriotic Union of Kurdistan—were too divided politically, and the city of Mosul lay too far outside the Kurdish sphere of influence.[32] Given the lack of choices, considerations regarding the ISF's specific capabilities were not relevant, and it was only the ISF's capacity and availability that led to the choice.

Mosul's social terrain in 2014. As mentioned above, social terrain is the combination of a city's physical terrain (both natural and man-made), the people that inhabit the city, and the city's social and economic infrastructure. Mosul, Iraq's second largest city, maintained a rich and extensive history that contributed greatly to Iraq's identity. Split in half by the Tigris River, Mosul also stands as one of Iraq's most diverse cities. Prior to the rise of the Baathist regime of Saddam Hussein, Mosul was home to diverse ethnic groups, including mostly Kurds, Turkmens, and a smaller Arab minority. In the early 1970s, the Baathist regime began to resettle Sunni Arabs into Mosul, increasing their numbers to exert greater influence over northern Iraq.[33]

In Chapter 2, Wendy Pullan proposes that "urban order can be understood as the relationships embodied in the physical and spatial fabric and registered over time to support and articulate broad forms of civic life." She also notes that cities that predate the modern era tend to be perceived as "particularly chaotic and incomprehensible for strangers." As an endorsement of these ideas, Mosul stands as a sprawling and complex city. Home to over 1.3 million people, Mosul has eight administrative sectors, 251 neighborhoods—91 on the right bank of the Tigris and 160 on the left bank. The left bank exceeds the living standards of the right bank, providing better security, more access to services, and more newer buildings. Eighty percent of the city's residents are Sunni Muslims, with the remaining 20 percent consisting of Kurds, Christians, Turkomans, Shabak, and Yazidis. The majority of minorities lived on the left bank, but the majority of Christians inhabited the old city on the right bank.[34]

The distribution of power in Iraq shifted substantially after the fall of the Baathist regime in 2003. Sunni domination came to an abrupt end with the ousting of Saddam Hussein's regime. Shia and Kurdish influence increased, but the Shia in Iraq represented an overwhelming majority of the population. De-Baathification pushed thousands of Sunni Arabs into the swelling insurgency and eradicated numerous Iraqi institutions. The United States and its coalition partners failed to separate the Iraqi identity from the Baathist identity. As a result, hundreds of thousands of military-trained men became disenfranchised and marginalized in their ability to secure a better future for Sunnis under al-Maliki's regime.

The sectarian violence that ravaged southern areas in Iraq were more muted in Mosul. Instead, residents of Mosul saw an influx of Shia soldiers that bore the face of al-Maliki's regime and the growing domination of Shia over Sunni.[35] Prior to 2003, Mosul had always been a consistent source of senior leaders in Iraq's Baathist-led military, creating a strong bond between the city and the Baathist regime in Baghdad. Further, reports suggest that "Mosul contributed over 300,000 people to the military, security, and intelligence services."[36] In Mosul, Sunni Arabs felt compelled to join ranks with AQI and later IS, despite significant disagreement with the motives and methods those organizations pursued.[37]

Ethnic and religious competition drastically altered Mosul's social terrain between 2004 and 2014. Armed groups filled the vacuum left after the fall of the Baathist regime. Widespread corruption and a lack of support or management of Mosul's infrastructure decimated the city's economy. The minority groups living in Mosul were displaced as individuals working with AQI seized control of Mosul's administrative positions and created Mosul's "shadow government," allowing AQI to seize control of the city's administration and economy.[38]

Between 2006 and 2008, an escalation of internecine violence led thousands of Mosul's minorities to flee the city in search of a safer place to live. At the same time, a large number of Sunni Muslims from surrounding areas sought refuge in Mosul.[39] Mosul's massive housing shortage created an opportunity for newly placed and corrupt politicians to allow wealthier people to purchase property and build homes while pushing the poorer population to seek housing in the cramped, old city or in the southwestern sector of the city.[40]

Prime Minister al-Maliki's response to the Arab Spring in 2011 involved an increase in the use of state-sanctioned violence to quiet or quell protests. Violent crackdowns spurred further civil unrest and caused disenfranchised Sunnis to join armed opposition groups. The most notable of these was IS, led by Abu Bakr al-Baghdadi. Sunnis in northern Iraq feared the politicization of the ISF.[41] As unrest increased in early 2013, the ISF responded to an armed protest in Hawija, killing forty-four unarmed Sunni protesters and causing a spike in Sunni outrage and armed-Sunni protests.[42] The presence of IS, a Sunni movement offering an alternative form of governance and an outlet for Sunni-based grievances, undermined the already weakened Shia-dominated ISF in Mosul.[43] The proximity of Mosul to Syria's border and al-Maliki's support of the Bashar al-Assad regime's efforts to break Sunni resistance in Syria may have also pushed the majority Sunni population in Mosul to protest Shia control.[44]

Prime Minister al-Maliki's efforts to consolidate power led to the corruption of the ISF at the highest levels. Senior military officers were reported to have accepted bribes from soldiers to look the other way when they deserted their posts. In Nineveh province, where Mosul is the capital, the ISF commander there had a reputation for using repressive counterinsurgency tactics and torturing prisoners. Under the ISF, Sunni areas grew more insecure and the city's economy plummeted. The Shia-dominated government turned a blind eye to the citizens of Mosul, contributing to the inhabitants' belief that al-Maliki's government had no intention of looking out for their interests.[45] Such practices erode the public's confidence and destroy intelligence operations. The social terrain of the city blinded the ISF to the intentions and operations of IS in the spring and early summer of 2014.[46]

How the ISF helped the United States and other coalition members. What help did the ISF offer? In two words: not much. As mentioned above, the ISF division tasked to defend Mosul reportedly contained 25,000 to 30,000 soldiers. Although it lacked the capability of a more experienced and professional army, the ISF had access to significant manpower and resources (capacity) to hold a city from a much smaller force. Further, the ISF was arrayed inside the city and had access to the structural defenses offered in any city.

Properly fielded and commanded, the ISF should have been able to hold Mosul. The shift in the social terrain between 2004 and 2014, however, robbed the ISF from holding any form of initiative. The ISF's access to information, one of the most important contributions in terms of capability purportedly offered by a surrogate, was nonexistent. The ISF could not read the social terrain due to its mismanagement and poor administration of the city. The al-Maliki regime and the military commanders in the Governate of Nineveh did not perceive the threat of a Sunni-dominated shadow government in Mosul. Prioritizing the desire to stave off a coup, al-Maliki's government did not exert the control or attention necessary to assure that the ISF could do what it was tasked. Reports suggest that the ISF was underequipped, even after having received advanced warning, because it had moved infantry, armor, and tanks to Anbar province to stem violence there. In addition, it appears the ISF only had 10,000 soldiers available out of the 25,000–30,000 posted in Mosul.[47]

How the ISF hindered the United States and other coalition members. Most significantly, the ISF did not attempt to engender support from the inhabitants of the city. Quite the opposite, in fact. The ISF's intelligence capability was essentially nonexistent. The ISF did not perceive the threat from allowing AQI-turned-IS to alter the city's ethnic and ideological landscape. Mosul's social terrain was being manipulated by an extremist Sunni movement led by IS, a fact the ISF appears to have either ignored or missed altogether. Despite the fact that the IS-led shadow government was corrupt and primarily used violence to intimidate the minority inhabitants of Mosul, a larger number of people living in Mosul felt oppressed by the ISF, especially given the use of violence to quell social unrest among Sunnis in northern Iraq. Many IS fighters were from the city and exerted tremendous influence there. As a result, the social terrain in Mosul greatly favored IS. The ISF had little to no ability to collect valuable intelligence. Lacking both intelligence and local legitimacy, the ISF provided no help to Iraq's government or to the coalition of states supporting it.

The ISF as tinder. Going beyond the ways ISF hindered efforts to hold IS forces from taking Mosul, it was the ISF's lack of commitment (moral hazard) to the task that created the tinder leading to the rapid and total loss of the city. The ISF's lack of commitment, however, had several sources. First, the soldiers and commanders of the ISF witnessed significant losses to its forces in 2013 when fighting IS in Fallujah and Ramadi. Added to this, horrific rumors circulated about how IS forces massacred the ISF soldiers it captured.[48] Second, the ISF was plagued by "ghost soldiers," who paid half their salary to their commanders to avoid having to show up for duty. Third, ISF's frontline soldiers had to purchase a significant chunk of the resources needed to survive from the local economy in Mosul because senior leaders pocketed the money intended to support the ISF's soldiers. The ISF's soldiers would have been well aware of the corruption among the ISF leadership and would have certainly raised doubts about their commanders' commitment to defending Mosul.[49] Fourth, the military leaders were Shia and likely felt no allegiance to the Mosul area, compromising their commitment to fight IS for Mosul.[50] Given advanced notice of the impending attack on Mosul, the ISF did not move to strike IS forces. Instead, the commander of the ISF in the region went on holiday only a few days before the predicted date of the attack. Frontline ISF soldiers, the majority of whom were also Shia, would have likely been aware of this, explaining why they followed suit and "simply dropped their weapons, took off their uniforms, and drove off."[51]

The conclusion here is that the ISF did not feel the city was worth defending when pressed by a significantly smaller force of IS fighters bolstered by a reputation for brutality and savagery, evidenced by the fact that the ISF did not really attempt to fight IS in Mosul.[52] Instead, two divisions of the ISF basically walked away from Mosul.[53] Although al-Maliki managed to hold on to power, his efforts decreased the ISF's military capability and demoralized its soldiers, obliterating the ISF's willingness to fight for the city of Mosul.[54] Added to this, the international community remained committed to supporting al-Maliki's regime, further feeding Sunni disenfranchisement.[55] Rather than being an instrument to defend all of Iraq's people, the ISF was perceived as an instrument of the al-Maliki government bent on marginalizing Sunnis and propping up Shias.[56]

The Battle for Mosul, 2017

The capture and occupation of Mosul by IS shook coalition forces. The hands-off approach, allowing Iraq to stand on its own against IS, had failed. Four months later, the United States along with seventy-eight states and five international organizations launched Operation Inherent Resolve (OIR) under the auspices of Article 51 of the UN Charter.[57] To put it in perspective, the effort to recapture Mosul was "the first sustained urban operation involving US forces since the 1968 Battle of Hue." It was also the "largest conventional land battle since the capture of Baghdad in 2003."[58] This case study looks specifically at the use of the ISF, the same force that simply abandoned the city in 2014, to recapture Mosul only two years later.

Background. After its fall, Mosul became the economic and political capital of IS. Inside the city, IS maintained 3,000 to 12,000 fighters armed with "significant numbers of heavy machine guns, rocket-propelled grenade launchers, recoilless rifles, mortars, and rockets. IS forces constructed an elaborate series of defensive works inside the city, fortifying buildings, blocking avenues of approach, creating obstacles, and constructing underground shelters and tunnels." IS also made effective use of vehicle-borne improvised explosive devices and off-the-shelf unmanned aerial vehicles. Lastly, IS prevented many inhabitants of Mosul from leaving the city so they could be used as human-shields.[59]

Phase II of OIR began at the end of 2015 and sought to support the ISF and partner forces in Syria in their efforts to take back the territory IS had captured as well as liberate the people living under IS control.[60] OIR's mission continued to be "to militarily defeat [IS] in the Combined Joint Operations Area by, with, and through regional partners in order to enable whole-of-governmental actions to increase stability."[61]

Operation Inherent Resolve sought to take the city of Mosul back from IS, reassert the Iraqi government's sovereignty in the region, degrade IS military capability, and set the stage for the defeat of IS as a protostate.[62] The US 101st Airborne Division led the nineteen-nation coalition and served as the primary interface with ISF. "The ISF consisted of a combination of Iraqi army, air force, special operations forces, and police who together provided the essential and decisive but finite ground-maneuver component."[63] In light of the lessons learned after losing Mosul, it is important to reflect on the notion that despite the impressive and overwhelming support for the ISF, it still took 10 months for the ISF to achieve what IS accomplished in only a handful of days.[64]

Mosul's social terrain in 2016. After IS captured Mosul, the city's already frail economy collapsed. Under the administration of IS, Mosul's infrastructure declined due to mismanagement and intentional destruction. Public institutions disintegrated and the city's cultural heritage suffered calculated efforts to erase things unrelated to a very narrow and extremist interpretation of Sunni Islam. Basic needs and services required by Mosul's inhabitants became significantly limited, including food and water. Electricity-generating stations stopped operating for want of fuel. Hydroelectric production continued at a reduced rate, providing electricity for a few hours a day on a rotational basis. IS took control of the city's resources, such as the central bank, and began taxing private businesses to fund military operations. Lastly, the vulnerable people who had escaped to Mosul from surrounding areas during the previous 8 years in order to find refuge from Shia domination were now "easy prey for IS recruitment."[65]

Shortly after gaining control of Mosul, IS published the "City Constitution," dictating the ways IS intended to influence "all aspects of social life and city administration and planning." IS instituted dress codes and restricted movement inside the city. Essential services were offered only for those who complied with IS dictates. IS appointed "emirs" to administrate hospitals, schools, and commerce. IS decreed that property ownership transfers were null and void after June 10, 2014.

IS control inspired a mass exodus of the city's religious and ethnic minorities, including those Sunni Muslims affiliated with the Iraqi government or those "deemed disloyal" to IS.[66] The population of Mosul, however, remained roughly the same. Those forced out were replaced with rural immigrants and other displaced people in the surrounding region, as well as Syrian refugees and IS fighters and their families.[67]

In the early weeks of the occupation, IS went to considerable lengths to immediately exert its authority in Mosul. As expected, the city's social terrain resisted. Iraqi nationalists and moderate Sunnis living in Mosul began pushing back against IS. Saddam-era banners and songs infiltrated city streets to the ire of IS leaders. Resistance, however, all but evaporated when al-Baghdadi declared himself caliph of the new Islamic State from Mosul's Hadba' Mosque just a few weeks after the city was captured. Some suggest that those resisting IS control were either sufficiently intimidated to yield or they decided to bide their time and wait for coalition forces to break IS's grip on Mosul.[68] From another perspective, it is possible many Sunnis abandoned their resistance to IS control of Mosul based on the pragmatic choice of avoiding a return to Shia control. Based on this notion, experts suggested that IS would have to be defeated politically, rather than militarily, to recapture Mosul and destroy IS's proclaimed caliphate.[69]

From a military perspective, Mosul's social terrain was manipulated to support IS's ability to hold the city. IS forces had over 2 years to alter the social terrain of the city, creating barriers and underground bunkers, changing ethnic and ideological aspects of the city, and planning for the defense of the city. Mosul was the center of IS's power, which is why it was the last IS urban stronghold targeted for liberation.[70]

Why use the ISF as a surrogate (again)? Due to imposed political constraints, coalition forces "had neither the force nor the authority" to fight IS forces. Instead, the leadership of OIR focused on advising and assisting ISF, concentrating on developing and facilitating ISF's ability to "grow and field the levels of organizational confidence, trust, and respect necessary to win against" IS on the ground.[71] Experts suggested that it would take months to recapture Mosul, but they stressed the need to lay the foundation for both "security and political elements of liberation" in advance of beginning the campaign.[72]

As one would expect, many of the groups involved in OIR's campaign to retake Mosul had different ideas about what a postconflict Mosul would look like.[73] Having a better understanding of the social terrain in and around Mosul, the leaders of OIR diluted the appearance and influence of ISF forces by adding other elements. The 40,000 ISF fighters were augmented by 30,000 members of Iraqi federal police brigades; specialized units from Iraq's Emergency Response Division under the Ministry of the Interior and Iraq's Counter-Terrorism Service (CTS; an organization that operated independently from the ministries of defense and interior); approximately 40,000 Peshmerga soldiers along with some Kurdish militias; and various ethnic and religious minority militias made up of Turkmen, Christian, Sunni, and Shia fighters. In addition to all that, OIR provided 500 personnel along with air assets to directly support the campaign.[74]

How the ISF helped the United States and other coalition members. Despite its failure in 2014 to hold Mosul, the ISF remained the most viable surrogate to retake Mosul. Looking at how the ISF helped the coalition supporting OIR, the primary benefit stemmed from the manpower it offered. Unlike in the previous case, coalition forces embedded deeply into the organization to provide an "advise and assist network" all the way down to the tactical level. In addition, coalition members provided fire support, ISR capabilities, and logistical support. Lastly, coalition support provided significant assistance in preparing, training, and equipping ISF forces.[75] It is obvious that the leaders of OIR realized that a hands-off approach would not work, but the entire enterprise reflects the deep political restraints of participating states that required the ISF to serve as a surrogate. In this particular campaign, the ISF proved a pliant and willing surrogate in pursuit of ridding Mosul of its IS occupiers.

ISF forces worked cooperatively with other Iraqi government forces. Special operators from Iraq's CTS led ISF forces during tactical missions to capture neighborhoods one at a time. After the ISF finished liberating East Mosul in January of 2017, police and militias were put in place to promote security and governance while reconstruction efforts took place.[76] CTS forces continued to lead the ISF, as well as Iraqi police from the Emergency Response Division and "vetted militia units," during operations to recapture West Mosul.[77]

Retaking Mosul took three times as long as anticipated. The ISF and its partners, however, exhibited superb tactical and operational patience. The benefits of a surrogate in urban warfare, apart from the numbers required to physically occupy the terrain that had been captured, never materialized. In the end, the ISF recaptured Mosul through attrition warfare, rather than achieving information superiority and an adroit ability to wield the city's social terrain.

How the ISF hindered the United States and other coalition members. Shortfalls in the ISF's ability to conduct deliberate and sophisticated urban operations meant that coalition members (primarily the United States) had to provide the capabilities needed to enable the ISF and its partners to operate in a dense urban environment against a deeply entrenched, motivated, and adaptive adversary. The professionalism and competency of coalition advisers closely controlled the pace of operations. A tactical pause after capturing East Mosul focused on learning from that part of the campaign to prepare for the challenges of taking West Mosul.[78]

There was no illusion about the capability of the ISF to serve as a surrogate in Mosul. The leaders of OIR minimized the negative effects of adverse selection through intense participation and the use of its vast resources to provide the intelligence needed to overcome the challenges created by Mosul's social terrain. In the end, it was the depth of the coalition force's participation in ISF operations that provided the main impetus for success. That depth, however, was sustained by the political will of the international community, buoyed by the widely shared fear of IS gaining more territory and influence. The ISF was incapable of providing the surgical precision required to succeed in urban warfare. That shortfall in capability nearly exhausted the US stockpile of precision munitions.[79]

Tinder. For IS, the positional advantage offered by the city of Mosul was amplified by the dense urban terrain and the city's abundant supply of noncombatants. More specifically, IS had the opportunity to push the costs of retaking Mosul beyond what its coalition partners could accept. The resource superiority of the ISF and the members of OIR won a battle of attrition against IS. The potential tinder associated with using the ISF and its potential moral hazard was mitigated by the shared aim to defeat and destroy IS. Minus that overarching objective, the ISF and its partners would have likely failed to retake Mosul in this campaign.[80]

From another perspective, the ISF and its partners managed risk through the use of precision strikes. How would the Battle for Mosul have evolved if the United States had exhausted its stockpile of weapons? As it was, estimates suggest that approximately 10,000 civilians died in the campaign, many crushed to death in buildings destroyed by precision munitions. The negative effects (the tinder) associated with using the ISF as the ground force were also negated by managing the loss of ISF fighters in difficult situations brought on by the dense, urban environment through the use of precision weapons. It is plausible that the atrocities that would have resulted from the ethnic and ideological difference between the ISF and IS fighters were stemmed because ISF advances were tightly controlled. The ISF was allowed to manage the risk to its forces rather than minimizing the death of noncombatants and the destruction of Mosul's infrastructure.

Conclusion

The size of the force on its own may not matter very much. These two cases show that a force's size and capability are necessary but not sufficient to determine the outcome of an urban conflict. In more direct conflict with accepted understandings of urban warfare, neither offense nor defense should be seen as a singular determinant of advantage or disadvantage. In a battle where the adversary appears to have no understanding of a city's social terrain, size appears to have been irrelevant. In the first battle for Mosul, the ISF's inability to wield the social terrain had a significant factor in its undoing. The ISF's occupation did not reflect its connection to the city of Mosul. The horizontal arrangement of the city became the relevant terrain. The ISF, however, ignored rather than misinterpreted the complex and layered vertical (political, economic, and social) arrangement of Mosul (as explained in Chapter 2). Absent this view, the ISF saw no value in staying to fight IS; it was as if the ISF, stationed inside the walls, was like a liter of water resting in a 5-liter bucket. When IS tipped the bucket, the ISF simply poured out.

In the beginning of this chapter, I argued that a surrogate's ability to control and wield a city's social terrain was a necessity. Although I stand by the theoretical value of the notion, the second battle for Mosul challenges the idea. Under circumstances when the available surrogate does not have the ability to wield a city's social terrain, then one of the most valuable qualities a professional army should seek in a surrogate (in addition to availability, of course) is a political/ethnic/ideological distance from its own people and government. Under such circumstances, attrition will become the required strategy. In such a situation, a professional army contracting the services of a surrogate must be able to sustain its operations in spite of surrogate losses.

An additional thought is that the sufficient condition for taking back Mosul came from the enabling functions provided by coalition partners. The initiative was regained only through attrition. Will professional armies be permitted to use attrition warfare to fight in urban settings in the future? Perhaps the greatest contribution of a surrogate has nothing to do with capability. The surrogate's capacity combined with a nearly complete penetration of operations from the strategic level down to the tactical level (including the organization, training, and equipping of the surrogate force) means urban warfare fought via surrogate can significantly minimize the risks of adverse selection and moral hazard.

Although the loss of Mosul in 2014 represents a case where the use of a surrogate greatly hindered efforts to engage in urban warfare, there are some lessons to learn from the ease with which IS took the city. Imagine if coalition forces wanted to take the city of Mosul from the Iraqi government. IS would have provided a strong surrogate in many ways, but we should not lose sight of the fact that victory is not always, or even often, the objective in urban warfare. When attempting to hold a city, the notion should be to exhaust an adversary's efforts to take the city. This means a surrogate must be connected to the city. Evaluating the social terrain should offer some ideas about how invested the surrogate is in the city. The existence of a shadow government exerting significant influence in the city of Mosul should have provided a very real warning sign that the ISF was not poised to stand its ground. In addition, the cultural divide between the leaders and soldiers of the ISF and the majority of the inhabitants inside the city of Mosul should have served as another warning sign. Lastly, perhaps more effort should be given to how to employ and enable a surrogate in shaping the battlespace. I understand the foolishness of suggesting that professional armies and their civilian leaders could conduct shaping operations for years prior to taking a city. Accepting that limitation, perhaps this is where a whole-of-government effort and covert operations may take center stage.[81]

Another lesson that can be learned regards how training a surrogate in a postoccupation scenario has to focus on more than providing the fighting and policing skills needed to maintain order. When the city seizes the initiative, it will be very difficult for a sponsor to provide the resources necessary to recapture or control the initiative through its surrogate. Politically, ethnically, and ideologically disconnected surrogates do not appear to have the ability to hold a city. Under different conditions, such a surrogate could hold territory, but the dense social terrain of a city will negate the influence of overwhelming force supplied by the surrogate's sponsors.

Reflecting on this research, I find it interesting that the US Army concludes that a "city's narrow streets and corridors, rubble, power lines, and unforeseen environmental hazards negatively impacted mobility and the ability to maneuver." Although that appears accurate, the conclusion drawn from that point is that "dense urban terrain aids in the conduct of the defense against a superior force."[82] It almost seems the US Army forgot that IS took the city from a far superior force just 2 years earlier.

In the end, it seems we have more to learn from the way the ISF lost Mosul. The Taliban seems to have inculcated the positive lessons offered by the way IS took Mosul. Evaluating a force's commitment to hold a city trumps its ability to wield social terrain effectively. The lightning speed with which the Taliban took Mazar e Sharif, Herat, and Kabul demonstrates that a more sophisticated understanding of a city's social terrain is the key to urban warfare strategy beyond the use of attrition.

Surrogate warfare in cities is not going away in the future. If anything, it will likely increase. Although I support additional work on improving considerations about how to enable a surrogate through the use of logistical, intelligence, operational, tactical, and precision strike capabilities, all of that commits future urban conflicts to a battle of attrition.[83] I agree with Amos Fox in his assessment that the US Army should focus more on developing a better theory of proxy/surrogate warfare.[84] I would extend that advice to any professional army considering how to prepare better for urban warfare. Focusing on how to enhance the control of a proxy or surrogate and how to sustain the coherence of a policy from planning cells all the way to the battlespace will provide the doctrinal foundation needed to achieve political objectives in an urban setting, regardless of the surrogate's capability and intentions.[85]

Notes

1. Von Clausewitz, *On War*, 86–87.
2. Helmuth von Moltke, *Oxford Essential Quotations*, 5th edition, edited by Susan Ratcliffe (Oxford University Press, 2017), https://www.oxfordreference.com/view/10.1093/acref/9780191843730.001.0001/q-oro-ed5-00011287.
3. US Joint Chiefs of Staff, "Joint Urban Operations," Joint Publication 3-06 (November 2013), x.
4. Krieg and Rickli, *Surrogate Warfare*, 4–5.
5. Spencer, "City Is Not Neutral."
6. Reynolds, "British Army and Mass in Urban Warfare," 52–53.
7. Quinn and Hilmer, "Strategic Outsourcing."
8. A strong theme that consistently runs through many paradigms in the field of international relations posits that decisionmakers require that the perceived benefits must outweigh the perceived costs of engaging in war. Most notably, see Fearon, "Rationalist Explanations for War."
9. "Know the enemy and know thyself; in a hundred battles you will never be defeated" is an oft quoted maxim from Sun Tzu highlighting the importance of intelligence in war. "Sun Tzu," in *Oxford Essential Quotations*, 5th edition, edited by Susan Ratcliffe (Oxford University Press, 2017), https://www.oxfordreference.com/view/10.1093/acref/9780191843730.001.0001/q-oro-ed5-00010536.
10. Krieg and Rickli, *Surrogate Warfare*, 58.
11. One example of this comes from US Department of Defense joint publication "Joint Intelligence Preparation of the Operational Environment," Joint Publication 2-01.3 (May 21, 2014), xi–xiii.
12. Adverse selection and moral hazard are two agency costs associated with using a proxy or surrogate in war. See Groh, *War on the Cheap*, 27.
13. Groh, *Proxy War*, 32–34.
14. Groh, *Proxy War*, 84–85.
15. Chapman, *Jungle Is Neutral*.
16. Groh, *War on the Cheap*, 7.
17. Krieg and Rickli, *Surrogate Warfare*, 69.
18. Groh, *War on the Cheap*, 27.

19. Ibid., 27–29.

20. Abdulrazaq and Stansfield, "Enemy Within," 530; O'Driscoll, "Liberating Mosul," 61, 64.

21. Saouli, "Sectarianism and Political Order," 70.

22. Al-Qarawee, *Imagining the Nation*, 127.

23. Al-Qarawee, *Imagining the Nation*, 3.

24. Saouli, "Sectarianism and Political Order," 81.

25. UN Habitat, *City Profile of Mosul*, 11.

26. Hoekstra, "How Mosul Fell," 684.

27. Strachan, *Factors Behind the Fall of Mosul.*

28. Committee on Financial Services, US House of Representatives, 113th Congress, Second Session, November 13, 2014 (Washington, DC: US Government Publishing Office, 2015).

29. Hoekstra, "How Mosul Fell," 685.

30. *New York Times*, "Premier of Iraq Accedes to Calls to Give Up Power," August 15, 2014, A1.

31. Hoekstra, "How Mosul Fell," 685.

32. Van Wilgenburg and Fumerton, *Kurdistan's Political Armies*, 1–2.

33. "Mosul," *Encyclopedia Britannica*, April 21, 2021, https://www.britannica.com/place/Mosul.

34. UN Habitat, *City Profile of Mosul*, 13.

35. Abdulrazaq and Stansfield, "Enemy Within," 535.

36. O'Driscoll, *Future of Mosul*, 16.

37. Abdulrazaq and Stansfield, "Enemy Within," 540–542.

38. UN Habitat, *City Profile of Mosul*, 3, 11, 29.

39. Ibid., 11.

40. Ibid., 4–5.

41. Hoekstra, "How Mosul Fell," 695.

42. Abdulrazaq and Stansfield, "Enemy Within," 527–529.

43. Saouli, "Sectarianism and Political Order," 82.

44. Abdulrazaq and Stansfield, "Enemy Within," 538.

45. Craig Charney, "Polls Key to Understanding—and Defeating—ISIS," International Peace Institute Global Observatory, September 9, 2014, http://theglobalobservatory.org/2014/09/polls-understanding- solving-isis-crisis/.

46. Hoekstra, "How Mosul Fell," 695–696.

47. Parker, Coles, and Salman, "How Mosul Fell: An Iraqi General Disputes Baghdad's Story."

48. Dodge, "Can Iraq Be Saved?" 12.

49. Ibid.

50. O'Driscoll, *Future of Mosul.*

51. Hoekstra, "How Mosul Fell," 697.

52. Ottaway, *ISIS.*

53. Abdulrazaq and Stansfield, "Enemy Within," 539.

54. Hoekstra, "How Mosul Fell," 684.

55. Strachan, *Factors Behind the Fall of Mosul.*

56. International Crisis Group, *Iraq's Jihadi Jack-in-the-Box.*

57. Combined Joint Task Force Operation Inherent Resolve, "Our Mission," Fact sheet, n.d., https://www.inherentresolve.mil/Portals/14/Documents/Mission/20210915%20Updated%20Mission%20Statement%20Fact%20Sheet.pdf.pdf?ver=5OLdNQ7TrF7R4YjokCHosQ%3d%3d.

58. Mosul Study Group, *What the Battle for Mosul Teaches*, 1.

59. Mosul Study Group, *What the Battle for Mosul Teaches*, 4–5; Amos Fox, "What the Mosul Study Group Missed," Modern War Institute, October 22, 2019, https://mwi.usma.edu/mosul-study-group-missed/.

60. Combined Joint Task Force Operation Inherent Resolve, "The Campaign," n.d., https://www.inherentresolve.mil/campaign/.

61. Ibid.

62. Arnold and Fiore, "Five Operational Lessons," 58.

63. Volesky and Noble, "Theater Land Operations," 2.

64. Although combat operations officially concluded after 9 months, it took an additional month to complete mop-up operations in Mosul. See Fox, "What the Mosul Study Group Missed" (cited here in note 59).

65. UN Habitat, *City Profile of Mosul*, 3–5.

66. Ibid., 21.

67. Ibid., 3.

68. Abdulrazaq and Stansfield, "Enemy Within," 540–541.

69. O'Driscoll, "Liberating Mosul," 62.

70. John Spencer and Jayson Geroux, "Urban Warfare Project Case Study Series: Case Study #2—Mosul," Modern War Institute, September 15, 2021, https://mwi.usma.edu/urban-warfare-project-case-study-2-battle-of-mosul/.

71. Volesky and Noble, "Theater Land Operations," 3.

72. O'Driscoll, "Liberating Mosul," 61.

73. Spencer and Geroux, "Case Study #2—Mosul" (cited here in note 70).

74. Ibid.

75. Mosul Study Group, *What the Battle for Mosul Teaches*, 1.

76. Ibid., 7.

77. Ibid., 7–8.

78. Ibid., 9–23.

79. Fox, "What the Mosul Study Group Missed" (cited here in note 59).

80. Ibid.

81. This concept is not new. For example, the US government attempted to protect Laos in this way. Although the effort eventually failed, the methods could be studied on a smaller scale, focused on a specific city rather than an entire state. See Groh, *Proxy War*, chap. 5.

82. Mosul Study Group, *What the Battle for Mosul Teaches*, 26.

83. Fox, "What the Mosul Study Group Missed" (cited here in note 59).

84. Ibid.

85. For a detailed theory of proxy war, see Groh, *Proxy War*, chap. 4.

12

Mass and Maneuvers

Louise A. Tumchewics

THE EASTERN MEDITERRANEAN CITY OF JAFFA, NOW PART OF TEL AVIV-Yafo, has a long and storied past. Established in the Bronze Age, it has been the site of myths, miracles, and urban conflict for millennia. The Greek myths of Perseus and Andromeda took place in Jaffa, and the biblical stories of Jonah, Solomon, and Saint Peter are all believed to have taken place in Jaffa.[1] The city has witnessed waves of invasion, occupation, and conflict over the centuries. Alexander the Great's armies were stationed in Jaffa, the Romans occupied it, and the city changed hands numerous times during the Crusades. Jaffa was later part of the Ottoman Empire, before being invaded by Napoleon during his Egyptian campaign at the very end of the eighteenth century.[2] In 1947, it was the site of a battle between Arab and Jewish forces in the Arab-Israeli civil war, in the final weeks of the British Mandate Administration of Palestine.

In the space of six days, a small group (approximately 850) Irgun fighters gained control of the city of Jaffa, defeating a numerically superior force of Arab defenders. It was also the first—and only—time in the months of the Arab-Israeli civil war that the British Army intervened directly in the intercommunal strife.[3] Over seventy years have passed since the battles took place, yet it still offers pertinent lessons about urban operations to contemporary armed forces. The Arab-Israeli war was a civil war, and the events of this period bear tragedy and controversy for both sides. This chapter does not aim to engage in these controversies, nor discuss the wider Palestinian-Israeli conflict that has endured since the civil war. Rather, it focuses on the events of the battle of Jaffa, and the lessons smaller contemporary armies can draw from these events. It does not intend to be a complete history of Jaffa nor the Arab-Israeli conflict.

A wide range of primary sources are available on the events that unfolded in Jaffa. The British Army and British Mandate Administration left primary documents in the National Archives. The records of the British Ministry of Defence, Foreign and Commonwealth Office, and Cabinet Office provide insight into the British perspective on events in Jaffa and throughout Palestine at the close of the Mandate Administration. The records of the Haganah and Israel Defense Forces also contain material on the battle of Jaffa. A language barrier and travel restrictions due to the Covid-19 pandemic meant Hebrew archives and sources were regrettably not part of the research for this chapter.

There is a rich historiography of the Arab-Israeli civil war, and the battle of Jaffa has been well documented by other historians. Adam LeBor's *City of Oranges* presents a social history of Jaffa, from 1920 to present day, tracing the history of the city through the profiles of Jewish and Arab individuals and families that lived through the city's tumultuous history. His chapter on the months of the Arab-Israeli civil war's effect on Jaffa, and the battle itself, provide insight into the impact on civilians, both Arab and Jewish, caught up in the current of conflict, chaos, and confusion of urban warfare.[4]

Benjamin Runkle's study of the battle of Jaffa, a chapter in the anthology *City Fights*, focuses on the tactical details of the battle and provides a summary of relevant observations for contemporary urban conflict.[5] Arnon Golan's informative article "The Battle for Jaffa, 1948" presents the battle of Jaffa within the wider context of Irgun-Haganah relationships and expounds on the complexity and intricacies of the local Arab defense organizations. He demonstrates how the battle of Jaffa was a battle for the Irgun's political aspirations as well as a military effort.[6] Golan, Yaakov Peleg, Alon Kaddish, and Yona Bandman have used Haganah archival sources to present a comprehensive study of the battle of Jaffa in their work *The Campaign for Jaffa, 1948*.[7] In their histories of the Arab-Israeli civil war, Ilan Pappe and Benny Morris discuss Jaffa as a notable event in the trajectory of the civil war, where fighting contributed to the flight of the Arab civilian population, as part of the wider chronology of the Arab-Israeli civil war. Itamar Radai's *Palestinians in Jerusalem and Jaffa, 1948*, draws on extensive Arabic sources to present an insightful account of the Palestinian experience of the battle, and the factors within Palestinian Arab society that contributed to societal breakdown and battlefield defeat.[8]

Jaffa the City

By the mid-nineteenth century, Jaffa had grown from its ancient roots into a thriving, prosperous city, second only in size to Jerusalem. Jaffa's sunny climate was ideal for the cultivation of citrus fruits, and Jaffa gained international recognition for its famous oranges, the city's most lucrative crop.[9] Urban growth expanded with the removal of Jaffa's ancient stone walls in the 1880s, and the town's thriving economy attracted newcomers from the nearby countryside, as well as Lebanon, Syria, and elsewhere in the Eastern Mediterranean basin. The city expanded and Muslim, Jewish, and Christian neighborhoods developed around the docks and old town center. In 1909, a new garden suburb of Jaffa, named Tel Aviv, was established by newly arrived Jewish settlers, many arriving from Eastern Europe.[10]

The end of World War I brought the collapse of the Ottoman Empire, which had governed Jaffa, with a brief upset by Napoleon, for centuries. In its wake, Palestine was governed instead by a League of Nations Mandate that gave oversight powers to Britain and promised the eventual creation of a Jewish homeland.[11] As Ottoman rule imploded and the mandate was established, tensions between the Arab and Jewish populations rose, and there was a brief outbreak of violence and rioting in the early 1920s, leading many of the city's Jewish population to relocate to nearby Tel Aviv.[12]

During the years of the British Mandate, Jaffa modernized, and its role in national life changed. Its importance as a port was superseded by the city of Haifa, with its modern, purpose-built deepwater port and strategic location as an oil pipeline terminus.[13] Although not as economically important as Haifa, nor as religiously significant as Jerusalem, Jaffa was a thriving center of Arab cultural life, with numerous newspapers, a radio station, public libraries, sports clubs, a cinema, and schools. It was a busy cosmopolitan center, connected to the rest of Palestine and other parts of the Middle East by road, railway, and regular sea travel.[14] Neighboring Tel Aviv had grown from a mere suburb to Palestine's largest urban area in two decades. The two cities were largely coexistent, with integration of businesses and populations and with goods and people crisscrossing the municipal boundary between the two cities daily.[15] Jaffa's modernization and urbanization attracted many newcomers from the countryside, and the population continued to grow, totaling 104,000 by 1947. Most of the new arrivals lived in the poor, working-class neighborhoods of Manishiya and Abu Kabir, on the border between Jaffa and Tel Aviv.[16]

In the years of the mandate, Britain had been unable to resolve intercommunal claims to Palestine satisfactorily. By the time World War II ended in 1945, Britain was unable to contain rising sectarian violence, and the security situation was steadily worsening. Jewish paramilitary groups attacked British forces, and tensions and violence escalated between communities. Unable to contain or defuse the situation, Britain referred the matter to the newly formed United Nations.[17] On November 29, 1947, the UN General Assembly voted in favor of the partition of Palestine and the creation of a Jewish state. Jewish jubilation over the creation of a new Jewish state was countered almost immediately by Arab anger and anxiety. The UN resolution was rejected by the Arab League and the Arab High Command (AHC). After the partition announcement, fighting broke out between Arab and Jewish communities, including the seam between Jaffa and Tel Aviv. Over the next 5 months, Jewish and Arab forces engaged in a fierce civil war. These armed clashes would determine the strategic position of the opposing sides when the British Mandate expired on May 15, 1948.[18]

Jaffa in the Arab-Israeli Civil War

The civil war was fought by a number of militia groups on both sides. The main Jewish militia, the Haganah, was established in 1920 to protect the Jewish community from outbreaks of Arab violence at the end of the Ottoman Empire and the beginning of the British Mandate. It was aligned with David Ben-Gurion's Mapai Party and more open to accommodation with British and Arabs than other Jewish paramilitary groups. By 1947, it had 35,000 members, and during the first few months of the civil war, the Haganah expanded in size and reorganized, shifting from a territorially based amateur militia into a relatively professionalized army of twelve brigades.[19]

The second right-wing paramilitary organization was the Irgun Zvai Levi (also known as IZL or by its Hebrew acronym, Etzel), with 2,000–3,000 members. Established in the 1930s, the Irgun maintained that violent action against the British presence in Palestine was necessary to deliver the Jewish homeland; it had targeted the British for years with random attacks, including the bombing of British military headquarters at the King David Hotel in 1946. The UK security service MI5 monitored the Irgun, as it classified the group's activities as terrorism.[20] Another, smaller group, the Lohamei Herut Israel (LHI, or Lehi), known to the British as the Stern Gang, after their leader, Avram Stern, was formed of a splinter faction from the IZL and comprised 200–500 members. It too espoused violence as a means of delivering Jewish independence from the British Mandate.[21]

The Palestinian Arab community had no organization similar to the Haganah. During the brief life of the Palestinian nationalist movement a decade prior, attempts had been made to establish two paramilitary youth organizations, the Najada and Futuwwa, but neither of these developed into a coherent national militia organization. It was only in 1947 that these two competing organizations merged into National Youth Liberation Army, but this never became a force equivalent to the Jewish paramilitary organizations.[22]

Rather than the central organization of their Jewish counterparts, Palestinian forces were more ad hoc. In villages, groups of 10, 50, or 100 or so able-bodied men, armed with whatever weapons they had available, formed their own local self-defense units.[23] In Jaffa, these bands comprised mainly young men from various neighborhoods. In addition to these very localized defense forces, a number of large organized local bands or gangs sprung up in December 1947 in Judea, Samaria, and Galilee, just as had happened in the Arab revolt a decade prior. Each band had 200–500 members and, unlike village militias, moved throughout the countryside. Some villages were reluctant to assist them, fearful of Jewish retaliation. In other villages, smaller local militias would join forces with these larger bands for short, specific objectives.[24]

The Arab League raised the Arab Liberation Army (ALA), which consisted mainly of foreign fighters—volunteers from Syria, Iraq, Lebanon, and Bosnia, with officers seconded or retired from Jordanian and Iraqi forces. Syria provided the ALA's training, money, and arms (mostly light arms and an assortment of mortars). Despite this external support and military experience, however, the ALA was not a pan-Arab fighting force nor a professional army.[25] Furthermore, the ALA lacked centralized coordination of existing Palestinian bands and local militias. The lack of centralized coordination of effort among Arab forces severely undermined their effectiveness.[26]

Recurrent supply issues—primarily guns, ammunition, and fuel—further stymied Palestinian forces. They relied on the support of neighboring states for crucial supplies, but many of these states were themselves poor, and the exiled AHC in Egypt was unable to fundraise for the Palestinian cause or purchase or organize arms shipments. Furthermore, there was no central coordination of supplies either inside or outside of Palestine.[27]

The third and largest military force in Palestine was the British Army. In 1947, it was about 100,000 strong. In addition, there were naval assets in the Eastern Mediterranean, Royal Air Force (RAF), and army units stationed in Cyprus and Egypt. Britain in 1947 was still severely weakened by World War II, facing economic crisis and commodity shortages, and grappling with the emergence of a new global security paradigm: the Cold War. The government of Prime Minister Clement Attlee wished to dispense with costly imperial commitments. Facing political concerns, frustrated diplomacy, and fear of unknown damage to Britain's already perilous economy, it sought to extract itself from Palestine without damaging its reputation among Arab states. As the British Chiefs of Staff Committee emphasized, good relationships with the Arab states was Britain's "major strategic requirement," lest the Arab states drifted into the Soviet orbit, thereby denying Britain—and other Western nations—access to oil, "setting us at a great disadvantage in the event of future war." Managing the British departure from Palestine was of critical importance as "the final effect of our position in the Middle East will depend on the success with which our withdrawal can be made acceptable to the Arab world."[28]

British authorities anticipated violence—particularly from Jewish paramilitary organizations—in the period of transition. But by late 1947, there was "a very serious deterioration in security," noted the British Inspector

General of Police in Palestine, as intercommunal violence increased throughout the country.

With less than 6 months before the departure of the army and Mandate Administration from Palestine, there was no intention or desire to intervene in fighting between the Jewish and Arab populations.[29] Though they were not about to defend the city for the Arab population, British commanders were aware of Jaffa's strategic position and stationed an infantry battalion on the highway to Jerusalem. Two infantry companies were assigned to protect the British security zone in the northwest of the city, and also patrolled the neighborhoods of Abu Kabir and Manishiya.[30]

In the opening days of the civil war, municipal leaders formed an Arab National Committee (ANC) in Jaffa. Contrary to its title, this was a regional emergency committee rather than a national body. Security of the city was among the top concerns of the committee, with a specific subcommittee headed by Mohamed al-Huari, the Najada commander for Jaffa. But the coordination of an effective military response was complicated by intra-Palestinian rivalries. Some members of the committee supported Haj Mohammed Amin al-Husseini, the mufti of Jerusalem and head of the AHC. Al-Husseini had played a leading role in the Arab revolt of 1936–1939 and had been duly exiled to Egypt by the British. From Cairo, he continued to influence Palestinian Arab politics and fighting efforts through his envoys and supporters, including those on the Jaffa ANC.[31] Al-Huari, the military commander, and Yousef Heikal, the mayor, advocated for a policy of restraint toward the Jewish population and opposed al-Husseini's more militant stance, causing friction with other committee members.[32] The divisions in Arab leadership meant that Jaffa did not have a coherent defensive plan.[33]

The fighting between the two sides in the initial weeks of the civil war was primarily sniper and machine gun fire between Jaffa and Tel Aviv, and some small attacks and local raiding. Two weeks after hostilities commenced, Jewish and Arab leaders met to discuss a possible cease-fire arrangement.[34] The arrangement was agreed to but never implemented. Supporters of al-Husseini were insistent that fighting should continue, and al-Huari was summoned to Cairo, where al-Husseini ordered him to remain. Less than a month into the civil war, the Jaffa ANC was without a military commander. Al-Huari was not replaced, and defense became the responsibility of local neighborhood organizations.[35]

An unofficial cease-fire was observed in and around Jaffa in December at the start of the citrus harvest season. Fighting dwindled as the world-famous fruit was picked and packed at its peak, ready for export. The Arab and Jewish communities agreed to a tacit modus vivendi and refrained from fighting in and around the orange groves that surrounded the city, but this was not a firm cease-fire agreement.[36]

In January 1948, the Irgun attacked Jaffa with a truck bomb in the central square, outside the New Saraia Hall, a municipal government building housing offices and a charity kitchen. The explosions killed twenty-six people and injured hundreds of civilians, shattering the confidence of many Jaffans.[37] Many members of the city's middle class fled for Lebanon, Egypt, and elsewhere in the Middle East. Fearful municipal leadership, shaken by the attack and anxious about Jaffa's evident vulnerability, sought assistance from al-Husseini and the ALA. Yet ALA divisions in Jaffa proved unruly and struggled to find effective military leadership, cycling through three commanders in less than 2 months.[38]

By April 1948, the situation in Jaffa was increasingly bleak. The economy was crumbling, as the abrupt severing of close ties between Jewish and Arab communities prompted a sudden spike in unemployment. As elsewhere in the country, commodity prices rose as wages evaporated. Criminality and homelessness were on the rise, and civil order was unraveling. Many more residents left the city, fleeing by sea or by land.[39]

By this time, international support for the proposed partition plan was wavering, and Egypt threatened to invade Palestine as soon as the British Mandate Administration officially expired.[40] For the first months of the civil war, Haganah policy had been to defend existing Jewish settlements and protect the vital supply convoys moving between towns. In April 1948, as the date of partition drew closer, the Haganah shifted from their defensive posture to the offensive, with the objective of defeating Palestinian militias in significant towns, including Haifa and Jaffa, on main supply routes through the country and along the borders of Palestine, in order to secure as much of the country as they could to withstand an anticipated invasion by neighboring Arab states and to prove to the international community the viability of the Jewish state.[41] The Haganah leadership judged that British military presence in Palestine would be so sufficiently depleted by the scheduled reduction of forces and the departure date that was close at hand that British forces would be unlikely to intervene in any Jewish-Arab confrontation.[42]

Jaffa's coastal location gave it strategic importance in an impending conventional war between the nascent state of Israel and the surrounding Arab states. Jewish leaders feared that Arab armies would try to drive a wedge from Jaffa to Jerusalem, cutting off the Negev desert, fracturing the Jewish territory in two, and thereby defeating the hopes of a Jewish state. Egyptians could easily land at Jaffa to launch an assault on Tel Aviv. Arab control of Jaffa was therefore a considerable vulnerability.[43]

As the city fell within Arab-designated territory, Jewish leadership was reluctant to engage in peremptory action that would violate the partition agreement and therefore might antagonize international support at a critical juncture. Nor did they wish to antagonize the British, who maintained a garrison in Jaffa's German colony in the northwest of the city, even as troop numbers were dwindling throughout the country. Instead, the Haganah chose to concentrate on gaining control of the urban periphery and surrounding villages, in hopes that the city would eventually fall to Jewish control without direct military action.[44]

The Battle for Jaffa

On April 21, 1948, the Haganah seized control of Haifa, a city of 167,000, in a surprising and swift battle. Most of the city's Arab population fled via the British-controlled port.[45] The Irgun were concerned about the implications of the Haganah's increased prestige and prominence in the wake of this victory and wanted to ensure that they too had a stake in the formation of the new Jewish state's defense forces. As a port city next to the country's largest Jewish urban center, Jaffa had undeniable military significance to the new Jewish state. It was also laden with political significance for the future ambitions of the Irgun and ideally located along the southern edge of Tel Aviv, home to the Irgun's main support base.[46]

The Irgun was less concerned than the Haganah about potential international diplomatic fallout from the seizure of Jaffa. Thus, on April 21, the Irgun decided to launch its first above-ground operation since its formation in 1938.[47] The Irgun high command met to determine a plan of attack. The Irgun's leader, Menachem Begin, appointed Amichai "Gidi" Paglin, Irgun's head of operations, as the leader of the Jaffa operation. Paglin had a history of successful terrorist attacks, including the notorious King David Hotel bombing and the Goldschmidt House officers' club bombing. A skilled operative and charismatic leader, Paglin's abilities impressed Begin and won the loyalty of his juniors.[48]

Paglin developed a two-phase plan of attack against Jaffa. The first phase would be an infantry attack westward to the sea, at the narrowest point connecting Manishiya, the northernmost Arab neighborhood, to the rest of Jaffa. This would cut Manishiya off from Jaffa. Then, two motorized formations, using requisitioned, stolen, or confiscated trucks and cars, would race through Manishiya and into the center of the city. Only the first phase was planned in detail as the Irgun command believed that the attack on Manishiya would precipitate the collapse of Arab defenses throughout the rest of the city.[49]

At Paglin's disposal were approximately 850 fighters—600 in infantry companies, and the remaining 250 responsible for logistics tasks. Of the 600 fighters, only 250 were assigned to the initial attack, with the remainder to provide a reserve.[50] They were armed with an assortment of small arms, Bren guns, a small number of armor-piercing shells and two launchers, and two 3-inch mortars. Just weeks before, the Irgun had stolen 12 tons of mortar shells in a raid on a British ammunition train, and these would be liberally used against Jaffa.[51]

The Irgun fighters faced several thousand Arab militia, equipped with Spandau machine guns, among other weapons. In addition, there was the presence of the British garrison at Jaffa to contend with, including three infantry battalions from the 3 Brigade, a battery from 41st artillery regiment, and 4th/7th Dragoon Guards, part of the 1 Infantry Division. The British had no intention of fighting directly with Irgun forces but were under orders to shoot in self-defense if attacked.[52]

On the night of April 25, the Irgun gathered at Camp Dov, in Ramat Gan, a southern suburb of Tel Aviv. The original plan had outlined a night attack, but a series of delays pushed the start of the offensive to 8 a.m. on April 26, when a heavy mortar barrage began to rain down on Manishiya. The mortars had a devastating impact on civilian morale but had little impact on the fighting in Manishiya. The initial fighting did not go well for the Irgun. Accustomed to irregular attacks, their well-honed guerrilla skills did not translate to conventional fighting, and they were pushed to retreat.[53]

The following day, April 27, the Irgun renewed their attack with another massive mortar barrage on the city. The mortar fire was indiscriminate, hitting the main post office, coffee shops, and market stalls and stopping the movement of buses and food supplies. The constant bombardment prompted thousands of civilians to flee their homes and rush to the port in an attempt to leave. Adding to their panic was the knowledge that the attackers were the Irgun, the masterminds of the December café bombings, the January bombing of the city's main square, and of a recent massacre of Arab civilians in the village of Deir Yassin.[54]

In a telegram, British authorities informed London that "the situation is rapidly deteriorating, and Arabs already beginning to evacuate town by road and sea."

Learning from the experience of the previous day, the Irgun fighters changed tactics slightly, with the attacking formation reduced to one company. One assault platoon led the charge, while another provided covering fire, and the third held off the Arabs. The Irgun made small advances, fighting Arab forces from house to house and room to room, in the fierce close-quarters fighting that typifies urban warfare. But still, Manishiya did not fall. The Arab defenses held, despite the intense fighting, and Irgun forces were bogged down, with little appearance of a breakthrough. To add to their difficulties, Irgun logistics were scrambled due to inexperience, and they struggled to maintain supplies of food, water, and ammunition for their fighting forces.[55] By the end of the second day of fighting, the outlook for the Irgun was not promising. Begin declared that his fighters should hold the line and withdraw.[56]

From the teeth of an apparent military failure, the Irgun seized a measure of political success. That evening, Begin met with Haganah commanders, who asked him to sign a cooperation agreement between the two paramilitary forces. The agreement granted formal recognition of the Irgun, confirming its integration into the Israeli Defense Forces and autonomous military status. The Haganah and Jewish civilian leadership, the Jewish Agency, sought to restrain the Irgun from further action to avoid the risk of direct British involvement.[57]

The Irgun's junior officers and men urged Begin to reconsider and permit them to reprise the attack. Paglin made a reconnaissance tour of the front line, seeking out any possible weakness in the Arab defenses, crawling on his knees. He was convinced he had found some vulnerable points. Impressed by the fighting spirit of his forces, Begin agreed, and gave them 24 hours to reach the sea. He was aware this would be their third and final attempt.[58]

Paglin and his officers set about creating a new plan. While the Irgun had limited experience with conventional fighting, it did have considerable experience with explosives. Instead of fighting through the streets, fighters applied this know-how to blow a route through the inside of houses and construct sandbag walls across the streets. With picks and shovels, the men would "mousehole" their way through the houses, with the outer walls providing cover. Mouseholing was not a new invention—it was first mentioned by Thucydides in the Peloponnesian War—but was used to effect in Jaffa.[59] Once they had reached the Arab lines, they would place explosives nearby and blow them up. This would afford them the crucial element of surprise that had been missing from their earlier daylight attacks.[60]

The renewed operation began at 4 p.m. on April 28. The work was backbreaking and slow, "the work of ants," as one fighter later recalled.[61] Three hours later, Irgun forces reached a row of houses just 64 meters from Arab forward positions. However, a road ran in between, and Arab pillboxes provided cover for the defenders. To neutralize these strong points, Irgun forces waited until darkness and then blew up a three-story building across from the Arab positions. The falling debris destroyed the pillboxes, just as intended. At the same time, other Irgun sappers rushed forward, planting explosives on other buildings. Groups of Irgun fighters took advantage of the chaos to cross the road, while others hurriedly built sandbag walls for cover. Once across the road, they continued to burrow their way through houses to the sea, reaching their objective around 7 a.m.[62] Within 24 hours of beginning their attack, they had achieved their objective. By the late morning, Arab fighters had capitulated, and the Zionist blue and white flag flew over the Hassan Bek Mosque at the center of Manishiya.[63]

As the Irgun were fighting their way through the walls of Manishiya, the Haganah commenced Operation Hametz to seize the Arab villages of

Khiriya, Saqia, and Tel Arish on the outskirts of Jaffa. The concomitant operation absorbed most of a battalion of ALA reinforcements in a counterattack for Tel Arish, with only one platoon dispatched to assist the defenders of Manishiya, thereby assisting the Irgun's seizure of the neighborhood. By April 30, the Irgun controlled Jaffa's railway station, and the route into the center of Jaffa was open.[64]

The Arab defenders had held firm for the first days of the Irgun onslaught but were riven with internal frictions and fissures. In the months preceding the battle, the military command had cycled through a number of leaders, and foreign fighters with little trust or support of the local citizenry had assumed the defense of the city.[65] The arrival of the additional reinforcements on the night of April 28 triggered internal disputes over command. Angered at the suggestion of being replaced, one senior ALA commander departed the city on April 30, taking his foreign fighters with him, at a crucial juncture in the fighting.[66]

On April 26, William Fuller, British Lydda District commander, had urged the mayor of Tel Aviv to order the Irgun to cease their attack. He persisted in his request for the next two days. By the 28th, the British issued an ultimatum: either the Irgun cease and withdraw, or the British would bomb from the land, sea, and air.[67]

A "most immediate" telegram sent from General A. Cunningham, high commissioner of Palestine, warned that Irgun success in Jaffa would have a more severe effect on Arab States than the seizure of Haifa just a week prior.[68] The attack on an Arab-designated city could deteriorate regional stability and could "inflame" neighboring Arab states.[69]

When news of the Irgun's surprise seizure of Manishiya reached London, Foreign Secretary Ernest Bevin was angered and troubled by the fighting in Jaffa. Arab states had already accused Britain of allowing Haifa to fall to the Jews just a week before, and now another city, "the most important Arab town in Palestine," located outside of formally agreed partition boundaries, had fallen to Jewish paramilitaries, potentially damaging the future of Anglo-Arab relations and regional stability.[70] There were also concerns about the safety of British troops and civilians within Palestine and the safe completion of the troop withdrawal from the country with the seizure of Haifa and the ongoing attack on Jaffa.[71] Anxious to regain control over the unfolding situation, Bevin circumvented conventional protocol to communicate directly with British authorities in Palestine, urging them to take direct action in Jaffa.[72] Reinforcements were duly dispatched from Cyprus, Libya, Egypt, and Iraq.

At 11 a.m. on April 28, just after the Irgun's successful advance to the sea, three formations of British tanks rolled out into the main sector of Jaffa, advancing toward Manishiya. This caught the Irgun by surprise, but they swiftly fired one of their few antitank rounds, disabling the first tank.

Things went better for the second armor formation, but the third turned around, not willing to risk becoming disabled in Manishiya.[73] Artillery fire, mortar, and machine guns raked the Irgun positions. Meanwhile, a Royal Navy destroyer maneuvered off the coast, and RAF Spitfires strafed Haganah in Tel Aviv, signaling the severity of British intentions.

The next day, April 29, British tanks once again advanced on Manishiya. With no antitank rounds left, Irgun fighters resorted to dropping the facade of a nearby house on the column of tanks, disabling one and halting the progress of the rest of the column. Irgun resistance stayed strong, and conscious of the risk of becoming trapped in the rubble of Manishiya, British troops retreated into central Jaffa.[74]

British forces were reluctant to continue fighting the Irgun. The losses they risked suffering in urban combat were disproportionate to the temporary gains to be made, as the mandate had only a little over a fortnight to live. The British abandoned their demands that Jaffa be returned to Arab control. With the rapid flight of Jaffa's Arab population, there was no clear authority to whom the British could return control, despite Bevin's orders to do so. Both the ALA commander and the mayor had fled the city by boat.[75] Rather than continue fighting, the British attempted to reach a settlement with the Jewish forces.

Refusing to negotiate with the Irgun, which was deemed a terrorist organization, the British referred the matter to the Haganah and the Jewish Administration Executive.[76] The British would stop their attacks if the Haganah stopped Operation Hametz on the outskirts of the city and did not attack Jaffa until the end of the mandate, and if the Irgun would leave Manishiya with Haganah troops replacing them. The Irgun agreed to these terms and blew up the police station on their way out of Manishiya.[77]

On May 13, the last British troops departed Jaffa. On May 14, the State of Israel was declared. The very next day, armies of the surrounding Arab states invaded Israel. With Jaffa under Jewish control, Egyptian forces were unable to use it as a landing site.[78] The Irgun's action had achieved its political and military objectives.

Conclusion

The battle of Jaffa has enduring relevance for small armies considering operations in urban environments. Jaffa might not have had the same religious and cultural significance as Jerusalem, but as a port city, Jaffa was crucial to the economic well-being of Palestine and the nascent state of Israel for trade and connections with the wider world. The legitimacy and viability of the nascent state of Israel were determined by controlling this important economic center. Militarily, they were critically important as potential landing sites for invading forces and hubs for naval operations.

The outgoing British army and mandatory administration relied on Haifa as its primary exit point. The rapid fall of Haifa caught the British government and army by surprise. The collapse of Jaffa just five days later prompted the British to intervene directly—something they had avoided for the previous 5 months of the civil war. There was little appetite for what Winston Churchill, then in opposition, called a "senseless, squalid war."[79] But the swift fall of two important cities had long-term implications for Britain's strategic relationship with Arab states, far beyond immediate Arab-Jewish relations in Palestine.

Jaffa was an important target for the Irgun, but it was Haganah's seizure of Haifa that increased impetus for the Irgun's action in late April 1948. The seizure of Haifa put Britain in a quandary as it tried to balance its desire to preserve its reputation with the Arab states with its wish to respect the international intention to create a Jewish state. The British army's restraint in Haifa was not replicated in the subsequent battle for Jaffa. And after the seizure of Haifa, the British government was willing to have the army intervene in the intercommunal conflict. The seizure of Jaffa demonstrates how events in one city radiate outward to others in the country. The decision to attack Jaffa was influenced by the seizure of Tiberias and Haifa days prior. The Irgun's capture of Manishiya contributed to the Haganah's defeat of the surrounding villages, and likewise the outcome of Operation Hametz further demoralized the Arab defenders of the city.

What we see in Haifa and Jaffa and the surrounding area is that one battle and the rapid fall of one city influenced subsequent actions and reactions. Cities were, as Lieutenant General Andrew Graham notes in Chapter 13, "knots in a net," with events in one influencing subsequent actions and reactions in another. When considering urban operations, it will be important to understand and anticipate the interconnectedness and potential effects on other municipalities within the same region or country.

The Palestinian Arab forces' strengths and weaknesses had suited the earlier part of the civil war, with localized raids and smaller-scale engagements suiting their forces. They were not trained or equipped for conventional, large-scale offensive operations in urban areas, though they initially fought fiercely in Jaffa. The lack of command and control hampered efforts to mount a credible and cohesive defense, even though Arab forces outnumbered Jewish forces in Jaffa.

The Irgun were adept in the terrorist techniques they had used to effect against both the British and the Arab population but were ill-prepared for conventional combat and the intensity of urban combat. The skills honed in raids, bombings, and ambushes did not prepare them for the close-quarters fighting through the streets and alleyways of Jaffa. Moreover, they had little logistical experience and struggled to maintain adequate supplies to their frontline fighters, despite the proximity of their support base in Tel Aviv. The initial plan of attack did not survive contact with the adversary and was hastily readjusted—not once, but twice—to make the most of their existing skill set and limited ammunition. Although conventional armed forces would embark on an urban operation with considerably more training than this particular paramilitary force, the ability to adapt and think creatively in contact is essential.

What facilitated the Irgun's adaptation in Jaffa was the relative freedom of the tactical commander, Paglin, to develop, execute, and adjust his plans to achieve the Irgun's intended objective. The Irgun benefited from Paglin's strong tactical leadership and his willingness to lead from the front, which contributed to the trust and the high level of motivation within the organization's ranks.

As Runkle notes in his study of Jaffa, the Irgun made good use of its (improvised) engineering assets for mobility and countermobility.[80] Irgun sappers created above-ground tunnels and mouseholes in walls, allowing Irgun fighters to move in relative cover, undetected by Arab fighters in the streets. They used explosives expertise to demolish buildings, thwarting the British advance, and dropped buildings on British and Arab strong points in lieu of antitank rounds. Combat engineers were invaluable in using the characteristics of the urban environment—narrow streets, densely built-up areas—to the Irgun's advantage. Rather than seeing the urban landscape of Manishiya as an obstacle, the Irgun found a way to work with—and through—it.

The Irgun and the Haganah were paramilitary organizations fighting a civil war. They were not bound by the law of armed conflict or normative behaviors that mitigate the behaviors of contemporary Western professional, conventional armed forces. This gave them a latitude in the treatment of enemy combatants and civilians that contemporary armies could not—and would not want to—replicate. Although some elements of their experience, such as tactical adaptation, combat engineering, and leadership, are relevant to contemporary armies, others are practices professional armies would strive to avoid.

One example of this was the widespread use of indiscriminate mortar fires in both Haifa and Jaffa. Mortars were suited to the urban environment due to their high parabolic trajectory and, for the Irgun, their availability. But they were imprecise weapons that brought about collateral damage, particularly in the densely built-up areas of Manishiya and central Jaffa. The bombardments, criticized by British and Red Cross observers at the time, contributed to the panicked exodus of thousands of Palestinian civilians and the demoralization of Arab forces.[81] Whereas precision fires on military targets have utility in urban combat, the indiscriminate use of fires against civilians violates customary international humanitarian law (IHL) and thus contravenes conventional rules of engagement.[82] That said, an adversary, less concerned with the observance of IHL, might employ weapons to similar physical and psychological effect, as has been demonstrated in recent conflicts in Syria and Ukraine.

Thousands of Arab civilians fled the fighting in Jaffa, abandoning their homes and crowding onto boats in the harbor, many never to return. Future urban conflicts will likely see similar mass movement of the civilian population—either within a city or fleeing the city entirely, as in Marawi and Groningen (discussed in Chapter 5 and Chapter 7), presenting an enduring humanitarian as well as security challenge to forces.

The Irgun sought to claim Jaffa as Jewish territory and to achieve greater recognition and status as Jewish paramilitary groups grappled for power on the eve of the creation of the State of Israel. The Irgun sought to seize Manishiya and then the rest of Jaffa to demonstrate its capability and legitimacy. They were able to achieve part of this goal—the seizure of Manishiya—by capitalizing on the relative weakness of Arab defenses and social fabric, the existing fears of the Arab civilian population, and the rapid adaptation of their own combat plans. Rather than attempt to permanently control or occupy the city, which was beyond their scope and interest, they ceded control once they achieved their political aspiration and negotiated satisfactory terms.

The battle of Jaffa could be interpreted as a tale of the triumph of a small army against not only one, but two, larger adversaries. Tactical innovation, leadership, and persistence undoubtedly played a role in the Irgun's success, but their adversaries' lack of coordination, cohesion, and willingness to fight gave the Irgun an advantage, conditions that might not occur in other contexts.

Notes

1. LeBor, *City of Oranges*, 13, 14.
2. Ibid., 14.
3. Runkle, "Jaffa 1948."
4. LeBor, *City of Oranges*.
5. Runkle, "Jaffa 1948," 290.
6. Golan, "Battle for Jaffa, 1948."
7. Golan et al., *Campaign for Jaffa, 1948*.
8. Pappe, *Making of the Arab-Israeli Conflict*; Morris, *1948: The First Arab-Israeli War*; Radai, *Palestinians in Jerusalem and Jaffa*.
9. LeBor, *City of Oranges*, 2.
10. Golan, "Battle for Jaffa, 1948," 998.
11. Black, *Enemies and Neighbours*, 43, 44.
12. Golan, "Battle for Jaffa, 1948," 998.
13. Black, *Enemies and Neighbours*, 43–44.

14. LeBor, *City of Oranges*, 12, 13.
15. Golan, "The City and War," 200.
16. Golan, "Battle for Jaffa, 1948," 998.
17. Beckerman, *Unexpected State*, 157.
18. Runkle, "Jaffa 1948," 291.
19. Pappe, *Making of the Arab-Israeli Conflict*, 49.
20. Ibid., 50–51; Hoffman, *Anonymous Soldiers*, 63, 251.
21. Black, *Enemies and Neighbours*, 2, 95; Hoffman, *Anonymous Soldiers*, 104.
22. Morris, *1948: The First Arab-Israeli War*, 88, 89.
23. Morris, *1948: The First Arab-Israeli War*, 89; Radai, *Palestinians in Jerusalem and Jaffa*, 137.
24. Morris, *1948: The First Arab-Israeli War*, 89.
25. Ibid., 90, 93.
26. Ibid., 93.
27. Ibid., 91.
28. "Palestine Strategic Requirements," Chiefs of Staff Committee, Joint Planning Staff Memoranda, January–March 1947, Papers No. JP1-JP42, DEFE 6/1 12671342, The National Archives (TNA), Kew, London, United Kingdom.
29. Beckerman, *Unexpected State*, 157, 175.
30. LeBor, "Zion and the Arabs."
31. LeBor, *City of Oranges*, 110.
32. Radai, *Palestinians in Jerusalem and Jaffa*, 136–137.
33. Golan, "Battle for Jaffa, 1948," 999.
34. Radai, *Palestinians in Jerusalem and Jaffa*, 155.
35. Ibid., 150; Golan, "Battle for Jaffa, 1948," 1000.
36. Pappe, *Making of the Arab-Israeli Conflict*, 80; Lebor, *City of Oranges*, 107–108; Radai, *Palestinians in Jerusalem and Jaffa*, 150.
37. LeBor, "Zion and the Arabs," 66.
38. Golan, "Battle for Jaffa, 1948," 1000.
39. LeBor, "Zion and the Arabs," 67; Radai, *Palestinians in Jerusalem and Jaffa*, 172; Golan, "Battle for Jaffa, 1948," 1000.
40. Pappe, *Making of the Arab-Israeli Conflict*, 50–51.
41. Morris, *1948: The First Arab-Israeli War*, 161.
42. Ibid., 117.
43. Runkle, "Jaffa 1948," 292.
44. Golan, "Battle for Jaffa, 1948," 1001.
45. Khalidi, "Fall of Haifa Revisited," 39; Black, *Enemies and Neighbours*, 110.
46. Morris, *1948: The First Arab-Israeli War*, 148.
47. Golan, "Battle for Jaffa, 1948," 1002.
48. LeBor, "Zion and the Arabs," 69; Runkle, "Jaffa 1948," 293.
49. Golan, "Battle for Jaffa, 1948," 1002.
50. Ibid., 1002.
51. Runkle, "Jaffa 1948," 293.
52. Golan, "Battle for Jaffa, 1948," 1005.
53. LeBor, *City of Oranges*, 115.
54. Pappe, *Making of the Arab-Israeli Conflict*, 85.
55. Golan, "Battle for Jaffa, 1948," 1003.
56. Runkle, "Jaffa 1948," 297.
57. Golan, "Battle for Jaffa, 1948," 1003.
58. Runkle, "Jaffa 1948," 297.
59. Lehr, *Counterterrorism Technologies*, 52.

60. Golan, "Battle for Jaffa, 1948," 1003.
61. LeBor, "Zion and the Arabs," 70.
62. Runkle, "Jaffa 1948," 299.
63. LeBor, "Zion and the Arabs," 70; LeBor, *City of Oranges*, 119.
64. Runkle, "Jaffa 1948," 299.
65. Radai, *Palestinians in Jerusalem and Jaffa*, 169.
66. Ibid., 174–175.
67. Ibid., 151.
68. Telegram from General A. Cunningham to Secretary of State, April 28, 1948, CO 537/3861, TNA, Kew, London, UK.
69. Ibid.
70. Telegram from Commander in Chief Middle East Land Forces to War Office, London. April 29, 1948, CO 537/3861, TNA, Kew, London, UK.
71. "Irgun Fails to Seize Jaffa," *Manchester Guardian*, April 26, 1948, https://www.theguardian.com/theguardian/2011/apr/26/archive-irgun-fails-to-seize-jaffa-1948.
72. Morris, *1948: The First Arab-Israeli War*, 151.
73. Runkle, "Jaffa 1948," 300.
74. Ibid., 300–301.
75. Morris, *1948: The First Arab-Israeli War*, 151.
76. Golan, "Battle for Jaffa, 1948," 1006.
77. Runkle, "Jaffa 1948," 302.
78. Ibid., 303.
79. Black, *Enemies and Neighbours*, 106.
80. Runkle, "Jaffa 1948," 303.
81. Morris, *1948: The First Arab-Israeli War*, 150.
82. International Committee of the Red Cross, "Practice Relating to Rule 11."

13

Conducting Urban Warfare in the Twenty-First Century

Andrew Graham

Just as the Cold War placed security studies at the centre of the intellectual and political challenges confronting the West, so urbanisation and demographic change may mean that urban operations represent a critical issue in the 21st century.

—Alice Hills[1]

HISTORY DEMONSTRATES REPEATEDLY WHAT RECENT EXPERIENCE CONfirms starkly, that from the decision to commit and deploy military forces to the moment of extraction, strategies and campaigns that commit the military to operations in urban environments are as easy to get wrong as to get right. Military operations in urban and densely inhabited areas are logistically ravenous, unreasonably destructive in human and physical terms, and manpower voracious. By virtue of their duration, the related financial and human cost, and the potential damage to reputation and impact on public opinion, they are also unpalatable politically; they outlast and thus exhaust political patience.

Previous chapters have explored a range of historical, human, social, tactical, and other aspects and challenges for military operations in urban environments. Wendy Pullan's chapter emphasized that traditional armed forces are faced with the uncertainties of urban life and guerrilla responses and that state-of-the-art military technology is often less effective in an environment where "the overwhelming reality of urban warfare is civilian lives." In the contemporary crucible of the "battle of the narrative," the combined glare of instant communications and social media, the agility of unconventional adversaries in the contested media and information arenas, and the immediacy of urban networks make the timelines of "media action and institutional reaction increasingly out of sync."[2]

The insights and conclusions from those chapters combine to reinforce the notion that throughout history urban operations tended to be seen by land force commanders[3] as a necessary aberration from open warfare in the field because "urban combat . . . disrupts unit cohesion, complicates control, blunts offensive momentum and causes casualties to soar on both sides."[4] Paradoxically, and inconveniently given their unpalatability, the insights and conclusions of those same chapters also confirm the likely centrality and growing influence of urban areas and their populations on national and international politics, economics, and dynamics and thus on security and the conduct and outcome of campaigns. While Paul Latawski cautions in Chapter 3 that "war in urban environments may or may not be inevitable . . . armies may or may not have a choice as to where they fight," it seems inevitable that politics, cities, violence, and warfare will be ever more inextricably linked[5] and that political leaders supported by their military and civilian strategists and planners need to develop a deeper and more coherent understanding of the potential of cities as future strategic-operational battlespaces.[6]

Alice Hills has made the case that because cities are likely to be of increasing strategic significance, priority should be given in future campaigns to the "exploitation of the complex systems and high population densities characteristics [*sic*] of cities."[7] She adds that if cities are to be exploited and managed to advantage then "it is strategies (rather than tactics)" that need to be considered. Hills's observation cues this chapter to veer away from tactical considerations and place future military operations, and specifically land operations, in urban environments into their strategic and operational (theater and campaign) context and offer principles and ideas as catalysts for future thinking and planning.

Strategic Context and Developing Trends

National or grand strategy is the art and science of developing and using the diplomatic, economic, and informational powers of a nation, together with its armed forces, during peace and war to secure national objectives. Effective national strategy making is the responsibility of national leadership; it is a political activity that is informed by the intelligence, military, and security communities and others but is not led by them. The object and art of effective national strategy making is to interweave the diplomatic, information, military, and economic levers of power and influence, including the private sector and wider community,[8] to achieve a desired outcome arrived at by taking a long and a broad view forward of prevailing strategic trends and of the strategic context in which they are nested, balanced against a more parochial (and political) view inside, down, and out.

An Evolving Security Environment

Taking 2020 as a baseline, the strategic security environment is already underpinned and shaped by a combination of long-standing and fresh or emerging trends. Long-standing trends are driven by such things as globalization (and the backlash against it), the rise of nonstate actors with capabilities comparable to some nation-states, US conventional military superiority that forces most opponents to avoid its strengths and migrate toward unconventional approaches, and the implications of a global information environment. As David Kilcullen pointed out as far back as 2007, these trends would endure "even if al-Qaida [*sic*] disappeared tomorrow" and that "we may look back on al-Qaida as being the harbinger of a new era of conflict."[9]

Looking forward from a 2020 context to define social, economic, and security trends should include the pace and scale of urbanization, demographic and migration pressures, an emergent Russia and a more assertive and polydimensionally muscular China, the impact of and global responses to the pandemic and climate change, and the revolution in communications and interconnectedness. These combine to challenge contemporary Western strategy makers. They also intensify the complexity and scale of the campaign environments within which military commanders are likely to operate, and which are increasingly shaped by the style and nature of the activities, attitudes, perceptions, and behaviors of national, regional, and local civil society. These in turn influence decisively the activities, attitudes, approaches, and behaviors of protagonists and adversaries alike.

Wars Among the People

Since the end of the Cold War, academics and strategists around the world have grappled with the challenge of defining the character of future conflict and its implications not only for military forces but for interagency, multidimensional strategic- and operational-level planning and action. General Rupert Smith has argued that, although it is unwise to have a "preconceived model of the battlefield since adversaries seek to choose and shape a field which allows the greatest exploitation of their strengths and guards their weaknesses," nevertheless, from the perspective of deploying and employing the military instrument, "the era of war—defined as industrial, interstate warfare between armies, where the clash of arms decides the outcome—may no longer exist." Instead the trend is toward timeless operations conducted "among the people." He argues that as much as anything else the utility of military forces depends on their ability to adapt to complex political contexts and engage nonstate opponents under the critical gaze of public opinion, where the emphasis is on precision and localized effect to avoid unintended consequences.[10]

Iraq in 2003 is a case in point. The absolute ease and speed with which the conventional Iraqi Army was defeated in the field and the consequent speed of collapse of the regime exceeded the expectations of operational and tactical commanders both military and nonmilitary. However, from the moment that the clash of conventional armed forces failed to produce a decisive campaign outcome, the distinct objectives of interstate war were replaced by complex and malleable objectives aimed at establishing a condition in which a "decision" of some kind might be found or arrived at somehow rather than deciding the matter. The pace of events outpaced the planning, mobilization, deployment, and shift in mindset and approach required for what Paddy Ashdown calls "the demi-monde between hot war and a stable peace"[11] and especially the filling of the governance and security vacuum.[12] As early as May 2004, the head of the Coalition Provisional Authority, Paul Bremer, was to conclude that "the message to most Iraqis is that the Coalition can't provide them the most basic government service: security. We've become the worst of all things—an ineffective occupier."[13] Nowhere was this more apparent or consequential than in the all-important urban areas, home to 68 percent of the Iraqi population.

It is certainly arguable that the baseline for the years of political hiatus, violence, near civil war, insurgency, loss of government authority, and disruption that have afflicted Iraq since 2003 was set in part by the absence from the outset of an effective and agreed set of long-term strategic goals, those goals to be met by fusing political, civic, public, military, interagency, coalition, and especially, local capabilities at every level from the strategic to the tactical grassroots to "win and stay winning in" the key cities or towns. The ensuing years of violence and instability have demonstrated starkly the challenges for conventional military forces when forced to adapt to "complex political contexts" and engage "non-state opponents," especially when it is without the whole-hearted commitment of politicians, advisers, and enablers to a coherent and well-integrated ("comprehensive" in the UK Whitehall jargon of the early twenty-first century), cross-government and intergovernmental, multiagency approach that provides the unifying thread for activity from the strategic to the tactical level.

Urbanization

The facts of rising population growth (and where it is growing fastest), the acceleration of urbanization (and where it is taking place), and increasing connectedness challenge the ability of urban governance, public security, economies, and public services to keep pace with public demand and aspiration.[14] The inevitable deduction is that the future of global instability and violence at and below the threshold of warfighting is likely to be urban.[15]

In a twenty-first century characterized by an increasingly urbanized world, adversaries, whether conventional or irregular, are likely to seek out urban spaces both for their strategic importance and the advantages the urban environment offers, especially to the defender. Those advantages include cover, complexity, and concealment; the opportunity it gives to mitigate technological advantage;[16] and the complications and considerations that the presence of civilians brings into play. Not the least of these is that it opens the contest, in every domain where it is being staged, to extensive formal and informal media and public scrutiny. Operating in a multidomain manner in urban environments as one part of the security apparatus in a range of operational scenarios up to and including warfighting is therefore not a matter of choice for any military.

In an era of increasing urbanization, effective strategies therefore need increasingly to be informed by a comprehensive understanding of the nature, characteristics, and complexities of contemporary urban environments (including megacities) and of their populations. The imperative is to chart a fresh way of approaching planning, deploying, and employing the military instrument most effectively in urban environments as one element in a multiagency (or "comprehensive") approach, within the tolerances of national strategic appetites for risk, reward, cost and benefit, and local and international opinion. At the same time, strategic- and operational-level planning must mitigate the constraints of limited military capacity; the law (including the local jurisdiction, the law of armed conflict, responsibility to protect, and humanitarian and other law); the compromises that arise from a multiagency, "comprehensive" approach; and the strategic imperative to "not lose" any or all of the narrative battle, the physical or influence battles, or "the force."[17]

Hybrid and Compound Warfare

Among other things, today's battlefields occupy cognitive and cyber as well as physical space, and the traditional strategic reliance on mass has been twinned with the "subtle deceptions" of a more "hybrid" style of warfare.[18] Here conventional tactical military actions and irregular and "indirect" activity in the physical, cognitive, and cyber domains need to be blended effectively and persistently within multidisciplinary and overarching strategic- and operational-level thinking, with priorities and planning that flex both the formal[19] and informal[20] instruments of national power.

Writing in 2009, Robert M. Gates described the categories of warfare as "blurring meaning,"[21] meaning that since regular or conventional adversaries were blending the irregular and the unconventional into their strategic, operational, and tactical planning, the distinctions between conventional warfighting, insurgency, and terrorism were increasingly hazy. As the campaigns in Iraq and Afghanistan took their respective courses, the terms *hybrid warfare* and *compound warfare* emerged to describe the sort of nonlinear, polydimensional, "broken" battlefield being experienced by coalition and national forces.

Frank Hoffman defines the term *hybrid warfare* as "not a novel phenomenon but rather an enhanced and simultaneous fusion of different capabilities, tactics, methods that were previously separated in space and time." *Compound wars* on the other hand describe "those major wars that had significant regular and irregular components fighting simultaneously under unified direction . . . based on operationally separate forces."[22] The definition of *hybrid warfare* has been extensively scrutinized; however, in the context of operations in urban environments, the definition of *compound warfare* could and should be developed further given that *compound* has two subsidiary but active meanings. First, compounding increases the cumulative return on an investment over time (as in "compound interest"). Second, compounding enhances the negative impact on an individual or situation (as in "compounding the problem").

Compound warfare is therefore the actual or conceptual combining, sequencing, and concentration (in terms of one or several of a period of time, space, mindset, domain, activity, or impact) of conceptual or practical "dissimilars," "unlikes," or "opposites" (such as strategic-tactical, symmetric-asymmetric, state-nonstate, civil-military, social-political, conventional-unconventional, regular-irregular, technical-primitive, communications impact–physical force, etc.) in order to achieve compound benefits for the subject and compound impacts on the adversary. A compound or compounding activity in one domain, at one level of command, in one part of a city or theater, directed at one particular adversary audience is likely to have a variety of immediate or delayed, potentially polydimensional impacts on different adversaries and audiences, in several domains, at multiple levels of command (the "strategic corporal" is the often-quoted example) in different places.

Whether these and other assessments herald a new form of warfare matters less to the military practitioner than that they reset the terms of the timeless challenge of facing adversaries who seek to develop creative and disruptive uses for the means at their disposal. Such creativity includes blending conventional and unconventional, regular and irregular, overt and covert means, in any or all of the physical, cyber, and cognitive dimensions, to gain advantage, seize the initiative, and neutralize the strengths of the opponent.

Of course, adversaries have always sought to be both creative and disruptive, but three hard contemporary truths must be faced up to: First, there are now more dimensions to be exploited and wider, deeper, and increasingly nonmilitary battlefields upon which war can be waged. Second, the range of potential adversaries and the means at their disposal is certainly not limited to the conventional forces of a nation-state. Finally, urban areas and their populations are a fertile and accommodating environment for a blend of conventional and unconventional adversaries to exploit.

Evolving Threats

But what of the threats that are generated from these hard truths? Hoffman defines hybrid threats as incorporating "a full range of modes of warfare, including conventional capabilities, irregular tactics and formations, terrorist acts that include indiscriminate violence and coercion, and criminal disorder."[23] Russell Glenn's description of hybrid threats emphasizes the role of the adversary and distinguishes between means and methods: "an adversary that simultaneously and adaptively employs some combination of (1) political, military, economic, social and information means, and (2) conventional, irregular, catastrophic, terrorist and disruptive/criminal warfare methods."[24] Glenn's definition also points toward the ends-ways-means calculus that underpins the forging of effective strategic responses.

Faced with these hybrid or blended threats, effective polydimensional action—be it preemptive, disruptive, offensive, defensive, deterrent, or coercive—that has the potential to lead to a positive strategic outcome requires a blended and coordinated multinational, national, multiagency, not purely military, approach to conflict and campaigning. That approach must incorporate whole of government and multidomain planning, objective-setting, and action laterally across and vertically between the strategic, operational, and tactical levels.

Viewed through a Western lens, the experience of military forces operating in urban environments since the year 2000 has demonstrated that President John F. Kennedy's dictum about the need for absolute understanding by the military of "the limits of military power" should be extended to include the majority of recent and contemporary Western political leaders and policymakers.[25] As was also observed, Chinese thinking and the characteristics of Russian interventions in Georgia, Ukraine, and Syria are symptomatic of nations capitalizing on the nonmilitary tools of state power in what is described[26] as an era of perpetual conflict where states and nonstate actors seek marginal advantage in limited tactical and often nonmilitary engagements that cumulatively bring strategic benefit.

Beyond Limits Warfare: China

The concept of "beyond limits warfare" was first presented to Western audiences in 1999 in a book whose stated aspiration was to have the concept incorporated into "combat methods."[27] In broad terms, the concept envisages that any sphere of activity including politics, economics, psyche and culture, and military is a battlefield, with tangible and intangible resources such as culture, religion, ethnic identity, geography, and history as weapons in a new form of combat. Measures (where measures equal actions not resources) taken in any sphere of activity attack the adversary's weaknesses while seeking to achieve clear objectives. Commanders at all levels use effective and innovative "combinations" and "additions" across domains or spheres of activity to achieve asymmetry and grasp the initiative, reinforce impact, create dominance, and exploit advantage. The concept emphasizes that military forces have a unique and indispensable part to play in this blend of approaches (comprehensive, whole of government, multidomain) but stresses that "any idea that the military is supreme" must be resisted.[28]

The concept is polydimensional in its scope and presupposes a comprehensive understanding of the strategic, operational, and tactical environment of the adversary and of the conventional and unconventional measures, means, and tools available to be used as weapons alone or in combination, in one domain or several. "Beyond limits warfare" espouses principles that are particularly applicable to operations in urban operations at any level:

- synchrony (actions taken within the same time period in different domains)
- omnidirectionality (no blind spots; orientation with no blind angles; cross-domain action and thinking)
- multidimensional coordination achieved using combinations of measures in many domains
- minimal consumption (more measures in more dimensions to reduce the requirement for and enable the reasoned and sustainable use of combat resources)
- limited objectives coupled with unlimited measures (objectives smaller than the measures available, so you know you can achieve the objectives you have set yourself; limited objectives achieved by way of unlimited measures to overwhelm the adversary by the variety of measures pitched against him)
- asymmetry (seeking the opposite of symmetry; select battlelines and areas where adversaries do not expect to be hit)

The key deduction we might make when considering operating in large and complex urban environments is that joint, combined arms and interagency thinking and action, and moving the focus for battle away from the military, encourages combining and aggregating measures in many domains aimed at achieving a common set of objectives. The nature of the urban environment also blurs the traditional lines between the levels of war and significantly enhances the impact of small-unit operations in combination with the measures taken in other domains in the same period of time at the strategic, operational, and tactical levels. As John Spencer demonstrated in

Chapter 9, a small force at the tactical level can accomplish many missions in extremely large areas given the right circumstances, clear objectives, and accurate knowledge (understanding) of key physical and human terrain.

The Contours of Future War: Russia

Paralleling the thinking and practice of an increasingly assertive China, Russian views on the conduct of contemporary and future warfare are forged in practice in Chechnya, Georgia, Ukraine, Dagestan, and Syria. Russia appears to envisage the waging of contemporary war as blending and coordinating the use of military (including integrating proxies, militias, special forces [SF], and organized crime cartels alongside conventional forces) and nonmilitary measures with the aim of undermining and confusing political and public commitment, influencing behaviors, disorganizing the adversary's command and control options, and bringing recce (reconnaissance) and strike options to bear where the destruction of enemy forces will achieve tactical, operational, and strategic ends.[29]

Examining Russian participation in operations in urban environments in Syria in particular shows that the contours of war include the use of unmanned systems, automation and artificial intelligence, precision, C4 ISR (command, control, communications, computers along with intelligence, surveillance, and reconnaissance), the use of proxy forces, and perhaps especially, information operations. Russia treats information operations as a central tenet of contemporary conflicts. Information manipulation, propaganda, misinformation, and disinformation—collectively known as *maskirovka*[30]—are considered vital weapons at every level whose targets are military and civil,[31] at home and abroad. Described in UK Doctrine as "sharp power," the manipulation of information as part of the strategic communications effort gains strategic advantage outside the normal rules-based system and can significantly undermine a nation's soft power.[32]

Urban Environments as Complex and Adaptive Systems

Urban environments encapsulate and exemplify two of the key Properties of Complexity identified by Simon Atkinson and James Moffat.[33] The first is the property of decentralized control, where emergent behavior is generated through local interactions rather than centralized control. The second is the combination of the three properties of coevolution, collectivist dynamics, and nonlinear interaction, which means that clusters or avalanches of local interaction are constantly being created across the system, those interactions ripple through the system, and small changes in the nature of the interaction can have large, potentially convulsive effects.

Conceptually, therefore, a city or town can be likened to a "knot" in a national, and indeed international, net. Each "knot" is made up of a number of smaller district, community, social, economic, or behavioral "subknots" that bunch together to make the whole. Each subknot gets its form, strength, and energy from a weaving together of, and the interaction between, a plethora of local strands or filaments—physical, cultural, economic, political, tribal, ethnic, social, religious, age group, and so on. The larger the city, the more subknots there are, so the less it is desirable or possible to see a city as a single entity, and the more important it is to really understand the dynamics, interactions, attitudes, and behaviors within and between localities and their populations.

The overall tension and energy applied to each large urban knot is set by a combination of national and international dynamics, influences, and flows and the energy fields created by the interactions within and between subknots (internal) and other large urban knots (external). In turn each town or city exerts its own particular pressure and influence outward across the full national and international net along physical, virtual, cultural, economic, or other filaments or strings. That pressure and influence is generated by the city's own unique range and combination of circumstances, challenges, factors, influences, and social, political, and economic dynamics.

Again and again in Iraq in 2003–2004, as well as in the example described in Chapter 9 when activity in Basra sparked the outbreak of violence in Sadr City 500 kilometers away, the combination of pressures and influences generated across Iraq ("the national net") by the interactions taking place in and between towns and cities across the country had the potential to pose the coalition and the Iraqi government a geographically dispersed range of testing, tactical, and operational political, military, and interagency challenges—a potential that was fulfilled.

The Security Landscape

Consideration of how military forces are to be prepared, deployed, and employed to operate most effectively in pursuit of strategic and operational rather than tactical objectives in complex urban environments in a changed and changing strategic context is a crucial issue for any analysis of the future security landscape. The transformation required first in strategic thinking and then in whole of government, interagency, cross-departmental, multidomain and multidisciplinary, interallied planning, alignment, and operations from the strategic to tactical level is potentially profound.

Winning the Street Fight

As General Mark Milley noted, "If war is about politics it is going to be fought where people live. It will be fought in my opinion in urban areas."[34]

There is much written about the problems and difficulties of conducting tactical urban military operations in the future, including the challenges of developing forces and concepts capable of conducting what Louis DiMarco terms systematic urban warfare,[35] what Michael Evans refers to as "a joint and interagency philosophical variant of Bugeaud's Military Operations as Urban Planning,"[36] what Paul Van Riper describes as "chameleon manoeuvre,"[37] what George-Henri Des Vallons expresses as "Metropolitical warfare,"[38] and Robert Leonhard discusses as "Urban Warfare in the Information Age."[39]

Sun Tzu's observation that "to subdue the enemy without fighting is the acme of skill" notwithstanding,[40] all agree on one thing. On the one hand, the level of the urban fight and the manner in which it is prosecuted will be dictated increasingly by the strategic requirement for precision to minimize collateral damage and second-order effects, by the scale of available military capacity, and by the negative and costly implications of timeless campaigns among the people. But the imperative at the tactical level to be prepared to fight and to prevail in high-intensity combat at close quarters when and where the situation demands is not tradable.

The chapters describing the battles for Jaffa, Donetsk Airport, Groningen, and Sadr City are focused on the conduct of the battles themselves and assess the impact of tactical actions at the operational and strategic levels. We might look at the developing situation in Iraq in 2003–2004 as another example. From a Western perspective, the strategic, operational, and tactical considerations that played out in Iraq, from the initial invasion in March 2003 through the twenty months of "hot stabilization" to December 2004, test in a practical and appropriate manner Alice Hills's hypothesis that "urban operations are not only a hard security challenge but also offer a powerful conceptual means for generating new insights into military operations in an urbanising world."[41] That period of the campaign shows starkly that operating effectively in urban environments, especially in a fragmented and fractured security vacuum, toward any sort of positive and sustainable outcome beyond pacification or the taking, holding, or denial of a piece of terrain to an adversary asks much more than winning the close-quarter battles. Importantly, each of these examples also shows that, whatever else may have been at stake, it was a strategic as well as military imperative not to lose those close-quarter battles.

The coalition experience in the Iraq campaign offers the opportunity recommended by Evans among others[42] to draw together key insights into a practical strategic- and operational-level, generalist but joint operations framework outlining principles for the way in which armed forces in general, and smaller armed forces in particular, might plan for, prepare for, direct, and execute future military operations in urban environments. These insights range from understanding dynamics to using a comprehensive approach as outlined below.

Understanding. Urban environments are arguably the most significant, complex, symbolic and consequently influential elements in any theater or battlespace. Adversaries will likely choose to capitalize on them for their perceived advantages. Understanding the internal and external dynamics and flows and the way in which complex interactions within an urban system have impact and influence on the wider security situation, the popular mood, and on societal behaviors is a prerequisite for neutralizing those advantages, exerting influence appropriately, and developing opportunities leading to future success. Military actions across the spectrum of operations must be nested in a broad strategic and operational multiagency context that is itself rooted in a comprehensive, all-source understanding of the nature of the theater from international through national and regional to local. That understanding must be developed through persistent engagement, study, and sophisticated social science–based analysis over time, leveraging the analysis of big data and artificial intelligence to unravel, interpret, and understand the complex connections, flows, interactions, and dynamics that characterize a city.

Clarity of strategic intent. From the above context and understanding should cascade clear and unified strategic intent emanating from clarity of political will that recognizes and exercises its Clausewitzian responsibility: "The first, the supreme, the most far-reaching judgement that the statesman and the commander have to make is to establish . . . the kind of war on which they are embarking, neither mistaking it for, nor trying to turn it into, something which is alien to its nature."[43] Implicit in that judgment is an assessment of the strategic value and the symbolic importance of centers of population and the consequent campaign priority and weight to be attached to them and their populations, dynamics, and influences.[44]

Coalition and alliance working. The likelihood of not operating with at least a handful of partner forces and agencies as part of some sort of coalition or alliance is negligible. Conflicts in Iraq, Afghanistan, Syria, and Libya show that what could be called "host nation participation" of some kind is a certainty. More broadly, coalitions and alliances are political responses to strategic circumstances. Alliances can be and often are devised for one set of circumstances but may be amenable, with negotiation, to adapting to a changed set of circumstances. Coalitions are situation-based and the "sum of parts."[45] Since each "part" reflects the national mood and appetite for risk of each contributing nation at a particular moment, there is an ab initio weakness for achieving an agreed strategy and campaign plan around which to coalesce unity of effort and assurance of commitment. On the other hand, there is a case to be made that the adaptable, more loose-leaf and relatively unconstrained "come as you are" nature of coalitions could create the sort of agile construct that suits the fluid and more enduring nature of operations in urban environments in the twenty-first century.

Unity of purpose. History teaches that unity of purpose is the vital ingredient in the success of coalition or alliance war at the strategic level.[46] A coalition that lacks the necessary international resolve and the legitimacy that mass involvement under an internationally recognized mandate implies is likely to become increasingly fragile, however effective it may appear in military terms and however successfully it may perform tactically. At the strategic and operational levels, achieving unity of effort is as or more important than unity of command. The building of trust ("the Coin of the Realm in coalition operations"[47]) in the political and military spheres of action, and working through and around national caveats and potential fault lines, is a—and potentially *the*—principal task for commanders at all levels in pursuit of that goal.[48]

Comprehensive approach. The characteristics of complex environments, especially larger centers of population, demand a proactive, multidomain, polydimensional, multiagency approach to achieve compound effects. Commanders must shape, influence, and capitalize on opportunities to stimulate operational and tactical interagency as well as military combined maneuver; gain understanding; seize initiative; persuade and reassure the neutral, uncommitted, and undecided; and attack the cohesion and will of the hostile in the cognitive as much as or more than in the physical domain.

Emerging Framework Principles

Extrapolating from those insights and from observations from the operations of other nations in theaters such as Ukraine and Syria, four framework principles and a number of consequent strategic, operational, and tactical considerations emerge.

Framework Principle 1: Strategic Direction, Alignment, and Integration of Effort

Addressing any hard security challenge places a nonnegotiable imperative and premium on having clear political direction from which cascade clear strategic objectives and an overall plan agreeable to all participants. Achieving effective integration of nonmilitary and military activities toward an agreed outcome over a sustained period of time and effort requires completely aligned, collaborative, and persistent strategic, operational, and tactical intent, effort, and commitment not only from the military but, above all, from the national leadership and the political class of the nation or nations involved.

There is nothing new to this insight. Sir Robert Thompson's First and Third Principles drawn from his experience in Malaya and Vietnam cite "a clear political aim" and "an agreed overall plan" as fundamental to success.[49] More recently, Ashdown cites the recommendation in the Brahimi Report that an integrated Mission Task Force be created by the UN to achieve the "desirable but historically unachievable alignment and unity of effort not only between political, civil and military pillars but also across the international community."[50]

Without that political and strategic intent and commitment to cross-department collaboration "from Whitehall to the Basra/Kabul street" and consequent operational and tactical alignment, the ability to act effectively and consistently on such things as security, the rule of law, governance, economics, and the basic issues of survival (food, water, work, shelter, and power), and to address and influence the hopes, opinions, attitudes, and perspectives of the population will be compromised. An effective integrated plan for operating in urban environments requires what I term the polydimensional integration of "the uniquely relevant capabilities of all the instruments of national power,"[51] including intelligence gathering and sophisticated social science analysis to develop understanding, cue information operations, influence behaviors, and design and direct appropriate military activity.

Framework Principle 2: Winning in the Cities

As Thompson noted, "Secure base areas first."[52] Ally the trends of increasing urbanization and demographic change with the strategic and symbolic centrality of centers of population, and a key element of an integrated strategic plan needs to pay particular attention to doing what it takes to "win and keep on winning in cities and towns" and to accelerate action to gain, or at least not lose, traction, support, momentum, and credibility in the eyes of the key urban audiences.

In Iraq, a country where 68 percent of the population were urban dwellers, there was extraordinary value to be placed on understanding, interpreting, and influencing both the systemic nature of towns and cities and the urban-dwellers' local and broader opinions, attitudes, linkages, and perceptions. The increasingly adverse reaction to the coalition reflected the absence of that urban focus and of effective, visible government and interagency action with consequent impact on the inherently fractured but also unsplintered political and security climate inside towns and cities. The consequent levels of violence in towns and cities and their impact across the theater on the attitudes, activities, and opinions of the population in turn set the long-term tone, tempo, narrative, and complexion of the campaign.

Fighting for cities. Observations from Yemen, Syria, Chechnya, the Balkans, and elsewhere reinforce the importance of urban areas as key battlegrounds. The prize is arguably less about domination or ownership, with all the administrative, economic, and social demands and responsibilities that control and authority (i.e., "government"[53]) bring, than about neutralizing and challenging the authority, influence, control, and consequent freedom of action of an adversary. In March and April 2003, Iraq's major cities of Basrah, Baghdad, and Mosul were symbolically crucial to the authority of the regime and to the idea of Iraq as a state. Their seizure (liberation?) were important indicators of campaign progress and success in collapsing the authority of the regime. Paradoxically, V Corps intention regarding Baghdad in March–April 2003 was to "avoid owning the city but still to be able to control it from outside."[54]

There is certainly an argument to be made that in warfighting the first consideration of the operational commander should be to seek alternatives to entering the city at all.[55] As an example, in Chapter 8 Steve Tatham described Brigadier Mackay's determination that influence should form the backbone of the brigade's efforts to capture Musa Qala. A primary consideration should be whether military means are sufficient to meet national or coalition objectives —any sort of setback or heavy casualty loss could be strategically counterproductive[56]—and whether covert SF, precision strikes, local forces, blockade, and targeted information operations efforts might be capable of achieving the mission of defeating the enemy or wresting control without deploying large-scale ground forces into the town to fight for ground.

The conventional resistance expected in the taking of Baghdad, Basrah, and Mosul in March–April 2003 after the invasion did not materialize to test the proposition that urban areas are an aberration because "urban combat . . . disrupts unit cohesions, complicates control, blunts offensive momentum and causes casualties to soar on both sides."[57] However, the next 20 months of stabilization operations in cities and other urban areas were certainly to test that proposition to the hilt in combined arms combat operations inside cities and towns at every scale from division to platoon in Fallujah, Kerbala, Sadr City, the cemetery and around the Imam Ali Mosque in An Najaf, Tal Aafar, Mosul, Ramadi, Al Amarah, and Al Kut—to name but a few.

Operating in cities. The Russians in Grozny in 1995 and 1999, the Iraqi Army in Mosul in 2017, the Israeli Army in Jenin in 2007, the Syrian Armed Forces in Aleppo and Idlib, and coalition and national forces in Afghanistan and in Iraq shared a common experience. Each faced the perspective, approach, and tactical, ethical, and presentational challenges posed by an unconventional adversary confronting a technologically and conventionally superior enemy in and around cities and challenging for the tacit or active support of the population.

Evaluating the nature and motivations of an adversary in those circumstances is especially difficult and complex in urban environments. In any unstable situation then—and particularly in the places where people are concentrated and where security and the rule of law are fragile or challenged—political power, authority, and control are fragmented or diffused among rival constituencies, including nonstate actors. These in turn compete for control and people's loyalty and keep hostility, factionalism, and conflict alive, with the consequence that violence begets more violence. Without a deep, nuanced, but comprehensive understanding of the local and wider situation, dynamics, narratives, politics, culture, and behavior, it is next to impossible to design proactive military and other operations to achieve "compound" effects that are carefully calibrated, proportionate, and precise and executed with a clear understanding of the likely consequences.[58]

On the one hand, this is confirmation for the military of the growing interface between insurgents and potential adversaries and between towns and cities and their populations.[59] On the other hand, for conventional armed forces likely to be acting in some circumstances on the defensive (e.g., North Atlantic Treaty Organization [NATO] prevention and deterrence tasks in the Baltic States), the smaller unconventional adversary's example, perspective,[60] and exploitation of the physical, information, cognitive, and temporal domains in the urban battlespace for tactical, operational, and even strategic advantage are important to consider for lessons and insights in negating, blunting, or deflecting perceived advantages of scale and technology.

Framework Principle 3: An Unconventional, Polydimensional, and Multiagency Mindset

"Okay Bubba, we're here, now what?"[61] This question from the commanding general of the US V Corps in April 2003 was a good one. The bursting of the bubble of Baathist authority created a vacuum that was filled rapidly by the unexpected release of pent-up energy and pressure from the population. That release presented the coalition military in every town and city with local conundrums unrelated to warfighting but instead around such things as return on equity, the approach to be taken to looting, how to fill the vacuum in law enforcement, the securing of ammunition sites and key installations, the status of former regime officials, and the meeting of the expectations of the liberated.

At the operational level, operations in Iraq in mid-2003 showed that tempo and momentum in all activity across the PMESII-PT[62] spectrum had to shift from being relative to the conventional and/or unconventional adversary to being relative to the local political and security context and the population and their needs, expectations, and perceptions. This was especially so within cities and towns whose population and infrastructure were so often the targets of insurgent attacks or affected by coalition operations and whose lives, perceptions, attitudes, and support were the prize over which the campaign should have been being fought.[63]

The operational and tactical bellwethers of success for the people of the country and of the cities were the prevailing political and security situation, power, water, economic opportunity, and food. This required an operational-level blend of military and interagency command authority, backed by the necessary unity of political, financial, and civil affairs craft, heft, muscle, and effort at all levels "*to disassemble and reassemble while maintaining momentum* most crucially in the places where the majority of citizens live their lives, in the cities and towns."[64]

Among other things, threats lay in all directions, the enemy was not linear or conventional, and influences and connections mattered.[65] Without shifting from linear military considerations of conventional threat, adversary motivations, weaknesses, centers of gravity, popular perceptions, and so on to human intelligence, network-based analysis, cultural analysis, demographics, social science, and urban studies, it was hard to understand the source and nature of the resistance, especially in cities and towns; what to do about it; and how to influence behaviors.

The challenges for coalition military forces over the period exemplified a "Continuum of Operations" in which a range of high-intensity to lower intensity or counterinsurgency/armed policing tasks were conducted alongside stability operations, including humanitarian and nation-building tasks.[66] The engagement in Iraq, and more starkly that in Afghanistan, has demonstrated that on their own the military actions that this continuum represents can only go so far in terms of achieving enduring political, economic, and social outcomes.

Framework Principle 4: Understanding

"The Art of Command is to make choices in the midst of ambiguity," as Clausewitz noted.[67] British military doctrine defines understanding as "the perception and interpretation of a particular situation in order to provide the context, insight and foresight required for effective decision making."[68] The US-led coalition in 2003 faced a nation-state whose organization, capabilities, resources, lines of authority, and conventional armed forces were susceptible to the full and formidable range of conventional and technical intelligence assessment and analysis capabilities at its disposal. The coalition found applying the "perception and interpretation" elements of the doctrine in the context of the deployment and preparation of an expeditionary force (Phases I and II) and the subsequent warfighting phase (Phase III) to be relatively straightforward.

The complete collapse of government authority and the very rapid transition from conventional conflict to state-building and stabilization (Phase IV) resulted in an unanticipated and precipitate breakdown of law and order, particularly in towns and cities.[69] The coalition were to experience the painful impact of an apparently superficial understanding at any level of, among other things, Iraqi culture and the Arab way of thinking or the internal political, religious, or social dynamics of the country as a whole and those of each of its provinces, cities, and towns in detail.

Understanding at the strategic and operational levels must take a whole-of-theater perspective. Integral to theater-level understanding is an appreciation and understanding of strategic dynamics—national and international politics and attitudes, regional and local narratives, economic and information considerations—balanced by detailed understanding of regional and local attitudes, perceptions, priorities, and dynamics. Despite years of operations to impose sanctions and enforce a no-fly zone, there appeared to be a dearth of knowledge or understanding of the state of infrastructure and public services in Iraq;[70] of the scale and range of demographic, ethnic, and sectarian laydowns; and of the diverse general, sectarian, historical, and parochial expectations, attitudes, and perceptions of the population inside Iraq as a whole and in each town, city, and region. Such an understanding should have been a prerequisite for enabling the correct or appropriate poly-dimensional and multiagency planning, resource allocation, decisions, and preparations to have been completed before what should have been the decisive phase of the campaign[71] and for ensuring that the most effective strategic, operational, and tactical decisions would be taken thereafter.

In his evaluation of the challenges that urban operations pose for joint intelligence organizations, J. A. Steadman[72] observes that "nothing is more critical to accurate and relevant intelligence than independent analysis."[73] He concludes with the necessity to analyze interrelationships and interdependencies between all the factors by a combination of nodal analysis and analysis of the urban area as an integrated "system of systems," drawing on the expertise of civil engineers, city planners, and social scientists among others in order to understand what he calls the synergistic effects. By understanding those effects, maneuverist action and "compound" pressure can be combined to gain further understanding, seize initiative, generate tempo, and attack cohesion and will in the physical, cognitive, and virtual domains.

PMESII-PT and ASCOPE[74] are relevant frameworks for considering the factors at play in understanding complex urban environments in a theater context. Both need the input and expertise of civil affairs, urban planning, nongovernmental organization, humanitarian, legal, political, economic, social science, and cultural experts and practitioners to mitigate their susceptibility to conventional military analysis techniques that tend to focus on an individual factor to come to a plan of action for that particular line of activity.[75]

Translating the deductions into a coherent framework and executable plan for action, with capacity and resilience and resources committed in the places where they matter most on an outcome basis, requires operational-level military-civil and interagency coherence, collaboration, and unity of pre-agreed purpose, running from the strategic to the tactical level. That purpose must be nuanced to the prevailing circumstances in each town or city, with a premium placed on dissemination of information and intelligence to those who need it—at a pace and accuracy to be both timely and usable.

Strategic and Operational Considerations

At the military strategic level, the responsibility of military commanders is to prepare the armed forces for war and to advise the civilian authorities and policymakers on the application of military force to achieve the aims of policy. That combined responsibility requires senior commanders to provide policymakers with an honest estimate and analysis of the strategic probabilities for military success. When considering future urban operations, military strategic planning advice has two factors to consider. One is the preparedness, training, equipment, and design of military forces at any time for likely upcoming or future wars or operations. The other is the planning, direction, deployment, and employment of military forces in that war or operation, including as part of a coalition or multinational force.

Preparedness, Training, Equipment, and Design of Military Forces

The debate over the strategic advisability and necessity, or otherwise, for armed forces to operate in urban environments extends as far back as Sun Tzu. For all armies, but perhaps especially for smaller ones, the debate's conclusions tend toward support for what has been described as the "article of faith amongst military professionals that cities swallow armies."[76] However, global trends and almost every contemporary analyst, military commander, academic, and significant military power agree that operations in urban environments, while politically unpalatable and potentially costly in terms of blood and treasure, are likely to become not only unavoidable but also more significant and symbolic elements of any campaign. This poses a significant conceptual and practical force design challenge at the tactical level given the growth in urbanization, the concomitant growth in the scale of urban areas, and the reality of demography and other facts.

Armed forces with the luxury of scale can achieve a degree of tactical specialization and organization, including dedicated combined arms subunits and units grouping infantry, armor, electronic warfare and cyber, joint fires, and engineers. Smaller armed forces need to be ingenious in the extent to which they structure and prepare for conventional peer-to-peer high-intensity open warfare as part of a combined arms conventional joint force while being sufficiently prepared and trained to operate effectively in complex, "compound" urban environments across the continuum of operations, arguably most likely in defense when in a NATO Article V context.

Design

Drawing on national experience and learning from the experience of others, design parameters should include the following:

- Capitalize on the military's unique conventional capability and responsibility to use force in order to "compel the enemy to compliance, to force them to accept a solution of some difference against their will,"[77] while operating as part of an integrated multidisciplinary force of military and nonmilitary generalists and specialists and in accordance with humanitarian and legal norms in armed conflict.[78]
- Recognize that in the history of urban street fighting it is skill and experience in small unit tactics, leadership and resilience, and combined arms tactical operations that have been the basis upon which armies have more successfully faced adversaries of all kinds at very close quarters in the urban environment. However desirable it may be to aspire to win without fighting, this characteristic is not tradable.
- Adopt innovative thinking around a "new concept for weapons" rather than "new concept weapons,"[79] recognizing that "mankind is endowing virtually every space with battlefield significance"[80] and optimizing the opportunities that nonmilitary action as well as technology offers at every level for maximizing multidomain indirect and kinetic, lethal and nonlethal precision effect; leveraging dispersion, deception, disruption, and detection; achieving operational and tactical poly-domain A2/AD[81] (including physical blockade and isolation [land, maritime, air, information], offensive and defensive EW (electronic warfare), nonkinetic obstacles, persistent presence in all domains, strategic communications, and information operations); achieving enhanced force protection, precision, economy of force, concentration of effect, synchronized action in a 5G-enabled tactical environment; developing an effective approach to influence—the subtle art of persuasion—drawing on Steve Tatham's observations in Chapter 8 advocating a fresh approach to achieving

influence, in particular the need to understand the reasons for and limits of behavior and to recognize that influence is about much more than messaging and social media.

- Place a premium on a sophisticated, polydimensional, crosscutting, outcomes-based approach to deterrence and prevention, constraint, and coercion.
- Meld the tactical level continuities and human absolutes implicit in operations in an urban environment at any level of intensity with the strategic- and operational-level imperatives of information operations, interagency and multidomain joint action, integrated action doctrine,[82] influence, and strategic communications.
- Be relevant to the effective deployment and employment of the military instrument at any stage of a campaign, and be applicable, whatever the intensity of conflict or the scale of the military involvement.
- Consider all operations of war. A legacy of recent campaign experience is the tendency to think always of being operationally on the offensive. The NATO context suggests that holding and denying urban terrain to prevent, deter, and constrain may be a crucial part of a strategy preceding an operational-level campaign of defense and delay to buy strategic time and breathing space.
- Think and plan broadly, including the human factor, emerging technology, optimum use of local forces, task organization, combined arms (infantry, armor, aviation, engineers, fires, counter–improvised explosive devices, special operations forces, etc.), training, the air and maritime dimensions, electronic warfare, logistics and supply, understanding, intelligence, and influence.

The result of that strategic design should be a concept for developing, deploying, and employing 4G/5G enabled military forces optimized for a conventional operational setting but capable of adapting to be the military element of a technically enabled, combined arms, interagency force in urban settings. The defining battle-winning characteristics of such forces might be the following:

- compressed operations and levels of war;
- demassed forces, noncontiguous forces, and dispersed forces;
- a drive to fight first for information;
- deep sensor reach ("he who sees can strike"[83]);
- dynamic self-synchronization and high-quality shared awareness;
- speed of command; and
- ability to adapt to initial conditions altering at increased rates of change, with events becoming ever more closely coupled.

Planning, Direction, Deployment and Employment of Military Forces

The operational level of military command takes the strategic objectives and priorities for the use of armed forces in both a theater and a multi-domain context and translates them into a campaign plan directing and sequencing tactical military and interagency actions toward the achievement of broad operational goals. The operational level of command is theater-focused and must think to the finish in thinking through and influencing the international and national political, strategic, military, and multiagency context; stretching the practical, political, temporal, and resource parameters; and setting the optimum conditions for short-term tactical and long-term campaign success.

In a conventional war between states, the initial military strategic, operational, theater, and tactical-level military considerations are focused around the use of violent military means, conducting sequenced or simultaneous operations of war through maneuver, threatening or striking vulnerabilities, attacking cohesion and will, hurting, compelling, coercing, and ultimately bringing the enemy to the treaty table or to regime collapse.[84]

The transition from conventional operations to "wars among the people" places a premium on the effectiveness of the combined/collaborative (interagency and military) operational level of command as the demands on commanders rapidly broaden and deepen. As the operation develops, the environment evolves and the influence of urban areas and their populations on the campaign "climate" increases so the need for combining maneuver and influence grows.[85] That is not least because the expectations and perceptions of the population, the "battle of the narrative," the media in all its aspects, the demands of the law and relevant return on equity, humanitarian and restoration activity, security, the rule of law, criminality, and factional interests move to center stage.

Adding to these complications is the fact that it is impossible for a template approach or solution to be applied because the distinctive characteristics and dynamics of each urban area shape and color the operational and tactical context within which that urban environment rests. This amplifies the higher commander's absolute responsibility to understand and shape the operational context in all possible dimensions in a manner that not only gives the forces on the ground as good a chance as possible of achieving stated objectives in the circumstances they meet on the ground but follows through beyond the fighting as circumstances change.

Although some argue that the era when commanders can bypass pitched battles in cities is over,[86] the necessity or otherwise for military forces to operate in cities and towns will be determined as much by strategic political, cultural, economic, and symbolic imperatives as military ones. The efficient, effective, and economical[87] tactical commitment and employment of ground forces into urban areas must be set inside a theater-level, joint, multidomain operational concept and design that taps into maritime warfare concepts such as command of the sea and sea control, including blockade and aspects of anti-access/area denial (A2/AD) thinking for its approach to activity outside, around, and between cities and towns while neither tying up small forces in attritional fighting or a framework of ground-holding tasks nor ceding physical and cognitive advantage to adversaries and thus exacerbating the challenges of the conventional tactical urban fight.

In an urban environment, command often becomes fragmented, with control driven from the operational to lower tactical levels. Mission command comes into its own, and the ability to surge both responsively and preemptively within urban areas and between urban areas generates tempo and momentum relative to the situation and achieves surprise as well as dislocating and disruptive impact. The agility and speed at which, and the accuracy with which, joint and information effects are brought to bear and forces are deployed between and into urban areas within a theater is a key tool in the operational commander's maneuverist toolbox at every stage of the campaign.[88]

Whatever the intensity of conflict, all supported and supporting military actions must be alive to and appropriately synchronized and sequenced with local, government, and official civil activity across the spectrum of PMESII-PT and ASCOPE planning being applied up, down, and across the theater to ensure unity of effort and coherence of outcome.

Once the decision to commit forces is made and the operational conditions are set, then a number of characteristics pertain to tactical operations in urban environments, whatever the intensity:

- The impact of the physical environment. Although you cannot fight urban physics,[89] you can test and exploit the "chemistry" of the urban system and population, provided of course that you understand its constituent elements and the interactions between them.
- The need to think "outside the urban box." Cities and towns are more than separate or discrete pieces of terrain or geographic points on a map. Actions and influences outside the urban area bear on the situation and local perceptions, attitudes, and behaviors inside the urban area, and vice versa, for all sides.
- Opportunity for local initiative. Seizing the local initiative mitigates the risk of being ground down in a battle of attrition.
- The complexity and diversity of urban operations. These are facts of life that must be dealt with.
- Your own, and the adversary's, will. Focus on actions and measures directed at breaking the adversary's cohesion and will and protecting one's own cohesion and will through indirect and direct, physical and

nonlethal maneuver in the physical, information, and cognitive domains—including treating bases as "hives" (opportunity for occupation, effect focused, chance of surprise, aggression to prick/sting/entangle the adversary and grasp initiative) not "nests" (passive, static, ponderous, predictable, resource sapping). Plan raids and strikes by concentrating fires and multidomain effects, not massing forces.
- Who owns the city. The army does not own the city. Just as the city has an impact on operations, so operations will have an impact on the city and its people; life in the city, nefarious or otherwise, will continue and serve as the context within which civil decisions are made, and reactions and perceptions are colored by the nature of the urban multi-actor fight[90] and the audiences involved.[91]
- An array of tactical- and operational-level centers of gravity. Among others, consider political, social, economic, physical, and psychological tactics and operations.
- The level of the fight. The level of the fight in urban environments demands small unit leadership and action, resilience, initiative, mission command, and combined arms maneuver.

Conclusion

The campaign in Iraq in the years 2003–2004 and subsequent military involvement in operations in urban environments elsewhere in the world have demonstrated vividly the observation by Evans that "urban operations are frequently different in terms of their size, demographic composition, cultural character and physical layout."[92] In a twenty-first century characterized by an increasingly urbanized world, adversaries—whether conventional or irregular—are likely to seek out urban spaces both for their strategic importance and the advantages the urban environment offers, especially to the defender. Managing cities requires a strategic view to be taken, but the major problem facing strategic and operational thinkers is one of "coherently synthesising the sheer variance and divergence of urban environments."[93]

Recent Western experience of warfighting in and around urban areas is from the perspective of an intervention where the offensive is the dominant operation of war and the attacker has the option of establishing the timing and conditions for the fight while the defender imposes himself on the ground of his choosing. In a NATO context, the commitment of military forces to a deterrent or preventative, thus fundamentally defensive, posture seems likely, and the defender can elect and prepare for the urban fight. Deducing a tactical concept for the forces of smaller armies operating in the complex urban environments of mainland Europe against a peer adversary blending hybrid and conventional capabilities, most likely below the Article 5 threshold but with the clear potential to escalate, is an imperative.

Looking forward, Leonhard argues that "urban warfare is the fight of the future" and advocates optimizing US force structure for the city rather than the countryside.[94] The challenge and opportunity for smaller armed forces is to consider the application of the principles of "beyond limits warfare" not as force drivers that tend to meet the stumbling block of mass and affordability but as stimulants for conceiving the conduct of operations in urban and other environments in a different, system-of-systems way.

Iraq in 2003–2004 and thereafter demonstrated the necessity for an unconventional, polydimensional, multiagency strategic mindset and approach to set the context within which military operations in urban environments must be framed and integrated. That mindset recognizes that "winning in the cities" is operational vital ground and sets the conditions for a range of nuanced tactical military actions suited to the local threat situation, urban environment and population dynamics, aspirations, and perceptions.

The insights gained from almost two decades of experience and observation yield four framework principles, as explained above. The product of that thinking leads toward a range of fluid, adaptive, time-unbounded tactical military actions built on the following:

- Recognition that the foundation skill that is not tradable for any army is that of being prepared to fight and win the close-quarter tactical close-combat street battle whenever and wherever it has to be fought.
- Exploiting the manner and the domains in which the contest will be fought is both inevitable and desirable.
- A persistent foundation must be maintained of strategic, operational, and tactical understanding and intelligence preparation of the environment (to tip the balance of what Leonhard describes as the ignorance versus knowledge trade-off[95]) generated by a fusion of joint and interagency intelligence, net assessment and multiagency input and analysis, capitalizing on artificial intelligence, relationships (attaches, exchanges, politics, other agencies, allies), political and social science analysis, and collaboration (training events, alliances, special operations forces, short-term training teams, host nation capability for intelligence, logistics, key leader engagement, etc.).
- A timely committal must be made of an operational fusion of host-nation, interagency, and military (conventional forces, SF, joint, proxy) elements focused not on occupation and domination or ground-holding but on applying multispectrum, "compound" pressure (virtual or physical, kinetic or indirect, multispectrum/multibattlefield) to the pulses and flows of the urban system (and thus on the population and/or the adversary and their means of control, support, resilience, influence, and sustainment) aimed at the achievement of specific objectives.
- Scaled conventional tactical supporting operations across the theater of operations must be undertaken for demonstration, deterrence or

prevention, AAAD, consolidation, stabilization, assurance, internal security, counterinsurgency, and other tasks.
- Transition should occur simultaneously or sequentially as circumstances alter at and between any or all of the strategic, operational, and tactical levels of command and authority.
- There must be an understanding of the value of, and strains on, the human element in urban environments.

From this baseline can be developed the consequent organizational, equipment, training, and conditioning preparations needed to put soldiers in the best prepared position both to exercise their unique capability to close with and defeat a determined enemy in combat at close quarters in one of the most complex of tactical environments and to transition seamlessly from high-intensity combat to other tasks. As the third decade of the century looms, military forces—which can operate enabled by information age command and control[96] to gain advantage in the information and cognitive domains and adopt some or many of the principles of "beyond limits combined war"[97] at the operational and tactical levels—will find themselves better prepared to understand, operate, and thrive in urban environments as part of an integrated "practical, resourced and exceedingly well trained interagency fighting force, rather than a theoretical interagency concept or a military force expected to do it all."[98]

Notes

1. Hills, *Future War in Cities*, 260.
2. Gowing, *Skyful of Lies*.
3. Sun Tzu, *Art of War*, 78.
4. Collins, *Military Geography*, 195.
5. DiMarco, *Concrete Hell*, 16.
6. Evans, *City Without Joy*, 2.
7. Alice Hills, "Strategies in Cities," *British Army Review* 126 (December 2000). Quoted in British Army Review Special Report—Urban Operations (Winter 2018), 71.
8. Steve Eames's conceptual formulation quoted in David Kilcullen, "New Paradigms for 21st Century Conflict," *Small Wars Journal* (blog), July 2007.
9. Kilcullen, "New Paradigms for 21st Century Conflict" (cited here in note 8).
10. Smith, *Utility of Force*, 226.
11. Ashdown, *Swords and Ploughshares*, 75.
12. "In the history of post-conflict reconstruction, there are no mistakes more regularly made or more dearly paid for than failing to dominate the security space immediately after the fighting has stopped and failing to transition this to the rule of law in the months which follow." Ashdown, *Swords and Ploughshares*, 69.
13. Bremer, *My Year in Iraq*, 358.
14. Kilcullen, *Out of the Mountains* (2013).

15. Konaev and Spencer, "Era of Urban Warfare."
16. DiMarco, *Concrete Hell*, 26.
17. Smith, "Requirements for Effective Military Operations."
18. Fry, "Strategic Defence and Security Review."
19. Formal would be DIME—diplomatic, intelligence, military, and economic—activities (including aid).
20. Informal activity includes communications, big data, artificial intelligence, advertising, influence, and so on.
21. Gates, "Balanced Strategy."
22. Hoffman, "Hybrid Warfare and Challenges," quoted in Glenn, "Thoughts on Hybrid Conflict," 1–8.
23. Ibid.
24. Glenn, "Thoughts on Hybrid Conflict," 7.
25. "You (military professionals) must know something about tactics, strategy, logistics, but also economics, politics, diplomacy and history. You must know everything you can about military power, and you must also understand the limits of military power. You must understand that few of the important problems of our time have, in the final analysis, been solved by military power alone." President John F. Kennedy, Remarks at Annapolis to the Graduating Class of the United States Naval Academy, June 7, 1961.
26. UK Defence Doctrine, Joint Doctrine Publication 0-01 (6th edition).
27. Qiao Lan and Wang Xiansui, *Unrestricted Warfare*, 217.
28. Ibid., footnotes, last chapter.
29. Konaev and Bendett, "Russian AI Enabled Combat," 3.
30. *Maskirovka*—Russian military deception; a doctrine devised from the start of the twentieth century that encompasses a range of measures from camouflage to denial, disinformation, propaganda, and so on.
31. "Information work with the population in Aleppo liberated neighbourhoods without a fight." "False news spreads farther, faster, deeper and more broadly." Konaev and Bendett, "Russian AI Enabled Combat."
32. UK Defence Doctrine, Joint Doctrine Publication 0-01, 12.
33. Atkinson and Moffat, *Agile Organisation*, 35.
34. General Mark Milley CAS (US Army) quoted in *Defence News*, October 2016.
35. DiMarco, *Concrete Hell*.
36. Evans, *City Without Joy*, 28–31.
37. Van Riper, *Concept for Military Operations*.
38. Bricet des Vallons, *Strategic Globalization and Urban Warfare*, 58. Quoted in Evans, *City Without Joy*, 21.
39. Leonhard, *Principles of War*; Leonhard, "Urban Warfare in the Information Age."
40. Sun Tzu, *Art of War*, 77.
41. Hills, *Future War in Cities*, 260.
42. Evans, *City Without Joy*, 30.
43. Von Clausewitz, *On War*.
44. Alex Neads, *Of Cyborgs and Separatists: Making Symbolic Meaning During the Fighting for Donetsk Airport*. Chapter 4.
45. Graham, "Military Coalitions in War."
46. Senior US commander addressing the UK James Calvert Spence College, 2009.
47. Ibid.

48. Graham, “Military Coalitions in War,” 319–331.

49. Thompson, *Defeating Communist Insurgency*, 50–55.

50. Ashdown, *Swords and Ploughshares*, 164–165.

51. Intelligence agencies and departments, including treasury, agriculture, health, justice, foreign service, and overseas development. Jones, *Re-examining Tomorrow's Battlefield.*

52. Thompson, *Defeating Communist Insurgency*, 57.

53. Witness the difficulties experienced by the Taliban in August 2021 and thereafter—after their military success in defeating the Afghan National Army and taking power—transitioning from military campaign to government.

54. Rayburn and Sobchak, *US Army in the Iraq War.*

55. Cooling, *Shaping the Battlespace.*

56. General P. J. Schoomaker CAS (US Army) in a letter regarding the operational level considerations for urban conflict, October 7, 1999. Cited in Cooling, *Shaping the Battlespace.*

57. Collins, *Military Geography*, 195.

58. For a discussion and expansion of the expression “compound warfare,” see the discussion earlier in this chapter.

59. “From the Somalis, through the Chechens to the militias, the Former Regime Elements and *mujahideen* in Iraq and ISIS in Syria most forces seeking to fight in towns and cities have been irregulars or militias.” Evans, *City Without Joy*, 13–17. *The Economist*, April 18, 2020, describes the Rakane shifting the focus of their insurgency in Myanmar from the countryside to the city.

60. “Controlling a major population centre for any significant period of time can be strategically important for a weak adversary.” DiMarco, *Concrete Hell.*

61. Rayburn and Sobchak, *US Army in the Iraq War*, 111.

62. PMESII-PT is an acronym developed in the US military that stands for political, military, economic, social, information, infrastructure, physical environment, and time.

63. Nagl, *Counterinsurgency Lessons.*

64. General Charles Krulak quoted in US Marine Corps Publication *MCWP 3-10 MAGTF Ground Operations*, November 8, 2017, amended April 4, 2018 (emphasis added).

65. Rayburn and Sobchak, *US Army in the Iraq War*, 196.

66. Army Code 71844, *Stability Operations in Iraq—An Analysis from a Land Perspective*, p. 10.

67. Carl von Clausewitz quoted in Handel, *Masters of War.*

68. UK Joint Doctrine Publication 04, *Understanding and Decision-making*, p. 2–1.

69. Army Code 71844, p. 4.

70. Ibid.

71. Defeat of the Iraqi Armed Forces and overthrow of the government was, in the author's opinion, the “shaping” rather than “decisive phase” of the campaign and should have been understood and treated as such.

72. Steadman, *Understanding the Urban Battlespace.*

73. Statement of Dennis C. Blair before the Senate Select Committee on Intelligence, US Senate, January 22, 2009.

74. ASCOPE covers areas, structures, capabilities, organization, people, and events.

75. Ashdown, *Swords and Ploughshares*, 67–75.

76. Evans, *City Without Joy*, 10.

77. Richmond, *Naval Warfare*, 11.
78. Konaev and Spencer, "Era of Urban Warfare Is Already Here," 5.
79. Qiao Lan and Wang Xiansui, *Unrestricted Warfare*, 9–34.
80. Ibid., 34.
81. A2/AD is anti-access/area denial.
82. UK Joint Doctrine Publication 0-20, p. 52, para 4.15.
83. General Sir Rupert Smith. Author's notes from a lecture on I (UK) Division. Insights from Operation Granby.
84. "Do not fight battles for the sake of winning; fight battles of which the result is fruitful." Richmond, *Naval Warfare*, 13.
85. Alderson, "Influence," 39.
86. Evans, *City Without Joy*, vii.
87. *Efficient*—concentration of power; multidimensional blending of mass, fires, and indirect means including special operations forces, local forces, info ops, etc.; economy of effort; cooperation; surprise; offensive action; timeliness and synchronization; simultaneity. *Effective*—outcomes achieved; exploitation; speed of transition; momentum and tempo relative to the adversary; distraction, disruption, dislocation of enemy action. *Economy*—economy of effort; logistics, no wastage; simplicity; minimize impact of enemy distraction and deception.
88. The Future Operating Environment and Threat for Expeditionary Operations in the Land Environment, UK Army Code 71999, April 2013, Annex A to Part B, para AA01.
89. "Basra the place quickly ate up scarce resources and limited physical and conceptual freedom of maneuver [*sic*] and that was before the force faced *Basra the people*." Baker, "Inescapable Truths of Urban Warfare," 3.
90. Baker, "Inescapable Truths of Urban Warfare," 1–10.
91. Ibid.
92. Evans, "Case Against Megacities"; Evans, *City Without Joy*, 25.
93. Evans, "Case Against Megacities."
94. Leonhard, "Urban Warfare in the Information Age," 40–44.
95. Leonhard, "Urban Warfare in the Information Age."
96. Atkinson and Moffat, *Agile Organisation*, 159.
97. Qiao Lan and Wang Xiansui, *Unrestricted Warfare*, 217. Discussed on page 7.
98. Jones, *Re-examining Tomorrow's Battlefield*.

Bibliography

Abdulrazaq, Tallha, and Gareth Stansfield. "The Enemy Within: ISIS and the Conquest of Mosul." *Middle East Journal* 70, no. 4 (Autumn 2016): 525–542.

Abujidi, Nurhan. *Urbicide in Palestine: Spaces of Oppression and Resilience*. London: Routledge, 2014.

Adair, Shafto. "The Defence of London." *Journal of the Royal United Service Institution* 4, no. 15 (1860).

Adams, Bianka J. *From Crusade to Hazard: The Denazification of Bremen Germany*. Plymouth, UK: Scarecrow Press, 2009.

Agamben, Giorgio. *Homo Sacer. Sovereign Power and Bare Life*, translated by Daniel Heller-Roazen. Stanford, CA: Stanford University Press, 1998.

Ahad, Ghaith Abdul, and Martin Chulov. "Two Suicide Bombers Kill at Least 32 Shoppers in Baghdad." *Guardian*, January 21, 2021. https://www.theguardian.com/world/2021/jan/21/baghdad-suicide-bombers-market-three-years.

Alderson, Alex. "Influence, the Indirect Approach and Manoeuvre." *RUSI Journal* 157, no. 1 (2012): 39.

Al Jazeera. "Afghanistan: Dozens Killed in Suicide Bombing at Kunduz Mosque." October 8, 2021.

———. "Growing Slums and Misery in Iraq's Sadr City." April 13, 2019. https://www.youtube.com/watch?v=47begFzEZvw.

Al-Qarawee, Harith. *Imagining the Nation: Nationalism, Sectarianism and Socio-Political Conflict in Iraq*. Lancashire, UK: Rossendale Books, 2012.

Amnesty International. "'Battle of Marawi' Leaves Trail of Death and Destruction." November 17, 2017. https://www.amnesty.org/en/latest/news/2017/11/philippines-battle-of-marawi-leaves-trail-of-death-and-destruction/.

Anderson, James, and Liam O'Dowd. "Borders, Border Regions and Territoriality: Contradictory Meanings, Changing Significance." *Regional Studies* 33, no. 7 (1999): 593–604.

Antouly, Julien. "Urban Warfare: A Challenge for Humanitarian Law and Action." *Alternatives Humanitaires* (March 2019): 108–117. https://alternatives-humanitaires.org/wp-content/uploads/2019/03/AH_N10_3_Focus_8_Antouly_VEN.pdf.

Ardolino, Bill. "Panjwai District Chief of Police: 'Pakistan Has Been Trying to Bring Afghanistan Under Its Control.'" *Long War Journal* (Foundation for Defense of Democracies), May 11, 2013. https://www.longwarjournal.org/archives/2013/05/pakistan_has_been_tr.php.

Arnold, Thomas D., and Nicolas Fiore. "Five Operational Lessons from the Battle for Mosul." *Military Review*, January–February 2019, 56–71.

Arquilla, John, and David Ronfeldt. "The Advent of Network (Revisited)." In *The Advent of Netwar (Revisited)*, edited by John Arquilla and David Ronfeldt, 1–25. Santa Monica, CA: RAND, 2001.

Ashdown, Paddy. *Swords and Ploughshares—Building Peace in the 21st Century*. London: Weidenfeld & Nicolson, 2007.

Ashworth, Gregory John. *The City as Battlefield: The Liberation of Groningen, April 1945*. Groningen, Netherlands: Faculty of Spatial Sciences, University of Groningen, 1995.

———. *War and the City*. London: Routledge, 1991.

Atkinson, Simon Reay, and James Moffat. *The Agile Organisation*. Washington, DC: DoD Command and Control Research Programme, 2005.

Auditor, Katrina. "Marawi Rehabilitation Delays Could Cast a Dark Shadow over Duterte's Peace Legacy." *Diplomat*, December 7, 2020. https://thediplomat.com/2020/12/marawi-rehabilitation-delays-could-cast-a-dark-shadow-over-dutertes-peace-legacy/.

Bailey, J. B. A. *Field Artillery and Firepower*. Annapolis, MD: Naval Institute Press, 2004.

Bailey, Michael, and Daniel Hopkins. "Unresponsive and Unpersuaded: The Unintended Consequences of a Voter Persuasion Effort." *Political Behavior* 38 (2016): 713–746. https://doi.org/10.1007/s11109-016-9338-8.

Bainbridge, W. F. "Gold Medal Prize Essay 1913–1914." *Journal of the United Service Institution of India* 44, no. 198 (January 1915).

Bairoch, Paul, and Gary Goertz. "Factors of Urbanization in the Nineteenth Century Developed Countries: A Descriptive and Econometric Analysis." *Urban Studies* 23 (1986): 285–305.

Baker, Ben. "The Inescapable Truths of Urban Warfare: Five Lessons from Basra." Modern War Institute, June 6, 2019. https://mwi.usma.edu/inescapable-truths-urban-warfare-five-lessons-basra-2007/.

Baker, Wayne E., and Robert R. Faulkner. "The Social Organization of Conspiracy: Illegal Networks in the Heavy Electrical Equipment Industry." *American Sociological Review* 58, no. 6 (1993): 837–860.

Bakshi, Anita. *Topographies of Memories*. London: Palgrave Macmillan, 2017.

Banlaoi, Rommel C. *The Marawi Siege and Its Aftermath: The Continuing Terrorist Threat*. Newcastle upon Tyne, UK: Cambridge Scholars Publishing, 2020.

Barabasi, Albert-Lászloe. *Linked: How Everything Is Connected to Everything Else and What It Means*. New York: Plume, 2003.

Baron, Kevin, and Marcus Weisgerber. "New Tactics, Quicker Decisions Are Helping to 'Annihilate' ISIS, Pentagon Says." *Defense One*, May 19, 2017.

Bartlett, Jamie. "Psychographics. What Is It? Does It Work? And Cambridge Analytica Use It?" *Reaction,* March 27, 2018. https://reaction.life/psychographics-work-cambridge-analytica-use/.

———. *The People vs Tech: How the Internet Is Killing Democracy (and How to Save It*. London: Penguin Random House, 2018.

Batrinca, Bogdan, and Philip C. Treleaven. "Social Media Analytics: A Survey of Techniques, Tools and Platforms." *AI & Society* 30 (2015): 89–116. https://doi.org/10.1007/s00146-014-0549-4.

BBC News. "Donetsk Airport: Ukraine's Coveted Prize." January 22, 2015. https://www.bbc.com/news/world-europe-30931105.

———. "Heavy Fighting for Donetsk Airport." October 3, 2014. https://www.bbc.co.uk/news/world-europe-29481979.

———. "Military Plane Shot Down in Luhansk." June 14, 2014. https://www.bbc.co.uk/news/world-europe-27845313.

———. "New Battle Rages at Donetsk Airport." December 1, 2014. https://www.bbc.co.uk/news/world-europe-30275259.

———. "Turchynov Announces Anti-Terror Operation." April 13, 2014. https://www.bbc.com/news/av/world-europe-27013169.

———. "UK IRA Killer Laughs at Sentence." March 19, 1999. http://news.bbc.co.uk/1/hi/uk/299635.stm.

———. "Ukraine Conflict: Rebel Leader Givi Dies in Rocket Attack." February 8, 2017. https://www.bbc.co.uk/news/world-europe-38905110.

———. "Ukraine Profile—Timeline." September 10, 2019. https://www.bbc.co.uk/news/world-europe-18010123.

Beall, Jonathan A. "The United States Army and Urban Combat in the Nineteenth Century." *War in History* 16, no. 2 (2009).

Bean, Timothy. "'Come Ye Back to Mandalay': 14th Army's Battle for Mandalay, March 1945." In Fremont-Barnes, *A History of Modern Urban Operations*.

Beckerman, Carly. *Unexpected State: British Politics and the Creation of Israel*. Bloomington: Indiana University Press, 2020.

Beckhusen, Robert. "Why Ukrainian Troops Are Calling the Donetsk Airport Siege 'Stalingrad.'" *Medium,* January 19, 2015. https://medium.com/war-is-boring/why-ukrainian-troops-are-calling-the-donetsk-airport-siege-stalingrad-cef7c4b6a2.

Ben-Ari, Eyal, Zev Lehrer, Uzi Ben-Shalom, and Ariel Vainer. *Rethinking the Sociology of Combat: Israel's Combat Units in the Al-Aqsa Intifada*. Albany: State University of New York Press, 2010.

Bennett, Huw. "From Direct Rule to Motorman: Adjusting British Military Strategy for Northern Ireland in 1972." *Studies in Conflict & Terrorism* 33, no. 6 (2010): 511–532.

Bennett, Huw, and David French (eds.). *The Kenya Papers of General Sir George Erskine 1953–1955*. Stroud, UK: History Press for the Army Records Society, 2013.

Bercuson, David Jay. *Battalion of Heroes: The Calgary Highlanders in World War II*. Toronto: Penguin Books, 1994.

Berdal, Mats. "The New Wars' Thesis Revisited." In Strachan and Scheipers, *The Changing Character of War*, 109–133.

Berman, Eli, and David Lake (eds.). *Proxy Wars: Suppressing Violence Through Local Agents*. Ithaca: Cornell University Press, 2019.

Berry, Mary Elizabeth. *The Culture of Civil War in Kyoto*. Berkeley: University of California Press, 1994.

Betz, David, and Hugo Stanford-Tuck. "The City Is Neutral: On Urban Warfare in the 21st Century." Texas National Security Review 2, no. 4 (2019). https://tnsr.org/2019/10/the-city-is-neutral-on-urban-warfare-in-the-21st-century/.

Biersack, John, and Shannon O'Lear. "The Geopolitics of Russia's Annexation of Crimea: Narratives, Identity, Silences, and Energy." *Eurasian Geography and Economics* 55, no. 3 (2014): 247–269.

Black, Ian. *Enemies and Neighbours: Arabs and Jews in Palestine and Israel, 1917–2017*. London: Penguin Random House, 2018.

Black, Jeremy. *Fortifications and Siegecraft: Defence and Attack Through the Ages*. Lanham, MD: Rowman and Littlefield, 2018.

Blanqui, Auguste. "A Blueprint for Insurrection." In *The Guerrilla Reader: A Historical Anthology*, edited by Walter Laqueur. London: Wildwood House, 1978.

Bloom, Mia. "Constructing Expertise: Terrorist Recruitment and 'Talent Spotting' in the PIRA, Al Qaeda, and ISIS." *Studies in Conflict & Terrorism* 40, no. 7 (2017): 603–623.

Bogdanović, B. "The City and Death." In *Storm 6: Out of Yugoslavia*, edited by J. Labon, translated by M. H. Heim, 36–74. Manchester: Carcanet Press, 1994.

Borg, Stefan. "Below the Radar. Examining a Small State's Usage of Tactical Unmanned Aerial Vehicles." *Defence Studies* 20, no. 3 (2020): 185–201.

Bottomley, N. H. "Letter to Air Chief Marshal Sir Arthur Harris, 3 September 1943." In *The Strategic Air Offensive Against Germany 1939–1945, Volume IV, Annexes and Appendices*, by Charles Webster and Noble Frankland, 160. London: Her Majesty's Stationery Office, 1961.

Boucher, Francis H. "Artillery Attacks in Stone Villages." *Field Artillery Journal* 34, no. 10 (October 1944): 708–709.

Bowen, Andrew. "Coercive Diplomacy and the Donbas: Explaining Russian Strategy in Eastern Ukraine." Journal of Strategic Studies 42, no. 3-4 (2019): 312–334.

Bowers, Christopher O. "Future Megacity Operations: Lessons from Sadr City." *Military Review*, May–June 2015, 8–16. https://www.armyupress.army.mil/Portals/7/military-review/Archives/English/MilitaryReview_20150630_art006.pdf.

Bradbury, Jim. *The Medieval Siege*. Woodbridge, UK: Boydell Press, 1992.

Brathwaite, Kirstin J. H., and Margarita Konaev. "War in the City. Urban Ethnic Geography and Combat Effectiveness." *Journal of Strategic Studies* (2019): 33–68. https://doi.org/10.1080/01402390.2019.1672159.

Bremer, L. Paul, III. *My Year in Iraq*. New York: Simon & Schuster, 2006.

Bricet des Vallons, Georges-Henri. *Strategic Globalization and Urban Warfare*. Current Strategic Thinking, 2007.

Brimelow, Ben. "ISIS Wants to Be as Dangerous as the Taliban—But It's Not Even Close." *Business Insider*, February 11, 2018.

British Army Review. *Special Report—Urban Operations—Volume 1*. Warminster: British Army, 2018.

———. *Special Report—Urban Operations—Volume 2*. Warminster: British Army, 2019.

Brown, Kingsley, and Brereton Greenhous. *Semper Paratus: The History of the Royal Hamilton Light Infantry (Wentworth Regiment), 1862–1977*. Hamilton, ON: RHLI Historical Association, 1977.

Brown, Seyom. "Purposes and Pitfalls of War by Proxy: A Systemic Analysis." *Small Wars & Insurgencies* 27, no. 2 (2016): 243–257.

Brunstetter, Daniel. "Drones: The Future of Warfare." E-International Relations, April 10, 2012. https://www.e-ir.info/2012/04/10/drones-the-future-of-warfare/.

Bugeaud, Maréchal. *La Guerre des rues et des maisons*. Paris: Jean-Paul Éditeur, 1997.

Burke, Edward. *An Army of Tribes: British Army Cohesion, Deviancy and Murder in Northern Ireland*. Oxford: Oxford University Press, 2018.

———. "Counter-Insurgency Against 'Kith and Kin'? The British Army in Northern Ireland, 1970–76." *The Journal of Imperial and Commonwealth History* 43, no. 4 (2015): 658–677.

Bury, Patrick, and Anthony King. "A Profession of Love: Cohesion in a British Platoon in Afghanistan." In *Frontline: Combat and Cohesion in the 21st Century*, edited by Anthony King, 200–215. Oxford: Oxford University Press, 2015.

Byers, Daniel. "Operation 'Canada': 5th Canadian Armoured Division's Attack on Delfzijl, 23 April to 2 May 1945." *Canadian Military History* 7, no. 3 (1998): 35–46.

Callwell, Charles, and John Headlam. *The History of the Royal Artillery: From the Indian Mutiny to the Great War, Volume I (1860–1890)*. Woolwich: The Royal Artillery Institution, 1931.

Carden-Coyne, Ana. *The Politics of Wounds: Military Patients and Medical Power in the First World War*. Oxford: Oxford University Press, 2014.

Carl, Peter. "*Convivimus Ergo Sumus*." In *Phenomenologies of the City. Studies in the History and Philosophy of Architecture*, edited by Henriette Steiner and Maximilian Sternberg, 11–31. Farnham, UK: Ashgate, 2015.

Carroll, Oliver. "A Ukrainian Father's Desperate Search for His Son, Trapped in Donetsk Airport." *Newsweek*, February 10, 2015. https://www.newsweek.com/2015/02/20/ukrainian-fathers-desperate-search-his-son-305859.html.

———. "Inside the Bloody Battle for Ukraine's Donetsk Airport." Newsweek, February 3, 2015. https://www.newsweek.com/2015/02/13/inside-bloody-battle-ukraines-donetsk-airport-304115.html.

———. "Residents Flee Savagery." *Independent*, January 15, 2015. https://www.independent.co.uk/news/world/europe/ukraine-crisis-residents-flee-savagery-of-the-battle-for-donetsk-s-airport-9981531.html.

———. "Ukraine Crisis: Residents Flee Savagery of the Battle for Donetsk's Airport." *Independent*, January 15, 2015. http://www.independent.co.uk/news/world/europe/ukraine-crisis-residents-flee-savagery-of-the-battle-for-donetsks-airport-9981531.html.

Carter, Keith L. "Coming Soon to a Theater (of War) Near You: Drones of all Shapes and Sizes." In *Drones and Global Order*, edited by Paul Lushenko, Srinjoy Bose, and William Maley. London: Routledge, 2022.

Cavanagh, Matt. "Ministerial Decision-Making in the Run-up to the Helmand Deployment." *RUSI Journal* 157, no. 2 (2012): 48–54.

Chamayou, Gregoire. *A Theory of the Drone*. New York: New Press, 2014.

Chamberlain, Peter, and Chris Ellis. "Churchill and Sherman Specials." In *British and Commonwealth AFVs 1940–46*, edited by Duncan Crow. Windsor, UK: Profile Publications, 1971.

Chapman, Frederick Spencer. *The Jungle Is Neutral*. London: Chatto and Windus, 1952.

Charap, Samuel, and Keith Darden. "Russia and Ukraine." *Survival* 56, no. 2 (2014): 7–14.
Chavez, Kerry, and Ori Swed. "The Proliferation of Drones to Violent Nonstate Actors." *Defence Studies* 21, no. 1 (2021): 1–24.
Chief of the General Staff. *An Analysis of Military Operations in Northern Ireland. Army Code 71842*. UK Ministry of Defence, 2006.
Cialdini, Robert B. *Influence: The Psychology of Persuasion*. New York: Collins, 2007.
Cirone, Alexandra, and Arthur Spirling. "Turning History into Data: Data Collection, Measurement, and Inference in HPE." *Journal of Historical Political Economy* 1, no. 1 (2021): 127–154.
Clout, Hugh. "Place Annihilation and Urban Reconstruction: The Experience of Four Towns in Brittany, 1940 to 1960." *Geografiska Annaler* 82, no. 3 (2000): 165–180.
Cluséret, P. G. "Street Fighting." *Journal of the Royal United Services Institute for Defence Studies* 117, no. 665 (1972).
Collier, David. "Symposium: Case Selection, Case Studies, and Causal Inference." *Qualitative & Multi-Method Research* 6, no. 2 (2008): 2–4.
Collins, John M. *Military Geography for Professionals and the Public*. Washington, DC: National Defense University Press, 1998.
Collins, Tim. *Rules of Engagement: A Life in Conflict*. London: Headline, 2005.
Colson, Bruno. *Napoleon on War*, translated by Gregory Elliot. Oxford: Oxford University Press, 2015.
Conflict in Cities (CinC). *Coping with Conflict*. Briefing Paper 9. Cambridge, UK: CinC, November 2012. https://www.urbanconflicts.arct.cam.ac.uk/downloads/briefing-paper-9.
Connelly, Tony. "UK Proposes Customs Posts on Both Sides of Border in Bid to Replace Backstop." RTÉ News, October 1, 2019. https://www.rte.ie/news/brexit/2019/0930/1079253-border/.
Cook, Martin. "Drone Warfare and Military Ethics." In Cortright, Fairhurst, and Wall, *Drones and the Future of Armed Conflict*.
Cook, Tim. *Fight to the Finish: Canadians in the Second World War, 1944–1945*. Toronto: Allen Lane, 2015.
Cooling, Norman L. *Shaping the Battlespace to Win the Street Fight*. USMC Command and Staff College, 2000.
Copp, Terry. *The Brigade: The Fifth Canadian Infantry Brigade, 1939–1945*. Mechanicsburg, PA: Stackpole Books, 1992.
———. *Cinderella Army: The Canadians in Northwest Europe, 1944–1945*. Toronto: University of Toronto Press, 2006.
Cornish, Paul. Written evidence submitted to House of Commons Public Administration Select Committee Report, *Who Does UK National Strategy?* HC 435. London: The Stationery Office, October 18, 2010.
Cortright, David, Rachel Fairhurst, and Kristen Wall (eds.). *Drones and the Future of Armed Conflict: Ethical, Legal, and Strategic Implications*. Chicago: University of Chicago Press, 2015.
Coward, Martin. *Urbicide. The Politics of Urban Destruction*. London: Routledge, 2009.
Dagher, Sam. "Baghdad Is Safer, but It's a Life Behind Walls." *Christian Science Monitor*, December 19, 2007.
Daher, Joseph. "The Paradox of Syria's Reconstruction." Carnegie Middle East Center, September 4, 2019. https://carnegie-mec.org/2019/09/04/paradox-of-syria-s-reconstruction-pub-79773.
Daly, Gavin. "'The Sacking of a Town Is an Abomination': Siege, Sack and Violence to Civilians in British Officers' Writings on the Peninsular War—The Case of Badajoz." *Historical Research* 92, no. 255 (February 2019): 160–182.
Dastrup, Boyd L. *The Field Artillery: History and Sourcebook*. Westport, CT: Greenwood Press, 1994.
Davies, Nicholas. *Ten-Thirty-Three: The Inside Story of Britain's Secret Killing Machine in Northern Ireland*. New York: Random House, 2011.
Davis, Paul K. *Besieged: 100 Great Sieges from Jericho to Sarajevo*. Oxford: Oxford University Press, 2003.
Davis, Paul K., and Brian Michael Jenkins. "A System Approach to Deterring and Influencing Terrorists." *Conflict Management and Peace Science* 21, no. 1 (2004): 3–15.

Deeb, Lara, and Mona Harb. *Leisurely Islam: Negotiating Geography and Morality in Shi'ite South Beirut*. Princeton, NJ: Princeton University Press, 2013.

de Malortie, General. *A Treatise on the Attack and Defence of Fortified Places*. London: no pub., 1824.

Demmers, Jolle, and Lauren Gould. "The Remote Warfare Paradox: Democracies, Risk Aversion and Military Engagement." E-International Relations, June 20, 2020. https://www.e-ir.info/2020/06/20/the-remote-warfare-paradox-democracies-risk-aversion-and-military-engagement.

Dempsey, Martin E. "Introduction." In *Beauty Is the Beginning of Terror*, compiled by West Point Cadets. Newberg, NY: Thornwillow Press, 2017.

Department of the Army. *ADP 3-0 Operations*. Washington, DC, July 2019. https://armypubs.army.mil/epubs/DR_pubs/DR_a/ARN18010-ADP_3-0-000-WEB-2.pdf.

Dewar, Michael. *The British Army in Northern Ireland*. London: Arms & Armour Press, 1985.

Dill, Janina. *Legitimate Targets? Social Construction, International Law, and US Bombing*. Cambridge: Cambridge University Press, 2015.

Dill, Janina, and Livia Schubiger. "Attitudes Towards the Use of Force: Instrumental Imperatives, Moral Principles, and International Law." *American Journal of Political Science* 65, no. 3 (2021): 612–633.

DiMarco, Louis A. *Concrete Hell: Urban Warfare from Stalingrad to Iraq*. London: Osprey, 2012.

DiNapoli, Emma K. "Urbicide and Property Under Assad: Examining Reconstruction and Neoliberal Authoritarianism in a 'Postwar' Syria." *Columbia Human Rights Law Review* 51, no. 1 (2019): 253–312.

Divinagracia, Major Erwin C. "Urban Terrorism: the Siege of Marawi City. Master's thesis, US Army Command and General Staff College, Fort Leavenworth, Kansas, 2018.

Dixon, Paul. "'Hearts and Minds'? British Counter-insurgency Strategy in Northern Ireland." *Journal of Strategic Studies* 32, no. 3 (2009): 445–474.

Dodge, Toby. "Can Iraq Be Saved?" *Survival* 56, no. 5 (2014): 7–20.

Doubler, Michael D. *Closing with the Enemy: How GIs Fought the War in Europe 1944–1945*. Lawrence: University of Kansas Press, 1994.

Duffield, Mark. *Global Governance and the New Wars: The Merging of Development and Security*. London: Zed Books, 2014.

Duffy, Christopher. *Fire and Stone: The Science of Fortress Warfare 1660–1860*. Newton Abbot, UK: David and Charles, 1975.

———. *The Fortress in the Age of Vauban and Frederick the Great 1660-1789*. Abingdon: Routledge, 2016.

———. *Siege Warfare: The Fortress in the Early Modern World 1494-1660*. London: Routledge, 1996.

Duncan, Francis. *History of the Royal Regiment of Artillery Volume II*. London: John Murray, 1873.

Dykstra, Ralph. "The Occupation of Groningen, Netherland, September 1944–April 1945 and the Liberation of the City of Groningen by the 2nd Canadian Infantry Division, April 13–16, 1945." Master's thesis, Wilfrid Laurier University, 2002.

Ebel, Édouard. "Du maintien de l'ordre à la guerre des rues: la gestion des foules entre 1830 et 1871." In *La ville en ébullition: Sociétés urbaines à l'épreuve*, edited by Pierre Bergel and Vincent Milliot. Rennes: Presses universitaires de Rennes, 2014. https://books.openedition.org/pur/50273.

Echevarria, Antullio. "War and Politics: The Revolution in Military Affairs and the Continued Relevance of Clausewitz." *Joint Forces Quarterly* (Winter 1995–1996). http://www.clausewitz.com/readings/Echevarria/ECHJFQ.htm.

The Economist. "Two Arab Countries Fall Apart." June 12, 2014. https://www.economist.com/middle-east-and-africa/2014/06/12/two-arab-countries-fall-apart.

Elish, M. C. "Remote Split: A History of US Drone Operations and the Distributed Labor of War." *Science, Technology, & Human Values* 42, no. 6 (2017): 1100–1131.

Elsdale, H. "The Defence of London and of England Part II: The Successive Stages of Defence." *Journal of the Royal United Service Institution* 30, no. 135 (1886): 637–670.

———. "The Defence of London and of England Part I: The Security of London Against a Coup de Main." *Journal of the Royal United Service Institution* 30, no. 135 (1886): 601–636.

Enders, David. "Behind the Wall. Inside Baghdad's Sadr City." *The Virginia Quarterly Review* 85, no. 3 (2009).

"[Eng subs] Combat Footage: Fights for Donetsk Airport 30/09/14; One Day with Givi." Separatist propaganda footage, October 3, 2014. https://www.youtube.com/watch?v=xP_ozv0qgXU#t=1430.

Erickson, Jennifer L. "Saint or Sinner? Human Rights and U.S. Support for the Arms Trade Treaty." *Political Science Quarterly* 130, no. 3 (2015): 449–474.

Esdale, Charles. *The Peninsular War*. London: Allen Lane, 2002.

Evans, Michael. "The Case Against Megacities." *Parameters* 45, no. 1 (2015): 33–43.

———. *City Without Joy: Urban Military Operations into the 21st Century*. Australian Defence College Occasional Series No. 2. Canberra: Department of Defence, 2007.

———. "Future War in Cities: Urbanization's Challenge to Strategic Studies in the 21st Century." *International Review of the Red Cross* 98, no. 901 (2016): 37–51.

———. "Lethal Genes: The Urban Military Imperative and Western Strategy in the Early Twenty-First Century." *Journal of Strategic Studies* 32, no. 4 (2009): 515–552.

Ewing, David. *Vietnam Studies: Field Artillery 1954–1973*. Washington, DC: US Government Printing Office, 1975.

Fealy, Greg. "The Battle for Marawi and ISIS in Southeast Asia." *Strategist* (blog). Australian Strategic Policy Institute, August 23, 2017. https://www.aspistrategist.org.au/battle-marawi-isis-southeast-asia/.

Fearon, James. "Rationalist Explanations for War." *International Organization* 49, no. 3 (Summer 1995): 379–414.

Felix, Kevin M., and Frederick D. Wong. "The Case for Megacities." *Parameters* 45, no. 1 (2015): 19–32.

Ferguson, Jonathan, and N. R. Jenzen-Jones. "Raising Red Flags: An Examination of Arms & Munitions in the Ongoing Conflict in Ukraine." Armament Research Service Report 3, November 2014. http://armamentresearch.com/Uploads/Research%20Report%20No.%203%20-%20Raising%20Red%20Flags.pdf.

Field Marshall Lord Roberts of Kandahar. *Forty-One Years in India: From Subaltern to Commander-in-Chief*, Volume I. London: Richard Bentley and Son, 1897.

Finnegan, Patrick. "Professionalization of a Nonstate Actor: A Case Study of the Provisional IRA." *Armed Forces & Society* 45, no. 2 (2019): 349–367.

Flynn, Michael T., Rich Juergens, and Thomas L. Cantrell. "Employing ISR: SOF Best Practices." *Joint Forces Quarterly* 50 (July 2008): 56–61.

Forde, Susan. "The Bridge on the Neretva: Stari Most as a Stage of Memory in Post-Conflict Mostar, Bosnia-Herzegovina." *Cooperation and Conflict* 51, no. 4 (December 2016): 467–483.

Forest, James J. F. (ed). *Influence Warfare: How Terrorists and Governments Fight to Shape Perceptions in a War of Ideas*. Santa Barbara: ABC Clio, 2009.

Foucault, Michel. "Space Knowledge and Power." Interview by Paul Rabinow in *Skyline* (March 1982). https://foucault.info/documents/foucault.spaceKnowledgePower/.

Fox, Amos. "'Cyborgs at Little Stalingrad': A Brief History of the Battles of the Donetsk Airport, 26 May 2014 to 21 January 2015." Association of the United States Army Land Warfare Paper 125, May 2019. https://www.ausa.org/sites/default/files/publications/LWP-125-Cyborgs-at-Little-Stalingrad-A-Brief-History-of-the-Battle-of-the-Donetsk-Airport.pdf.

Fox, Amos, and Andrew Rossow. "Making Sense of Russian Hybrid Warfare: A Brief Assessment of the Russo–Ukrainian War." Association of the United States Army Land Warfare Paper 112, 2017.

Franco, Joseph. "Assessing the Feasibility of 'Wilayah Mindanao.'" *Perspectives on Terrorism* 11, no. 4 (2017): 29–38.

———. *The Battle for Marawi: Urban Warfare Lessons for the AFP*. Working paper. Quezon City, Philippines: Security Reform Initiative, October 4, 2017. https://www.researchgate.net/profile/Joseph-Franco-3/publication/337076300_The_Battle_for_Marawi_Urban_Warfare_Lessons_for_the_AFP/links/5dc3cd02a6fdcc2d2ff7e747/The-Battle-for-Marawi-Urban-Warfare-Lessons-for-the-AFP.pdf.

Frankel, Matt. "The ABCs of HVT: Key Lessons from High Value Targeting Campaigns Against Insurgents and Terrorists." *Studies in Conflict and Terrorism* 34, no. 1 (2011): 17–30.

Freedman, Lawrence. "Ukraine and the Art of Crisis Management." *Survival* 56, no. 3 (2014): 7–42.

———. "Ukraine and the Art of Limited War." *Survival* 56, no. 6 (2014): 7–38.

Fregonese, Sara. "The Urbicide of Beirut? Geopolitics and the Built Environment in the Lebanese Civil War (1975–1976)." *Political Geography* 28, no. 5 (2009): 309–318.

———. *War and the City. Urban Geopolitics in Lebanon*. London: IB Tauris, 2020.

Fremont-Barnes, Gregory (ed.). *A History of Modern Urban Operations*. Cham, Switzerland: Palgrave Macmillan, 2020.

Fry, Robert. "A Strategic Defence and Security Review That Really Matters." *The Article*, February 6, 2020. https://www.thearticle.com/a-strategic-defence-and-security-review-that-really-matters.

Fuller, J. F. C. *The Conduct of War 1789–1961: A Study of the Impact of the French, Industrial and Russian Revolutions on War and Its Conduct*. London: Eyre and Spottiswoode, 1961.

Fussell, Paul. *The Great War and Modern Memory*. Oxford: Oxford University Press, 1975.

Gabriel, Ralph H. "American Experience with Military Government." *American Political Science Review* 37, no. 3 (1943): 417–438.

Galeotti, Mark. *Armies of Russia's War in Ukraine*. Oxford: Osprey, 2019.

Gall, Carlotta. "Afghan Villagers Take on Taliban in Their Heartland." *New York Times*, March 20, 2013.

Galliott, Jai. *Military Robots: Mapping the Moral Landscape*. London: Routledge, 2015.

Garner, Ian. "The Myth of Stalingrad in Soviet Literature, 1942-1963." PhD dissertation, University of Toronto, 2018.

Gates, Robert M. "A Balanced Strategy: Reprogramming the Pentagon for a New Age." *Foreign Affairs* 88, no. 1 (January/February 2009): 28–40.

Gellman, Barton. "William McRaven: The Admiral." *Time*, December 14, 2011.

Geneva Conventions. *The Geneva Conventions of August 12, 1949*. Geneva: International Committee of the Red Cross, 2002.

———. *Protocols Additional to the Geneva Conventions of 12 August 1949*. Geneva: International Committee of the Red Cross, 1977.

Gibbons-Neff, Thomas. "At Point 18 in Eastern Ukraine, the War Grinds on, Night After Night." *Washington Post*, August 8, 2015. https://www.washingtonpost.com/world/national-security/at-point-18-in-eastern-ukraine-the-war-grinds-on-night-after-night/2015/08/15/fffcf2c0-405f-11e5-9561-4b3dc93e3b9a_story.html?utm_term=.156a09a2b2b4.

Gilbert, G. M. *Nuremburg Diary*. Boston, MA: DaCapo Press, 1995.

Giles, Lionel. "Sun Tzu on the Art of War: The Oldest Military Treatise in the World." 1910. Translated from the Chinese. http://www.hqco9thmarines.com/History/Sun_Tzu_on_the_Art_of_War.pdf.

Gill, Paul, and John Horgan. "Who Were the Volunteers? The Shifting Sociological and Operational Profile of 1240 Provisional Irish Republican Army Members." *Terrorism and Political Violence* 25, no. 3 (2013): 435–456.

Gilli, Andrew, and Mauro Gilli. "The Diffusion of Drone Warfare? Industrial, Organizational, and Infrastructural Constraints." *Security Studies* 25, no. 1 (2016): 50–84.

Glenn, Russell W. *Combat in Hell: A Consideration of Constrained Urban Warfare*. Santa Monica: RAND, 1996.

———. "Thoughts on Hybrid Conflict." *Small Wars Journal*, March 2, 2009, 1–8.

Głowacki, Ludwik. *Obrona Warszawy i Modlina na tle kampanii wrześniowej*. Warszawa: Ministerstwa Obrony Narodowej, 1985.

Golan, Arnon. "The Battle for Jaffa, 1948." *Middle Eastern Studies* 48, no. 6 (2012): 997–1011.

———. "The City and War: The Case of Tel Aviv and Jaffa." *Urban Studies and Public Administration* 3, no. 3 (2020).

Golan, Arnon, Yaakov Peleg, Alon Kaddish, and Yona Bandman. *The Campaign for Jaffa, 1948*. Beer Sheva: Ben Gurion University Press, 2018.

Goldsmith, Jett. "Were Chemical Weapons Used in Donetsk Airport's Last Stand?" *Bellingcat*, January 23, 2015. https://www.bellingcat.com/news/uk-and-europe/2015/01/23/were-chemical-weapons-used-in-donetsk-airports-last-stand/.

Goncharova, Olena. "At Least 1,373 Soldiers Killed in Russia's War Against Ukraine." *Kyiv Post,* January 22, 2015. https://www.kyivpost.com/kyiv-post-plus/at-least-1373-soldiers-killed-in-russias-war-against-ukraine-378134.html?cn-reloaded=1.

Goodspeed, D. J. *Battle Royal: A History of the Royal Regiment of Canada, 1862–1962.* Ontario: Royal Regiment of Canada Association, 1962.

Gowing, Nik. *Skyful of Lies and Black Swans.* Oxford: Reuters Institute for the Study of Journalism, University of Oxford, 2009.

Graham, A. "Military Coalitions in War." In *The Oxford Handbook of War*, edited by J. Lindley-French and Y. Boyer, 319–331. Oxford: Oxford University Press, 2012.

Graham, Stephen. "Bulldozers and Bombs: The Latest Palestinian-Israeli Conflict as Asymmetric Urbicide." *Antipode* 34, no. 4 (2002): 642–649.

———. "Cities as Strategic Sites: Place Annihilation and Urban Geopolitics." In *Cities, War, and Terrorism: Towards an Urban Geopolitics*, edited by Stephen Graham, 31–53. Oxford: Blackwell, 2004.

———. *Cities Under Siege: The New Military Urbanism.* London: Verso, 2010.

———. *Cities, War, and Terrorism.* Oxford: Blackwell, 2004.

———. "Lessons in Urbicide." *New Left Review* 19 (January/February 2003).

———. *Vertical: The City from Satellites to Bunkers.* London: Verso, 2018.

Graham, Stephen, and Derek Gregory. "Urbicide." In *The Dictionary of Human Geography*, 5th ed., edited by Derek Gregory, Ron Johnston, Geraldine Pratt, Michael J. Watts, and Sarah Whatmore. Chichester: Wiley-Blackwell, 2011.

Granovetter, Mark S. "The Strength of Weak Ties." *American Journal of Sociology* 78, no. 6 (1973): 1360–1380.

Grant, Ulysses S. *Personal Memoirs of U. S. Grant.* Lincoln: University of Nebraska Press, 1996.

Gregory, Derek. "The Biopolitics of Baghdad: Counterinsurgency and the Counter-city." *Human Geography* 1, no. 1 (2008): 1–21.

Griffiths, Charles John. *A Narrative of the Siege of Delhi with an Account of the Mutiny at Ferozepore in 1857.* London: John Murray, 1910.

Groh, Tyrone. *Proxy War: The Least Bad Option.* Palo Alto, CA: Stanford University Press, 2019.

———. *War on the Cheap? Assessing the Costs and Benefits of Proxy War.* Washington, DC: Georgetown University, 2010.

Groningen in Vuur en Puin. Groningen: J. Haan, 1945.

Grytsenko, Oksana. "'Cyborgs' Pledge to Defend Ruined Donetsk Airport." *Kyiv Post,* October 10, 2014. https://www.kyivpost.com/kyiv-post-plus/cyborgs-pledge-to-defend-ruined-donetsk-airport-367553.html?cn-reloaded=1.

Guardian. "Pro-Russia Rebels Attempt to Seize Donetsk Airport in Ukraine." October 3, 2014. https://www.theguardian.com/world/2014/oct/03/pro-russia-rebels-donetsk-airport-ukraine.

———. "Russia-Backed Separatists Seize Donetsk Airport in Ukraine." January 15, 2015. https://www.theguardian.com/world/2015/jan/15/russian-backed-separatists-seize-donetsk-airport-ukraine.

Gunaratna, Rohan. "Ending the Fight in Marawi." *Counter Terrorist Trends and Analysis* 9, no. 10 (2017): 1–5.

———. "The Siege of Marawi: A Game Changer." *Terrorism in Asia* 9, no. 7 (2017): 4–8.

Gusterson, Hugh. *Drone: Remote Control Warfare.* Cambridge, MA: MIT Press, 2015.

———. "Drone Warfare in Waziristan and the New Military Humanism." *Current Anthropology* 60, Supplement 19 (February 2019).

Gutfraind, Alexander. "Understanding Terrorist Organizations with a Dynamic Model." In *Mathematical Methods in Counterterrorism*, edited by Nasrullah Memon, J. David Farley, D. L. Hicks, and T. Rosenorn. Vienna: Springer-Verlag, 2009.

Gwynn, Charles W. *Imperial Policing.* London: Macmillan, 1934.

Hafner-Burton, Emilie M., Stephan Haggard, David A. Lake, and David G. Victor. "The Behavioral Revolution and International Relations." *International Organization* 71, Supplement S1 (2017): S1–S31.

Hamadeh, Noor, and Krystel Bassil. "Demolishing Human Rights in the Name of Reconstruction: Lessons Learned from Beirut's Solidere for Syria." Tahrir Institute for Middle East

Policy, September 16, 2020. https://timep.org/commentary/analysis/demolishing-human-rights-in-the-name-of-reconstruction-lessons-learned-from-beiruts-solidere-for-syria/.

Hamill, Desmond. *Pig in the Middle: The Army in Northern Ireland*. London: Methuen, 1986.

Hamley, Edward. *National Defence: Articles and Speeches*. Edinburgh: William Blackwood and Sons, 1889.

Handel, M. I. *Masters of War*. 3rd ed. London: Routledge, 2009.

Harb, Mona, and Laura Deeb. "Culture as History and Landscape: Hizbullah's Efforts to Shape an Islamic Milieu in Lebanon." *Arab Studies Journal* 19, no. 1 (2011): 12–45.

Hardy, John, and Paul Lushenko. "The High Value of Targeting: A Conceptual Model for Using HVT Against a Networked Enemy." *Defence Studies* 12, no. 3 (2012): 413–433.

Harnden, Toby. *Bandit Country: The IRA and South Armagh*. London: Hodder & Stoughton, 2000.

Harress, Christopher. "Pro-Russian Forces About to Take Donetsk Airport as Ukraine Peace Talks Called Off." *International Business Times*, January 13, 2015. https://www.ibtimes.com/pro-russian-forces-about-take-donetsk-airport-ukraine-peace-talks-called-1782026.

Harris, Marc, Robert Dixon, Nicholas Melin, Daniel Hendrex, Richard Russo, and Michael Bailey. *Megacities and the United States Army: Preparing for a Complex and Uncertain Future*. Office of the Chief of Staff of the Army, Strategic Studies Group, Megacities Concept Team, June 2014.

Hartman, Geoffrey. "Marawi Battle Highlights the Perils of a Stalled Peace Process in the Philippines." Centre for Strategic and International Studies, June 29, 2017. https://www.csis.org/analysis/marawi-battle-highlights-perils-stalled-peace-process-philippines.

Harvey, Dan. *Soldiering Against Subversion: The Irish Defence Forces and Internal Security During the Troubles, 1969–1998*. Kildare, Ireland: Merrion Press, 2018.

Hastings, Michael. *The Operators: The Wild and Terrifying Inside Story of America's War in Afghanistan*. New York: Blue Rider Press, 2012.

Hatherly, Owen. "Architects of Revolt: The Kiev Square that Sparked Ukraine's Insurrection." *Guardian*, April 8, 2014. https://www.theguardian.com/cities/2014/apr/08/architects-revolt-kiev-maidan-square-ukraine-insurrection.

Headquarters, Department of the US Army. *Combat Tactics*. Field Manual No. 3-90. Washington, DC, July 4, 2001, B-15.

———. *Combat Tactics*. Field Manual No. 3-90. Washington, DC, July 4, 2001, B-16-7.

Helmus, Todd C., and Russell W. Glenn. *Steeling the Mind: Combat Stress Reactions and their Implications for Urban Warfare*. Santa Monica: RAND, 2005.

Herscher, Andrew. "Warchitectural Theory." *Journal of Architectural Education* 61, no. 3 (2008): 32–43.

Hersh, Eitan D., and Brian F. Schaffner. "Targeted Campaign Appeals and the Value of Ambiguity." *Journal of Politics* 75, no. 2 (2013): 520–534.

Heydarian, Richard Javad. "Crisis in Mindanao: Duterte Battles ISIS." *Foreign Affairs*, June 28, 2017. https://www.foreignaffairs.com/articles/philippines/2017-06-28/crisis-mindanao.

Hillenga, Martin. *Groningen 40-45*. Zwolle, Netherlands: W Books, 2015.

Hills, Alice. "Deconstructing Cities: Military Operations in the Urban Era." *Journal of Conflict Studies* 22, no. 2 (2002): 99–117.

———. "Fear and Loathing in Fallujah." *Armed Forces and Society* 32, no. 4 (2006): 623–639.

———. *Future War in Cities: Rethinking a Liberal Dilemma*. London: Routledge, 2004.

Hinnant, Lori, and Maggie Michael. "An Undercover Blogger Lived a Double Life for 2 Years in ISIS-Occupied Iraq to Document the Group's Atrocities—Here's His Story." Business Insider, December 10, 2017. https://www.businessinsider.com/mosul-eye-documented-isis-atrocities-iraq-2017-12?r=US&IR=T.

Hix, Major General Bill (US Army). "Future Combined Arms Operations." RUSI Land Warfare Conference speech, June 28, 2017. https://www.youtube.com/watch?list=PLFAgO2TZWpwDKqsQVyBj37kbUizOkBsBE&time_continue=4383&v=_EcrrD1dBhg.

HM Government. *A Strong Britain in an Age of Uncertainty: The National Security Strategy*. London: Stationery Office, 2010.

Hockey, John C. *Squaddies: Portrait of a Subculture*. Liverpool, UK: Liverpool University Press, 2016.

Hoekstra, Quint. "How Mosul Fell: The Role of Coup-Proofing in the 2014 Partial Collapse of the Iraqi Security Forces." *International Politics* 57 (2020): 684–703.

Hoffman, Bruce. *Anonymous Soldiers: The Struggle for Israel, 1917–1947*. New York: Alfred A. Knopf, 2015.

Hoffman, Frank. "Hybrid Warfare and Challenges." *Joint Forces Quarterly* 52, 1st quarter (2009).

Hofman, Beno. *Groningen in de Tweede Wereldoorlog, Van Oorlogsdreiging tot Bevrijdingsroes*. Assen, Netherlands: In Boekvorm Uitgevers, 2015.

Holm, Frank P. *A Backwards Glance: The Personal Story of an Infantry Signaller with the Calgary Highlanders in World War II*. Sault Ste. Marie, ON: Frank P. Holm, 1989.

Horne, Alistair. *The Price of Glory: Verdun 1916*. London: Macmillan, 1975.

Horowitz, Michael C., Sarah E. Kreps, and Matthew Fuhrmann. "Separating Fact from Fiction in the Debate over Drone Proliferation." *International Security* 41, no. 2 (2016): 7–42.

Horowitz, Michael C., Joshua A. Schwartz, and Matthew Fuhrmann. "Who's Prone to Drone? A Global Time-Series Analysis of Armed Uninhabited Aerial Vehicle Proliferation." *Conflict Management and Peace Science* 39, no. 2 (November 2020): 1–24.

Hughes, Geraint Alun. "Syria and the Perils of Proxy Warfare." *Small Wars & Insurgencies* 25, no. 3 (2014): 522–538.

House, Jonathan M. *Controlling Paris: Armed Forces and Counter-Revolution, 1789–1848*. New York: New York University Press, 2014.

Howard, Michael. "The Lessons of History." *History Teacher* 15, no. 4 (1982): 491–492.

Hudson, Myles. "Battle of Alesia, Ancient Roman History (52 BCE)." *Encyclopedia Britannica*, 2019. https://www.britannica.com/event/Battle-of-Alesia-52-BCE.

Hughes, Ben. *The British Invasion of the River Plate 1806–1807: How the Redcoats Were Humbled and a Nation Was Born*. Barnsley, UK: Pen and Sword Military, 2013.

Huizinga, M. H. *Maple Leaf Up: De Canadese opmars in Noord-Nederland, April 1945*. Groningen, Netherlands: J. Niemeijer, 1980.

Huizinga, M. H., and B. van Leusen. *Vier Dagen in April*. Groningen, Netherlands: Reco Multi Media, 1999.

Huntington, Samuel. *The Soldier and the State*. Cambridge, MA: Harvard University Press, 1957.

Hurst, Steven. "US: Raid of Baghdad's Sadr City Kills 49." *Washington Post*, October 21, 2007. https://www.washingtonpost.com/wp-dyn/content/article/2007/10/21/AR2007102100159_pf.html.

Ibrahimi, Niamatullah, and Shahram Akbarzadeh. "Intra-Jihadist Conflict and Cooperation: Islamic State–Khorasan Province and the Taliban in Afghanistan." *Studies in Conflict and Terrorism* 43, no. 12 (2020): 1086–1107.

"In International Airport of 'Dnepropetrovsk' Commemorated the Cyborgs." Dnepropetrovsk International Airport, January 20, 2020. http://dnk.aero/en/news/150/.

Instructions for the Government of Armies of the United States in the Field (Lieber Code). April 24, 1863. International Committee of the Red Cross Treaties and Documents. https://ihl-databases.icrc.org/applic/ihl/ihl.nsf/Treaty.xsp?action=openDocument&documentId=A25AA5871A04919BC12563CD002D65C5.

International Committee of the Red Cross. "Convention (IV) Respecting the Laws and Customs of War on Land and Its Annex: Regulations Concerning the Laws and Customs of War on Land. The Hague, 18 October 1907." Customary international humanitarian law (IHL) database. https://ihl-databases.icrc.org/ihl/INTRO/195.

———. "Practice Relating to Rule 11: Indiscriminate Attacks." Customary international humanitarian law (IHL) database, 2016. https://ihl-databases.icrc.org/customary-ihl/eng/docs/v2_rul_rule11.

———. "Practice Relating to Rule 50. Destruction and Seizure of Property of an Adversary." Customary international humanitarian law (IHL) database. https://ihl-databases.icrc.org/customary-ihl/eng/docs/v2_rul_rule50.

International Crisis Group. *Iraq's Jihadi Jack-in-the-Box*. Middle East Briefing No. 38. Beirut and Brussels: International Crisis Group, June 20, 2014. https://d2071andvip0wj.cloudfront.net/iraq-s-jihadi-jack-in-the-box.pdf.

———. "Philippines: Addressing Islamist Militancy After the Battle for Marawi." July 17, 2018. https://www.crisisgroup.org/asia/south-east-asia/philippines/philippines-addressing-islamist-militancy-after-battle-marawi.

———. "The Philippines: Militancy and the New Bangasamoro." June 27, 2019. https://www.crisisgroup.org/asia/south-east-asia/philippines/301-philippines-militancy-and-new-bangsamoro.

———. "Rebels Without a Cause: Russia's Proxies in Eastern Ukraine." Europe Report 254, July 16, 2019.

"Interview: Ukrainian 'Cyborg' Describes Nine Days Defending Donetsk Airport." Radio Free Europe/Radio Liberty, November 15, 2014. https://www.rferl.org/a/donetsk-fighters-ukraine/26693593.html.

Isin, Engin F. *Being Political. Genealogies of Citizenship*. Minneapolis: University of Minnesota Press, 2002.

Izady, Michael M. R. "Urban Unplanning: How Violence, Walls, and Segregation Destroyed." *Journal of Planning History* 19, no. 1 (2020): 52–68.

Jackson, Mike. *Soldier: The Autobiography of General Sir Mike Jackson*. London: Random House, 2007.

Jadoon, Amira. *Allied and Lethal: Islamic State Khorasan's Network and Organizational Capacity in Afghanistan and Pakistan*. West Point, NY: Combating Terrorism Center, December 2018.

Jadoon, Amira, and Andrew Mines. *Broken, but Not Defeated: An Examination of State-Led Operations Against Islamic State Khorasan in Afghanistan and Pakistan (2015–2018)*. West Point, NY: Combating Terrorism Center, March 2020.

Jahanbani, Nakissa. *Beyond Soleimani: Implications for Iran's Proxy Network in Iraq and Syria*. West Point, NY: Combating Terrorism Center, January 10, 2020.

Jansiz, Ghassa. "Let's Ditch Homs and Build a New One." *RIBA Journal*, July 16, 2014. https://www.ribaj.com/culture/lets-build-a-new-homs.

Jeffrey, Keith. "The British Army and Internal Security 1919–1939." *Historical Journal* 24, no. 2 (1981): 377–397.

Jenkings, K. Neil, Nick Megoran, Rachel Woodward, and Daniel Bos. "Wootton Bassett and the Political Spaces of Remembrance and Mourning." *Area* 44, no. 3 (September 2012): 356–363.

Jocelyn, Julian R. J. *The History of the Royal Artillery (Crimean Period)*. London: John Murrat, 1911.

Johnson, David E., M. Wade Markel, and Brian Shannon. *The 2008 Battle of Sadr City: Reimagining Urban Combat*. Santa Monica, CA: RAND Corporation, 2013. https://www.rand.org/pubs/research_reports/RR160.html.

Jones, S. M. *Re-examining Tomorrow's Battlefield; Taking the Fight into the Cities*. Volume S03-01. Carlisle, PA: US Army War College, June 2003.

Jordan, Jenna. *Leadership Decapitation: Strategic Targeting of Terrorist Organizations*. Stanford, CA: Stanford University Press, 2019.

Jukes, Geoffrey. *Hitler's Stalingrad Decisions*. Berkeley: University of California Press, 1985.

Kaag, John, and Sarah Kreps. *Drone Warfare*. Cambridge, UK: Polity, 2012.

Kahn, Paul. "The Paradox of Riskless Warfare." *Philosophy & Public Policy Quarterly* 22, no. 3 (2002): 2–8.

Käihkö, Ilmari. "A Nation-in-the-Making, in Arms: Control of Force, Strategy and the Ukrainian Volunteer Battalions." *Defence Studies* 18, no. 2 (2018): 147–166.

Kaldor, Mary. "In Defence of New Wars." *Stability: International Journal of Security and Development* 2, no. 1 (2013).

———. *New and Old Wars: Organised Violence in a Global Era*. 3rd ed. Stanford, CA: Stanford University Press, 2012.

Kaplan, Caren. "Drone-o-Rama: Troubling the Temporal and Spatial Logics of Distance Warfare." In *Life in the Age of Drone Warfare*, edited by Lisa Parks and Caren Kaplan, 162–177. Durham, NC: Duke University Press, 2017.

Kappeler, Victor E., and Peter B. Kraska. "Normalising Police Militarisation, Living in Denial." *Policing and Society* 25, no. 3 (2015): 268–275.

Karagiannis, Emmanuel. "Ukrainian Volunteer Fighters in the Eastern Front: Ideas, Political-Social Norms and Emotions as Mobilization Mechanisms." *Southeast European and Black Sea Studies* 16, no. 1 (2016): 139–147.

Kates, Glenn, and Olena Removska, "Disturbing Videos Raise War-Crimes Concerns in Ukraine," Radio Free Europe/Radio Liberty, January 27, 2015. https://www.rferl.org/a/ukraine-prisoners-of-war-separatists-war-crimes-questions/26816285.html.

Katz, Irit. "Between *Bare Life* and *Everyday Life*: Spatializing Europe's Migrant Camps." *Architecture_MPS* 12, no. 1 (2017).

Katzman, Kenneth, and Clayton Thomas. *Afghanistan: Post-Taliban Governance, Security, and U.S. Policy*. Washington, DC: Congressional Research Service, December 13, 2017.

Kaufman, David, and Michiel Horn. *A Liberation Album: Canadians in the Netherlands, 1944–45*. Toronto: McGraw-Hill Ryerson, 1980.

Keegan, John. *The First World War*. London: Hutchinson, 1998.

Kelly, Terrence, John E. Peters, Eric Landree, Louis R. Moore, Randall Steeb, and Aaron L. Martin. *The U.S. Combat and Tactical Wheeled Vehicle Fleets: Issues and Suggestions for Congress*. Santa Monica, CA: RAND, 2011. https://www.rand.org/pubs/monographs/MG1093.html.

Kennedy-Pipe, Caroline, and Colin McInnes. "The British Army in Northern Ireland 1969–1972: From Policing to Counter-Terror." In *Modern Counter-Insurgency*, edited by Ian Beckett, 203–226. London: Routledge, 2017.

Kern, Paul Bentley. *Ancient Siege Warfare*. Bloomington: Indiana University Press, 1999.

Khalaf, Samir. *Heart of Beirut. Reclaiming the Bourj*. London: Saqi, 2006.

Khalidi, Walid. "The Fall of Haifa Revisited." *Journal of Palestine Studies* 37, no. 3 (2008): 30–58.

Kilcullen, David. *Out of the Mountains: The Coming Age of the Urban Guerrilla*. London: Hurst and Company, 2013. Reprinted Oxford: Oxford University Press, 2015.

———. "Urban Combat: 'Cities Are Sponges That Soak Up Troops.'" Cipher Brief, October 22, 2017. https://www.thecipherbrief.com/urban-combat-cities-are-sponges-that-soak-up-troops.

King, Anthony. "Close Quarters Battle: Urban Combat and 'Special Forcification.'" *Armed Forces & Society* 42, no. 2 (April 2016): 276–300.

———. *The Combat Soldier: Infantry Tactics and Cohesion in the Twentieth and Twenty-First Centuries*. Oxford: Oxford University Press, 2013.

———. *Urban Warfare in the 21st Century*. Cambridge: Polity, 2021.

Kipfer, Stefan, and Kanishka Goonewardena. "Colonization and the New Imperialism: On the Meaning of Urbicide Today." *Theory & Event* 10, no. 2 (2007).

Kitchen, Martin. *A Military History of Germany from the Eighteenth Century to the Present Day*. London: Weidenfeld and Nicolson, 1975.

Klein, Margarete. "Ukraine's Volunteer Battalions: Advantages and Challenges." Swedish Defence Research Agency (FOI) Memo 5312/RUFS Briefing No. 27, April 2015.

Knight, Charles. "Urban Warfare Capability: A Background to the Challenges and a Call for Professional Debate." *Australian Army Journal* 15, no. 1 (2019): 9–30.

Knight, Charles, and Katja Theodorakis. *The Marawi Crisis: Urban Conflict and Information Operations*. Australian Strategic Policy Institute, July 31, 2019. https://www.aspi.org.au/report/marawi-crisis-urban-conflict-and-information-operations.

Komorowski, Krzystof. *Bitwa o Warszawę: Militarne aspekty Powstania Warszawskiego*. Warsaw: Oficyna Wydawnicza RYTM, 2004.

Konaev, Margarita. *The Future of Urban Warfare in the Age of Megacities*. Paris: Institut français des relations internationales (IFRI), 2019.

Konaev, Margarita, and Samuel Bendett. "Russian AI Enabled Combat: Coming to a City near You." *War on the Rocks*, July 31, 2019.

Konaev, Margarita, and John Spencer. "The Era of Urban Warfare Is Already Here." Foreign Policy Research Institute, March 21, 2018. https://www.fpri.org/article/2018/03/the-era-of-urban-warfare-is-already-here/.

Kovalenko, Oksana, and Galina Titish. "Brigade Commander Yevgeny Moysyuk About Airport, Raid and Features of Enemy Action." *Ukrainian Pravda*, February 12, 2016. https://www.pravda.com.ua/articles/2016/02/12/7098744/.

Krebs, Valdis E. "Mapping Networks of Terrorist Cells." *Connections* 24, no. 3 (2002): 43–52.

Krepinevich, Andrew F., Jr. *The Army and Vietnam*. Baltimore, MD: Johns Hopkins University Press, 1986.

Kreps, Sarah. *Drones: What Everyone Needs to Know*. Oxford: Oxford University Press, 2016.

Kreps, Sarah, and Paul Lushenko. "What Happens Now to U.S. Counterterrorism Efforts in Afghanistan?" *Washington Post*, September 21, 2021.

Kreps, Sarah, and Sarah Maxey. "Context Matters: The Transformative Nature of Drones on the Battlefield." In *Technology and International Relations: The New Frontier in Global Power*, edited by Giampiero Giacomello, Francesco Niccolò Moro, and Marco Valigi, 68–89. Northampton, MA: Edward Elgar Publishing, 2021.

Kreps, Sarah, and Geoffrey P. R. Wallace. "International Law, Military Effectiveness, and Public Support for Drone Warfare." *Journal of Peace Research* 53, no. 6 (2016): 830–844.

Kreuzer, Michael P. *Drones and the Future of Air Warfare: The Evolution of Remotely Piloted Aircraft*. London: Routledge, 2016.

Krieg, Andreas, and Jean-Marc Rickli. *Surrogate Warfare: The Transformation of War in the Twenty-First Century*. Washington, DC: Georgetown University Press, 2019.

Kudrytski, Aliaksandr. "Ukraine Forces Retake Most of Donetsk Airport, Government Says." *Bloomberg*, January 18, 2015. http://www.bloomberg.com/news/2015-01-17/pro-russian-rebels-pound-ukrainian-positions-at-donetsk-airport.html.

Kupfer, Matthew. "Ukrainian Veterans Accuse Apple Ad of Supporting Russian Propaganda." *Hromadske International*, September 21, 2017. https://en.hromadske.ua/posts/ukrainian-veterans-accuse-apple-ad-of-supporting-russian-propaganda.

Kuran, Timur. "Now out of Never: The Element of Surprise in the East European Revolution of 1989." *World Politics* 44, no. 1 (1991): 7–48.

LAC. RG 24, vol. 10948, file 249C5. (D84), Lieutenant-General P. J. Montague, Chief of Staff, Canadian Military Headquarters, Canadian Operations—North-West Europe: Extracts from War Diaries and Memoranda (Series 25), July 17, 1945, 12.

Langston, Henry. "The Battle for Donetsk International Airport: Russian Roulette (Dispatch 44)." Vice News, May 28, 2014. https://www.youtube.com/watch?v=XtdVVFhjdqA.

———. "Defending the Ruined Village of Pisky: Russian Roulette (Dispatch 91)." Vice News, February 5, 2015. https://www.youtube.com/watch?v=-LzEx03UQFI&list=PLw613M86o5o5zqF6WJR8zuC7Uwyv76h7R&index=90.

Larson, Eric V., Richard E. Darilek, Daniel Gibran, Brian Nichiporuk, Amy Richardson, Lowell H. Schwartz, and Cathryn Quantic Thurston. *Foundations of Effective Influence Operations: A Framework for Enhancing Army Capabilities*. Santa Monica, CA: RAND. https://www.rand.org/content/dam/rand/pubs/monographs/2009/RAND_MG654.pdf.

Latané, Bibb. "Dynamic Social Impact: The Creation of Culture by Communication." *Journal of Communication* 46, no. 4 (Autumn 1996): 13–25. https://www2.psych.ubc.ca/~schaller/528Readings/Latane1996.pdf.

Laurie, Clayton D., and Ronald H. Cole. *The Role of Federal Military Forces in Domestic Disorders 1877–1945*. Washington, DC: Center of Military History, 1997.

LeBor, Adam. *City of Oranges*. London: Bloomsbury, 2006.

———. "Zion and the Arabs: Jaffa as a Metaphor." *World Policy Journal* 24, no. 5 (2008): 61–75.

Lecher, Colin. "Facebook Executive: We Got Trump Elected, and We Shouldn't Stop Him in 2020." *Verge*, January 7, 2020. https://www.theverge.com/2020/1/7/21055348/facebook-trump-election-2020-leaked-memo-bosworth.

Leeninga, J. J., and E. Jac Westra. *En T'och Staat de Martini, Groningen onder Duitsch Schrikbewind*. J. Netherlands: Niemeijer, 1945.

Lehr, Peter. *Counterterrorism Technologies*. Cham, Switzerland: Springer Inter-national, 2019.

Leonhard, Robert R. *Principles of War for the Information Age*. Novato: Presidio Press, 1998.

———. "Urban Warfare in the Information Age." *Army*, April 2003.

Lewis, James. "The Battle of Marawi: Small Teams Lessons Learned for the Close-Quarters Fight." The Cove (Australian Army), October 1, 2019. https://cove.army.gov.au/article/the-battle-marawi-small-team-lessons-learned-the-close-fight.

Lewis, Rob. *Fishers of Men: The Gripping True Story of a British Undercover Agent in Northern Ireland*. London: John Blake Publishing, 2017.

Lindelauf, Roy, Peter Borm, and Herbert Hames. *Understanding Terrorist Network Topologies and Their Resilience Against Disruption*. Discussion Paper No. 85. Tilburg, Netherlands: CentER, Tilburg University, November 4, 2009.

Llewellyn, Alexander. *The Siege of Delhi*. London: Macdonald and Jane's, 1977.

Loiko, Sergei. "How Ukraine's Outgunned 'Cyborgs' Lost Donetsk Airport." *Los Angeles Times,* January 28, 2015. https://www.latimes.com/world/la-fg-c1-ukraine-donetsk-airport-20150128-story.html#page=1.

———. "Ukraine Fighters, Surrounded at Wrecked Airport, Refuse to Give Up." *Los Angeles Times,* October 28, 2014. https://www.latimes.com/world/la-fg-c1-ukraine-airport-20141028-story.html#page=1.

Loshkariov, Ivan, and Andrey Sushentsov. "Radicalization of Russians in Ukraine: From 'Accidental' Diaspora to Rebel Movement." *Southeast European and Black Sea Studies* 16, no. 1 (2016): 71–90.

Lushenko, Paul. "Coining an Ethical Dilemma: The Impunity of Afghanistan's Indigenous Security Forces." *Journal of Military Ethics* 13, no. 3–4 (2015): 272–275.

———. "Reconsidering the Theory and Practice of High Value Targeting." *Counter Terrorist Trends and Analysis* 7, no. 7 (2015): 23–30.

———. "Review of the Special Issue 'Robotics Autonomous Systems and Warfare,' *Small Wars and Insurgencies* 31, 4 June 2020." *Small Wars & Insurgencies* 32, no. 3 (2021): 550–570.

———. "U.S. Presidents' Use of Drone Warfare." *Defense & Security Analysis* 38, no. 1 (2022): 31–52.

Lushenko, Paul, and John Hardy. "Panjwai: A Tale of Two COINs in Afghanistan." *Small Wars & Insurgencies* 27, no. 1 (2016): 106–132.

Lushenko, Paul, Lance Van Auken, and Garrett Stebbins. "ISIS-K: Deadly Nuisance or Strategic Threat?" *Small Wars & Insurgencies* 30, no. 2 (2019): 265–278.

Lushenko, Paul, Srinjoy Bose, and William Maley (eds.). *Drones and Global Order: Implications of Remote Warfare for International Society*. London: Routledge, 2022.

MacDonald, Darach. *The Chosen Fews: Exploding Myths in South Armagh*. Cork, Ireland: Mercier Press, 2000.

Mackay, Andrew. "The Art of Influence: How the Taliban Were Driven from a Strategic Afghan Town." BBC News, September 17, 2015. https://www.bbc.co.uk/news/resources/idt-ff9a9c01-faa4-4038-b4e9-83e619460e1f.

Mackay, Andrew, and Steve Tatham. *Behavioural Conflict*. Saffron Waldon, Essex, UK: Military Studies Press, 2011.

Magnusson, Warren. *Politics of Urbanism. Seeing Like a City*. Abingdon, UK: Routledge, 2011.

Mahečić, Darja Radović, "The Case of Dubrovnik: UNESCO World Heritage Site Under Siege, 1991-1992." In *Architecture and Armed Conflict: The Politics of Destruction,* edited by JoAnne Mancini and Keith Bresnahan, 27–41. Abingdon: Routledge, 2015.

Maley, William. "State Strength and Rule of Law." In *Afghanistan: Challenges and Prospects*, edited by Srinjoy Bose, Nishank Motwani, and William Maley, 63–77. London: Routledge, 2017.

———. "Talibanisation and Pakistan." In *Talibanisation: Extremism and Regional Instability in South and Central Asia*, edited by Denise Groups, 53–74. Berlin: Conflict Prevention Network–Stiftung Wissenschaft and Politik, 2001.

Mallett, Michael. "Siegecraft in Late Fifteenth-Century Italy." In *The Medieval City Under Siege*, edited by Ivy A. Corfis and Michael Wolfe. Woodbridge, UK: Boydell Press, 1995.

Malyarenko, Tetyana, and David J. Galbreath. "Paramilitary Motivation in Ukraine: Beyond Integration and Abolition." *Southeast European and Black Sea Studies* 16, no. 1 (2016): 113–138.

Mann, Christopher. "'The Rest of the Day Was Perfectly Bloody': The 51st Highland Division and Town Clearance in Germany, February–March 1945." In Fremont-Barnes, *A History of Modern Urban Operations*.

Marighella, Carlos. *Minimanual of the Urban Guerrilla*. In *Terror and Urban Guerrillas: A Study of Tactics and Documents,* edited by Jay Mallin. Coral Gables, FL: University of Miami Press, 1982.

Marine Corps Warfighting Publication (MCWP) 3-10. *MAGTF Ground Operations*. US Marine Corps Publication, November 8, 2017 (amended April 4, 2018).

Marshall, Alex. "From Civil War to Proxy War: Past History and Current Dilemmas." *Small Wars & Insurgencies* 27, no. 2 (2016): 183–195.

Martel, William C. *Grand Strategy in Theory and Practice: The Need for an Effective American Foreign Policy*. Cambridge: Cambridge University Press, 2015.

Matthews, William. "Megacity Warfare: Taking Urban Combat to a Whole New Level." *Army*, March 2015, 31–35.

Matveeva, Anna. "No Moscow Stooges: Identity Polarization and Guerrilla Movements in Donbass." *Southeast European and Black Sea Studies* 16, no. 1 (2016): 25–50.

McChrystal, Stanley A. "Becoming the Enemy: To Win in Afghanistan, We Need to Fight More Like the Taliban." *Foreign Policy* 185 (March–April 2011): 66–70.

McConaghy, Kieran. *Terrorism and the State: Intra-state Dynamics and the Response to Non-state Political Violence*. London: Springer, 2017.

McCormick, G. H., and G. Owen. "Security and Coordination in a Clandestine Organization." *Mathematical and Computer Modelling* 31, no. 6–7 (2000): 175–192.

McMaster, H. R. "Continuity and Change: The Army Operating Concept and Clear Thinking About Future War." *Military Review* 95, no. 2 (2015): 6–20.

Meadows, Donella. *Thinking in Systems: A Primer*. White River Junction, VT: Chelsea Green Publishing, 2008.

Medley, J. G. "The Siege of Delhi." *Journal of the United Services Institution of India, 1882–1883*.

Meisels, Tamar, and Jeremy Waldron. *Debating Targeted Killing: Counter-Terrorism or Extra-judicial Execution?* Oxford: Oxford University Press, 2020.

Memon, Nasrullah, and Henrik Larsen. "Practical Algorithms for Destabilizing Terrorist Networks." In *Intelligence and Security Informatics*, edited by Sharad Mehrotra, Daniel D. Zeng, Hsinchun Chen, Bhavani Thuraisingham, and Fei-Yue Wang, 389–400. Vol. 3975 of *Lecture Notes in Computer Science*. Berlin: Springer, 2006.

Mendel, Jonathan. "Urban Violence and Militarization." In *International Encyclopedia of the Social and Behavioral Sciences*. Amsterdam, Netherlands: Elsevier, 2015. http://ebookcentral.proquest.com/lib/cam/detail.action?docID=1963260.

Menshikov, Major G. "Artillery in Street Fighting." *Field Artillery Journal* 36, no. 8 (August 1946): 471–474.

Mercur, James. *Attack of Fortified Places Including Siege-Works, Mining and Demolitions*. New York: John Wiley and Sons, 1894.

Miksche, Ferdinand Otto. *Secret Forces: The Techniques of Underground Movements*. London: Faber, 1952.

Miller, Christopher. "Art of War: Ukraine Explores Donbas Conflict On Screen, In Books." Radio Free Europe/Radio Liberty, December 7, 2017. https://www.rferl.org/a/ukraine-film-cyborgs-donetsk-airport-battle-premiere/28903701.html.

Miller, James, and Pierre Vaux. "The Death of Ukraine's Cyborg Army." *Foreign Policy*, January 22, 2015. https://foreignpolicy.com/2015/01/22/the-death-of-ukraines-cyborg-army-ukraine-russia-donetsk-airport-shelling/.

Ministry of Defence (UK). *Operation Banner: An Analysis of Military Operations in Northern Ireland*. July 2006.

Ministry of Defence of Ukraine. "'Cyborgs' Commemoration Day: Ukraine Commemorates Defenders of Donetsk Airport." January 16, 2020. https://www.mil.gov.ua/en/news/2020/01/16/cyborgs%E2%80%9D-commemoration-day-ukraine-commemorates-defenders-of-donetsk-airport/.

———. "Dmytro Yarosh Appointed as Advisor to Chief of General Staff." April 5, 2015. https://www.mil.gov.ua/en/news/2015/04/05/dmytro-yarosh-appointed-as-advisor-to-chief-of-general-staff/.

———. "Raid Documentary." April 27, 2016. https://www.youtube.com/watch?v=co-AHRfrSro.

Mir, Asfandyar, and Dylan Moore. "Drones, Surveillance, and Violence: Theory and Evidence from a US Drone Program." *International Studies Quarterly* 63, no. 4 (December 2019).

Mistry, Dinshaw. *Containing Missile Proliferation: Strategic Technology, Security Regimes, and International Cooperation in Arms Control*. Seattle: University of Washington Press, 2003.

Molnar, Andrew R. *Undergrounds in Insurgent, Revolutionary, and Resistance Warfare*. Washington, DC: Special Operations Research Office, American University, 1963.

Moloney, Ed. *Voices from the Grave: Two Men's War in Ireland*. New York: PublicAffairs, 2010.

Morris, Benny. *1948: The First Arab-Israeli War*. New Haven, CT: Yale University Press, 2008.

Morselli, Carlo, Cynthia Giguere, and Katia Petit. "The Efficiency/Security Trade-Off in Criminal Networks." *Social Networks* 29, no. 1 (2007): 143–153.

Moser, Caroline, and Philipp Horn. *Understanding the Tipping Point of Urban Conflict: Conceptual Framework Paper*. Working Paper 1. Manchester, UK: Global Research Centre, University of Manchester, 2011.

Moss, Robert. *Urban Guerrillas: The New Face of Political Violence*. London: Temple Smith, 1972.

Mosse, George. *Fallen Soldiers: Reshaping the Memory of the World Wars*. Oxford: Oxford University Press 1990.

Mosul Study Group. *What the Battle for Mosul Teaches the Force*. No. 17-24U. Leavenworth, KS: Army University Press, September 2017.

Mueller, John. "War Has Almost Ceased to Exist: An Assessment." *Political Science Quarterly* 124, no. 2 (2009): 297–321.

Mulroe, Patrick. "'The Most Notorious Trouble Spot Along the Entire Border': Exploring the Dynamics of Political Violence in an Irish Border Town, 1971–1974." *Irish Political Studies* 35, no. 4 (2019): 566–584.

Münkler, Herfried. *The New Wars*, translated by Patrick Camiller. Cambridge, UK: Polity Press, 2005.

Murray, Robert K. *Red Scare: A Study in National Hysteria, 1919–1920*. Minneapolis: University of Minnesota Press, 1955.

Myatt, Frederick. *British Sieges of the Peninsular War*. Cheltenham, UK: The History Press, 2008.

Myers, David. *Psychology*, 9th revised edition. New York: W. H. Freeman & Co, 2009.

Nagel, Caroline. "Reconstructing Space, Re-creating Memory: Sectarian Politics and Urban Development in Post-war Beirut." *Political Geography* 21, no. 5 (2002): 717–725.

Nagl, John A. *Counterinsurgency Lessons from Malaya and Vietnam*. London: Praeger Publishers, 2002.

Neiman, Max. "Urban Operations: Social Meaning, the Urban Built Form, and Economic Function." In *Soldiers in Cities: Military Operations on Urban Terrain*, edited by Michael C. Desch, 139–148. Carlisle: US Army War College Strategic Studies Institute, 2001.

Neukirch, Claus. "The Special Monitoring Mission to Ukraine in Its Second Year: Ongoing OSCE Conflict Management in Ukraine." In *OSCE Yearbook 2015*, edited by the Institute for Peace Research and Security Policy at the University of Hamburg, 229–239. Nomos: Baden-Baden, 2016.

———. "The Special Monitoring Mission to Ukraine: Operational Challenges and New Horizons." In *OSCE Yearbook 2014*, edited by the Institute for Peace Research and Security Policy at the University of Hamburg, 183–197. Nomos: Baden-Baden, 2015.

Neumann, Peter, Ryan Evans, and Raffello Pantucci. "Locating Al Qaeda's Center of Gravity: The Role of Middle Managers." *Studies in Conflict and Terrorism* 34, no. 11 (November 2011): 825–842.

New York Times. "Major Describes Move." February 8, 1968. https://www.nytimes.com/1968/02/08/archives/major-describes-move.html.

Nickerson, David W., and Todd Rogers. "Political Campaigns and Big Data." *Journal of Economic Perspectives* 28, no. 2 (Spring 2014).

Noll, Jeffrey D. "Restraint in Urban Warfare: The Canadian Attack on Groningen, Netherlands, 13–16 April 1945." Master's thesis, US Army Command and General Staff College, 2013.

Noorbergen, Rene. *Shadow of Terror*. Washington, DC: Review and Herald Pub Association, 1990.

Norman, C. L. "2nd Gold Medal Prize Essay 1913–1914." *Journal of the United Service Institution of India* 44 (April 1915).

"Numerical Return of Killed, Wounded, and Missing in the Field Force Under the Command of Major-General A. Wilson, During Operations of 14th September 1857." *Selections from the Letters Despatches and Other State Papers Preserved by the Military Department of the Government of India 1857–1858*, Volume I, edited by George W. Forrest. Calcutta: Military Department Press, 1893.

OCHA Services, Relief Web. *Urban Baghdad: Impact of Conflict on Daily Life*. United Nations Office for the Coordination of Humanitarian Affairs, 2011. https://reliefweb.int/report/iraq/urban-baghdad-impact-conflict-daily-life.

O'Connor, Tim. "ISIS Takes Credit for Kabul Attack That Killed U.S. Marines, Taliban Vows to Fight 'Evil' Attackers." *Newsweek*, August 26, 2021.

O'Driscoll, Dylan. *The Future of Mosul: Before, During, and After the Liberation*. Middle East Research Institute, September 2016.

———. "Liberating Mosul." *Middle East Policy* 23, no. 4 (2016): 61–73.

Official account of Lieutenant General Sir John Whitelocke, *London Gazette*, September 13, 1807.

OSCE. "Information Received until 18:00 hrs (Kyiv time), 27 May 2014." Organization for Security and Co-operation in Europe. https://www.osce.org/ukraine-smm/119189.

———. "Information Received until 18:00 (Kyiv time), 4 June 2014." Organization for Security and Co-operation in Europe. https://www.osce.org/ukraine-smm/119565.

———. "Information Received until 18:00 hrs (Kyiv time), 19 June 2014." Organization for Security and Co-operation in Europe. https://www.osce.org/ukraine-smm/120059.

———. "Information Received as of 18:00 (Kyiv time), 12 August 2014." Organization for Security and Co-operation in Europe. https://www.osce.org/ukraine-smm/122607.

———. "Information Received as of 18:00 (Kyiv time), 19 August 2014." Organization for Security and Co-operation in Europe. https://www.osce.org/ukraine-smm/122847.

———. "Information Received as of 18:00 (Kyiv time), 20 August 2014." Organization for Security and Co-operation in Europe. https://www.osce.org/ukraine-smm/122908.

———. "Information Received as of 18:00 (Kyiv time), 22 August 2014." Organization for Security and Co-operation in Europe. https://www.osce.org/ukraine-smm/122939.

———. "Information Received as of 18:00 (Kyiv time), 6 November 2014." Organization for Security and Co-operation in Europe. https://www.osce.org/ukraine-smm/126480.

———. "Information Received as of 18:00 (Kyiv time), 21 December 2014." Organization for Security and Co-operation in Europe. https://www.osce.org/ukraine-smm/132881.

———. "Information Received as of 18:00 (Kyiv time), 26 December 2014." Organization for Security and Co-operation in Europe. https://www.osce.org/ukraine-smm/133021.

———. "Information Received as of 18:00 (Kyiv time), 30 December 2014." Organization for Security and Co-operation in Europe. https://www.osce.org/ukraine-smm/133156.

———. "Information Received as of 18:00 (Kyiv time), 2 January 2015." Organization for Security and Co-operation in Europe. https://www.osce.org/ukraine-smm/133341.

———. "Information Received as of 18:00 (Kyiv time), 9 January 2015." Organization for Security and Co-operation in Europe. https://www.osce.org/ukraine-smm/133771.

———. "Latest Report by the OSCE Special Monitoring Mission (SMM) to Ukraine, Based on Information Received as of 18:00 (Kyiv time), 26 May 2014." Organization for Security and Co-operation in Europe. https://www.osce.org/ukraine-smm/119166.

———. "Latest Report by the OSCE Special Monitoring Mission (SMM) to Ukraine, Based on Information Received as of 18:00 (Kyiv time), 12 December 2014." Organization for Security and Co-operation in Europe. https://www.osce.org/ukraine-smm/131766.

———. "Latest Report by the OSCE Special Monitoring Mission (SMM) to Ukraine, Based on Information Received as of 18:00 (Kyiv time), 20 January 2015." Organization for Security and Co-operation in Europe. https://www.osce.org/ukraine-smm/135671.

———. "Latest Report by the OSCE Special Monitoring Mission (SMM) to Ukraine, Based on Information Received as of 19:30 (Kyiv time), 12 December 2017." Organization for Security and Co-operation in Europe. https://www.osce.org/special-monitoring-mission-to-ukraine/362576.

———. "Latest Report by the OSCE Special Monitoring Mission (SMM) to Ukraine on Alleged Shelling of Donetsk Hospital, 7 August 2014." Organization for Security and Co-operation in Europe. https://www.osce.org/ukraine-smm/122468.

———."Situation Report: Information Received as of 18:00 (Kyiv time), 22 August 2014." Organization for Security and Co-operation in Europe. https://www.osce.org/ukraine-smm/122939.

———. "Situation Report: Information Received as of 18:00 (Kyiv time), 25 August 2014." Organization for Security and Co-operation in Europe. https://www.osce.org/ukraine-smm/122949.

———. "Situation Report: Information Received as of 18:00 (Kyiv time), 27 August 2014." Organization for Security and Co-operation in Europe. https://www.osce.org/ukraine-smm/123030.

———. "Spot Report by the Special Monitoring Mission to Ukraine: Damage to Donetsk City Water Supply. 15 June 2014." Organization for Security and Co-operation in Europe. https://www.osce.org/ukraine-smm/119818.

———. "Spot Report by the Special Monitoring Mission to Ukraine: Shelling of Residential Areas in Donetsk, the SMM Observes Civilian Fatalities. 24 August 2014." Organization for Security and Co-operation in Europe. https://www.osce.org/ukraine-smm/122940.

———. "Spot Report by the OSCE Special Monitoring Mission to Ukraine: The Situation at Donetsk Airport and Shchastya. 7 September 2014." Organization for Security and Co-operation in Europe. https://www.osce.org/ukraine-smm/123267.

———. "Spot Report by the Special Monitoring Mission to Ukraine: OSCE SMM Patrol Caught Up in Small Arms Fire. 16 December 2014." Organization for Security and Co-operation in Europe. https://www.osce.org/ukraine-smm/132096.

Ostrovsky, Simon. "Russian Roulette: Dispatch 84." *Vice News*, November 8, 2014. https://www.vice.com/en/article/gy94gw/life-in-a-bunker-in-eastern-ukraine-russian-roulette-dispatch-84.

———. "Ukraine Elects A New President: Russian Roulette (Dispatch 43)." *Vice News,* May 28, 2014. https://www.vice.com/en_us/article/ywj9pv/russian-roulette-dispatch-43.

Ottaway, Marina. *ISIS: Many Faces, Many Battles*. Middle East Program Occasional Paper Series. Washington, DC: Woodrow Wilson Center for International Scholars, 2015. https://www.wilsoncenter.org/publication/isis-many-faces-different-battles.

Ousby, Ian. *The Road to Verdun: France, Nationalism and the First World War.* London: Jonathan Cape, 2002.

Overy, Richard. *The Bombing War: Europe 1939–1945*. London: Allen Lane, 2013.

Pappe, Ilan. *The Making of the Arab-Israeli Conflict 1947–1951*. London: IB Tauris, 2014.

Parker, Ned, Isabel Coles, and Raheem Salman. "How Mosul Fell: An Iraqi General Disputes Baghdad's Story." Reuters, October 14, 2014. http://www.reuters.com/article/us-mideast-crisis-gharawi-special-report-idUSKCN0I30Z820141014.

Pasley, C. W. *Rules for Conducting the Practical Operations of a Siege*, Parts I and II, 2nd ed. London: John Weale, 1843.

Pearlman, Wendy. "Emotions and Microfoundations of the Arab Uprisings." *Perspectives on Politics* 11, no. 2 (2013): 387–409.

Peters, Gretchen. *Seeds of Terror: How Heroin Is Bankrolling the Taliban and al Qaeda.* Oxford: Oneworld Publications, 2009.

Piëst, Jan. *The True Watcher: A Memoir of Events Surrounding the Occupation and Liberation of Groningen, and Their Aftermath*. Groningen, Netherlands: Studio van Stralen BNO, 2005.

Police Ombudsman for Northern Ireland. "Report into a Complaint from Rita and Rohn Restorick Regarding the Circumstances of the Murder of Their Son Lance Bombardier Stephen Restorick on 12th February 1997." 2006.

Porteous, J. Douglas, and Sandra E. Smith. *Domicide: The Global Destruction of Home*. Montreal: McGill-Queen's University Press, 2001.

Posen, Barry R. "Urban Operations: Tactical Realities and Strategic Ambiguities." In Soldiers in Cities: Military Operations on Urban Terrain, edited by Michael C. Desch, 149–166. Carlisle: US Army War College Strategic Studies Institute, 2001.

Power, Susan. "Siege Warfare in Syria: Prosecuting the Starvation of Civilians." *Amsterdam Law Forum* 8, no. 2 (2016): 1–22.

Price, Bryan C. "Targeting Top Terrorists: How Leadership Decapitation Contributes to Counterterrorism." *International Security* 36, no. 4 (2012): 9–46.

Provan, Keith G., and Patrick Kenis. "Modes of Network Governance: Structure, Management, and Effectiveness." *Journal of Public Administration Research and Theory* 18, no. 2 (April 2008): 229–252.

Prykhodko, Ihor, Janina Matsehora, Ivan Lipatov, Ihor Tovma, and Ilona Kostikova. "Servicemen's Motivation in the National Guard of Ukraine: Transformation After the 'Revolution of Dignity.'" *Journal of Slavic Military Studies* 32, no. 3 (2019): 347–366.

Przeworski, Adam, and Henry Teune. *The Logic of Comparative Social Inquiry*. New York: John Wiley & Sons, 1970.

Puglisi, Rosaria. "General Zhukov and the Cyborgs: A Clash of Civilisation within the Ukrainian Armed Forces." *Istituto Affari Internazionali Working Papers* 15, no. 17 (2015): 1–22.

Pullan, Wendy. "The Conditions of Urbicide." In *Being Urban: Community, Conflict and Belonging in the Middle East*, edited by Simon Goldhill, 153–171. Abingdon, UK: Routledge, 2020.

———. "Frontier Urbanism: The Periphery at the Centre of Contested Cities." *Journal of Architecture* 16, no. 1 (2011): 15–35.

———. "Justice as the Urban Everyday." In *Philosophy and the City: Interdisciplinary and Transcultural Perspectives*, edited by Keith Jacobs and Jeff Malpas, 117–131. London: Rowman & Littlefield, 2019.

———. "Violent Infrastructures, Places of Conflict: Urban Order in Divided Cities." In *The Sage Handbook of Urban Sociology: New Approaches to the Twenty-First Century City*, edited by Ricky Burdett and Suzanne Hall, 376–396. London: Sage, 2018.

Pullan, Wendy, and Ammar Azzouz. *Destruction and Reconstruction*. Conflict in Cities Briefing Paper 11. Cambridge, UK: CinC, 2019. https://www.urbanconflicts.arct.cam.ac.uk/downloads/briefing-paper-11-2019.pdf.

Pullan, Wendy, and Britt Baillie (eds.). *Locating Urban Conflicts: Ethnicity, Nationalism, Everyday Life*. Basingstoke, UK: Palgrave Macmillan, 2013.

Qiao Lan and Wang Xiansui. *Unrestricted Warfare—Wake Up, America!* Lambertville, NJ: Shadow Lawn Press, 2017.

Quataert, Jean H. "War-Making and Restraint by Law: The Formative Years, 1864–1914." In *The Cambridge History of War, Volume IV, War in the Modern World*, edited by Roger Chickering, Dennis Showalter, and Hans van de Ven, 142–162. Cambridge: Cambridge University Press, 2012.

Quinn, James, and Frederick Hilmer. "Strategic Outsourcing." *MIT Sloan Management Review* 35, no. 4 (Summer 1994). https://sloanreview.mit.edu/article/strategic-outsourcing/.

Raab, Jorg, and H. Brinton Milward. "Dark Networks as Problems." *Journal of Public Administration Research and Theory* 13, no. 4 (2003): 413–439.

Radai, Itamar. *Palestinians in Jerusalem and Jaffa, 1948*. London: Routledge, 2015.

Radio Free Europe/Radio Liberty. "Nine Days Defending Donetsk Airport." November 15, 2014. https://www.rferl.org/a/donetsk-fighters-ukraine/26693593.html.

Randall, Nelson H. "The Battle for Manila." *Field Artillery Journal* 35, no. 8 (August 1945): 451–455.

Ratcliffe, Susan (ed). *Oxford Essential Quotations*, 5th edition. Oxford: Oxford University Press, 2017. https://www.oxfordreference.com/view/10.1093/acref/9780191843730.001.0001/q-oro-ed5-00011287.

Rauta, Vladimir. "Towards a Typology of Non-State Actors in 'Hybrid Warfare': Proxy, Auxiliary, Surrogate and Affiliated Forces." *Cambridge Review of International Affairs* 33, no. 6 (2020): 868–887.

Rayburn, Joel D., and Frank K. Sobchak (eds.). *The US Army in the Iraq War*. Volume 1, *Invasion—Insurgency—Civil War, 2003–2006*. Carlisle, PA: US Army War College Press, 2019.

REACH-UNITAR. *Syrian Cities Damage Atlas*. March 16, 2019. https://reliefweb.int/sites/reliefweb.int/files/resources/reach_thematic_assessment_syrian_cities_damage_atlas_march_2019_reduced_file_size_1.pdf.

"Recollections of the British Army in the Early Campaigns of the Revolutionary War Expedition to South America." *United Service Journal and Naval and Military Magazine*, December 1836.

Regencio, Ted. "Delay in Return 'Boosts ISIL Recruitment' in Philippines' Marawi." Al Jazeera, October 23, 2019. https://www.aljazeera.com/news/2019/10/23/delay-in-return-boosts-isil-recruitment-in-philippines-marawi.

Renic, Neil C. *Asymmetric Killing: Risk Avoidance, Just War, and the Warrior Ethos*. Oxford: Oxford University Press, 2020.

Report of Major General W. T. Sherman, January 1, 1865. In *The War of the Rebellion: A Compilation of the Official Records of the Union and Confederate Armies*, edited by Fred C. Ainsworth and Joseph W. Kirkley. Series I, Volume 39, Part I Correspondence. Washington, DC: Government Printing Office, 1902.

Report of the Special Rapporteur on Extrajudicial, Summary or Arbitrary Executions. "Targeted Killings Through Armed Drones and the Case of Iranian General Qassam Soleimani." June 29, 2020. https://www.ohchr.org/en/issues/executions/pages/srexecuti onsindex.aspx.

Reporters Without Borders. "2014 World Press Freedom Index." https://rsf.org/en/world-press-freedom-index-2014.

———. "2019 World Press Freedom Index." https://rsf.org/en/ranking.

Reynolds, Nick. "The British Army and Mass in Urban Warfare." *RUSI Journal* 166, no. 4 (2021): 52–65.

Richmond, H. W. *Naval Warfare*. London: E Benn, 1930.

Rogers, Steven. "Beyond Abu Sayyaf: The Lessons of Failure in the Philippines." *Foreign Affairs*, January/February 2004, 15–20.

Rómmel, Juliusz. *Za honor i ojczyznę*. Warsaw: Iskry, 1958.

Rosenau, William. "'Every Room Is a New Battle': The Lessons of Modern Urban Warfare." *Studies in Conflict & Terrorism* 20, no. 4 (October–December 1997): 371–394.

Rotter, Juian B. "Generalized Expectancies for Internal Versus External Control of Reinforcements." *Psychological Monographs* 80, no. 1 (1966): 1–28.

———. "Internal Versus External Control of Reinforcement: A Case History of a Variable." *American Psychologist* 45, no. 4 (1990): 489–493.

Rowecki, Stefan. *Walki Uliczne*. Warsaw: Wojskowy Instytut Naukowo-Wydawniczy, 1928.

Rubin, Alissa J. "'Our Patience Is Over': Why Iraqis Are Protesting." *New York Times*, November 20, 2019. https://www.nytimes.com/2019/11/20/world/middleeast/iraq-protests-sadr-city.html.

Runkle, Benjamin. "Jaffa 1948." In *City Fights: Selected Histories of Urban Combat from World War II to Vietnam*, edited by John Antal and Bradley Gericke. New York: Ballatine Books, 2003.

Russell, Shawn. "Mao Zedong's *On Guerrilla Warfare* and Joseph Kabila's *Lost Opportunity*." *Small Wars Journal*, October 7, 2010. https://smallwarsjournal.com/jrnl/art/mao-zedong's-on-guerrilla-warfare-and-joseph-kabila's-lost-opportunity.

Ryssdahl, Kai. "I Asked a Security Expert to Reveal How Cambridge Analytica Might Target Me Based on My Personality." *Marketplace*, March 26, 2018.

Salem, Harriet. "Let Them Eat Cake or: The Battle for Donetsk International Airport." Vice News, May 27, 2014. https://www.vice.com/en_us/article/43853b/let-them-eat-cake-or-the-battle-for-donetsk-international-airport.

Salter, Michael. "Toys for the Boys? Drones, Pleasure and Popular Culture in the Militarisation of Policing." *Critical Criminology* 22 (2014): 163–177.

Sanders, Andrew. "Operation Motorman (1972) and the Search for a Coherent British Counterinsurgency Strategy in Northern Ireland." *Small Wars & Insurgencies* 24, no. 3 (2013): 465–492.

Sanders, Deborah. "'The War We Want; The War That We Get': Ukraine's Military Reform and the Conflict in the East." *Journal of Slavic Military Studies* 30, no. 1 (2017): 30–49.

Saouli, Adham. "Sectarianism and Political Order in Iraq and Lebanon." *Studies in Ethnicity and Nationalism* 19, no. 1 (2019): 67–87.

Schnabel, James F. *United States Army in the Korean War, Policy and Direction: The First Year*. Washington, DC: US Government Printing Office, 1972.

Schogol, Jeff. "Drones Pose the Biggest Threat to US Troops in the Middle East Since IEDs, Top General Says." *Task and Purpose*, February 8, 2021. https://taskandpurpose.com/news/drone-threat-ieds-middle-east-mckenzie/.

Scott, James C. *Seeing Like a State: How Certain Schemes to Improve the Human Condition Have Failed*. New Haven, CT: Yale University Press, 1998.

———. *Weapons of the Weak: Everyday Forms of Peasant Resistance*. New Haven, CT: Yale University Press, 1985.

Senn, Martin, and Jodok Troy. "The Transformation of Targeted Killing and International Order." *Contemporary Security Policy* 38, no. 2 (2017): 175–211.

———(eds.). *The Transformation of Targeted Killing and International Order*. London: Routledge, 2019.

Sennema, Greg. "The Liberation of Groningen: The Diary Account of Theo Polman." *Canadian Military History* 26, no. 2 (2017).

Sennett, Richard. *The Uses of Disorder: Personal Identity and City Life*. London: Faber and Faber, 1996.

Shakirov, Mumin. "Interview: 'I Was a Separatist Fighter in Ukraine.'" *Guardian*, July 15, 2014. https://www.theguardian.com/world/2014/jul/15/separatist-fighter-russia-eastern-ukraine-interview.

Shapiro, Jacob N. "The Terrorist's Challenge: Security, Efficiency, Control." *Center for International Security and Cooperation*. Stanford, CA: Stanford University, April 26, 2007.

"Shelling Makes Donetsk Airport Defenders to Leave Old Terminal." *Ukrainian Mirror Weekly*, December 5, 2014. http://mw.ua/UKRAINE/shelling-makes-donetsk-airport-defenders-to-leave-old-terminal-619_.html.

Sherr, James. "Russia-Ukraine Rapprochement? The Black Sea Fleet Accords." *Survival* 39, no. 3 (1997): 33–50.

Shevchenko, Vitaly. "Ukraine Conflict: The 'Cyborg' Defenders of Donetsk Airport." BBC Monitoring, October 31, 2014. https://www.bbc.co.uk/news/world-europe-29793696.

Shramovych, Viacheslav. "Ukraine's Deadliest Day: The Battle of Ilovaisk, August 2014." BBC News, August 29, 2019. https://www.bbc.co.uk/news/world-europe-49426724.

Shtohrin, Irina, and Sergei Loiko. *AD 424*. Karkiv: Radio Svoboda, 2016.

Simmons, Erica S. "Market Reforms and Water Wars." *World Politics* 68, no. 1 (2016): 37–73.

Sims, Christopher. "Academics in Foxholes. The Life and Death of the Human Terrain System." *Foreign Affairs*, February 4, 2016.

Slater, Paul A. "Field Artillery and the Urban Battle." *Field Artillery Journal* 50, no. 1 (January–February 1982).

Smith, Daniel J., Kelley Jeter, and Odin Westgaard. "Three Approaches to Center of Gravity Analysis: The Islamic State of Iraq and the Levant." *Joint Force Quarterly* 78 (2015): 129–136.

Smith, Robert Ross. *United States Army in World War II, The War in the Pacific: Triumph in the Philippines*. Washington, DC: US Government Printing Office, 1973.

Smith, Rupert. "Requirements for Effective Military Operations in Globalization and Conflict." In *National Security in a 'New' Strategic Era*, edited by Robert G. Patman. London: Routledge, 2006.

———. *The Utility of Force*. London: Allen Lane, 2005.

Solis, Gary. "Targeted Killing and the Law of Armed Conflict." *Naval War College Review* 60, no. 2 (2007): 127–146.

Sotiriou, Stylianos. "The Irreversibility of History: The Case of the Ukrainian Crisis (2013–2015)." *Southeast European and Black Sea Studies* 16, no. 1 (2016): 51–70.

Spencer, John. "The City Is Not Neutral: Why Urban Warfare Is So Hard." Modern War Institute, March 4, 2020.

———. "The Eight Rules of Urban Warfare and Why We Must Work to Change Them." Modern War Institute, December 1, 2021. https://mwi.usma.edu/the-eight-rules-of-urban-warfare-and-why-we-must-work-to-change-them/.

———. "The Most Effective Weapon on the Modern Battlefield Is Concrete." Modern War Institute, November 14, 2016. https://mwi.usma.edu/effective-weapon-modern-battlefield-concrete/.

———. "Stealing the Enemy's Urban Advantage: The Battle of Sadr City." Modern War Institute, January 31, 2019. https://mwi.usma.edu/stealing-enemys-urban-advantage-battle-sadr-city/.

Spencer, John, and Harshana Ghoorhoo. "The Battle of Shusha City and the Missed Lessons of the 2020 Nagorno-Karabakh War." Modern War Institute, July 14, 2021. https://mwi.usma.edu/the-battle-of-shusha-city-and-the-missed-lessons-of-the-2020-nagorno-karabakh-war/.

Spencer, John, and Charles Knight. "The Battle of Marawi." Urban Warfare Project Podcast, Modern Warfare Institute, May 5, 2021. https://mwi.usma.edu/the-battle-of-marawi/.

Stacey, C. P. *The Victory Campaign: The Operations of North-West Europe 1944–1945*. Ottawa: Queen's Printer, 1960.

Stahl, Lesley. "Facebook 'Embeds,' Russia and the Trump Campaign's Secret Weapon." CBS News, October 17, 2017. https://www.cbsnews.com/news/facebook-embeds-russia-and-the-trump-campaigns-secret-weapon/.

Steadman, J. A. *Understanding the Urban Battlespace: An Intelligence Challenge*. Naval War College, February 5, 2001.

Stedman, Stephen John. "Spoiler Problems in the Peace Process." *International Security* 22, no. 2 (1997).

Stohl, Cynthia, and Michael Stohl. "Secret Agents: The Communicative Constitution of a Clandestine Organization." *Organization Studies* 32, no. 9 (2011): 1–19.

Stone, David R. "Stalingrad and the Evolution of Soviet Urban Warfare." *Journal of Slavic Military Studies* 22, no. 2 (2009): 195–207.

Strachan, Anna Loise. *Factors Behind the Fall of Mosul to ISIL (Daesh) in 2014*. K4D Helpdesk Report, January 1, 2017. Brighton, UK: Institute of Development Studies.

Strachan, Hew. "Introductory Essay: The Changing Character of War." In *Conceptualising Modern War*, edited by Erik Haug Kar and Ole Jorgen Maao. New York: Columbia University Press, 2011.

Strachan, Hew, and Sibylle Scheipers (eds.). *The Changing Character of War*. Oxford: Oxford University Press, 2011.

Stryzhova, Olha. "Why the Battle for Luhansk Airport Is Just as Important as the Battle for Donetsk Airport." Euromaidan Press, February 10, 2018. http://euromaidanpress.com/2018/02/10/117926/.

Sun Tzu. *The Art of War*, edited and translated by Samuel B. Griffith. Oxford: Oxford University Press, 1963.

Sutyagin, Igor. "Russian Forces in Ukraine." RUSI Briefing Paper, March 2015, 1–3.

Synge, Millington Henry. *The Defence of England: A Military Sketch*. London: Longmans, Green, Reader and Dyer, 1872.

Syrbanism. "Marota City: Is This the Type of Reconstruction Syrians Need?" Video, 2019. https://www.syrbanism.com/marotacity.

Tarafder, Md Kamrul, and Barbara Custodio. "Microfinance in a Conflict Zone: The Battle of Marawi." *OIDA International Journal of Sustainable Development* 11, no. 9 (2018): 11–24. https://ssrn.com/abstract=3286480.

Tarrow, Sidney. *Power in Movement: Social Movements and Contentious Politics*. Cambridge: Cambridge University Press, 1998.

Taw, Jennifer Morrison, and Bruce Hoffman. *The Urbanization of Insurgency: The Potential Challenge to U.S. Army Operations*. Santa Monica, CA: RAND, 1994.

Therborn, Göran. *Cities of Power: The Urban, the National, the Popular, the Global*. London: Verso, 2017.

Thompson, Robert. *Defeating Communist Insurgency*. London: Chatto & Windus, 1970.

Thorp, Gene, and Lazaro Gamio. "The Cease-Fire War in Donetsk." *Washington Post*, October 28, 2014. https://www.washingtonpost.com/wp-srv/special/world/ukraine-cease-fire-war/.

Torres, Wilfredo. *Rido: Clan Feuding and Conflict Management in Mindanao*. Makati City: The Asia Foundation, 2007.

Traugott, Mark. *The Insurgent Barricade*. Berkeley: University of California Press, 2010.

Triandis, Harry C. *Individualism and Collectivism*. Boulder, CO: Westview, 1995.

Tulaganova, Shahida. "Airport Donetsk." Current Time TV & Radio Free Europe/Radio Liberty, 2015. https://www.currenttime.tv/a/aiportdonetsk/27255834.html.

Tunstall, Graydon A. *Written in Blood: The Battles for Fortress Przemyśl in WWI*. Bloomington: Indiana University Press, 2016.

Turner, John C., and Katherine J. Reynolds. "The Story of Social Identity." In *Rediscovering Social Identity: Core Sources*, edited by Tom Postmes and Nyla R. Branscombe, 13–32. East Sussex, UK: Psychology Press, 2010.

UK Defence Doctrine. Joint Doctrine Publication 0-01 (6th edition).

UK Ministry of Defence. *UK Land Power*. Joint Doctrine Publication 0-20.

"Ukraine Live." *Interpretermag* rolling blog. https://www.interpretermag.com/ukraine-live-day-333/#6304.

"Ukrainian Post Service Released Stamp Dedicated to Defenders of Donetsk Airport." *Ukraine in Arabic,* January 22, 2020. http://arab.com.ua/en/ukrainian-post-service-released-stamp-dedicated-to-defenders-of-donetsk-airport.

Ungváry, Kristián. *Battle for Budapest: One Hundred Days in World War II.* London: IB Tauris, 2007.

UN Habitat. *City Profile of Mosul, Iraq: Multi-Sector Assessment of a City Under Siege.* United Nations Human Settlements Programme, October 2016.

UN High Commissioner for Refugees (UNHCR). "Syrian Refugee Crisis Explained." 2021. https://www.unrefugees.org/news/syria-refugee-crisis-explained/.

Urban, Mark. *Big Boys' Rules: The SAS and the Secret Struggle Against the IRA.* London: Faber and Faber, 2012.

van Ommen Kloeke, W. K. J. J. *De Bevrijding van Groningen.* Assen: Van Gorcum, 1947.

Van Riper, Paul K. *A Concept for Military Operations on Urbanized Terrain.* Department of the Navy Marine Corps Combat Development Command, July 1997.

Van Wilgenburg, Wladimir, and Mario Fumerton. *Kurdistan's Political Armies: The Challenge of Unifying the Peshmerga Forces.* Carnegie Middle East Center, December 16, 2015.

Vesely, Dalibor. *Architecture in the Age of Divided Representation. The Question of Creativity in the Shadow of Production.* Cambridge, MA: MIT Press, 2004.

Vibart, Edward. *The Sepoy Mutiny as Seen by a Subaltern: From Delhi to Lucknow.* London: Smith, Elder and Co., 1898.

Vice News. "The Ruins of Donetsk Airport: Dispatch 95." February 18, 2015. https://www.vice.com/en/article/d39bjk/the-ruins-of-donetsk-airport-russian-roulette-dispatch-95.

Vilmer, Baptiste Jeangène. "Not So Remote Drone Warfare." *International Politics* (July 2021): 1–22.

Volesky, Gary, and Roger Noble. "Theater Land Operations: Relevant Observations and Lessons from the Combined Joint Land Force Experience in Iraq." *Military Review* (Online Exclusive, June 2017). https://www.armyupress.army.mil/Portals/7/Army-Press-Online-Journal/documents/Volesky-v2.pdf.

Von Clausewitz, Carl. *On War*, edited and translated by Michael Howard and Peter Paret. Princeton, NJ: Princeton University Press, 1984.

Wahlman, Alec. *Storming the City: U.S. Military Performance in Urban Warfare from World War II to Vietnam.* Denton: University of North Texas Press, 2015.

Walker, Shaun. "Ukraine Says it Controls Donetsk Airport After Fighting Leaves Dozens Dead." *Guardian,* May 27, 2014. https://www.theguardian.com/world/2014/may/27/ukraine-rebel-30-dead-donetsk-airport-air-strikes.

———. "Ukrainian Forces and Pro-Russia Rebels Clash over Donetsk Airport." Guardian, October 1, 2014. https://www.theguardian.com/world/2014/oct/01/ukrainian-forces-pro-russia-rebels-donetsk-airport.

Walker, Shaun, and Oksana Grysenko. "Ukraine Forces Admit Loss of Donetsk Airport to Rebels." *Guardian,* January 21, 2015. https://www.theguardian.com/world/2015/jan/21/russia-ukraine-war-fighting-east.

Walsh, Stephen. "The Battle of Stalingrad, September–November 1942." In Fremont-Barnes, *A History of Modern Urban Operations.*

Watson, Alexander. *The Fortress: The Great Siege of Przemyśl.* London: Allen Lane, 2019.

Watts, Duncan J. "Networks, Dynamics, and the Small-World Phenomenon." *American Journal of Sociology* 105, no. 2 (1999): 493–527.

Weinberg, Gerhard L. "Stalingrad and Berlin: Fighting in Urban Terrain." In *Soldiers in Cities: Military Operations on Urban Terrain,* edited by Michael C. Desch, 17–28. Carlisle: US Army War College Strategic Studies Institute, 2001.

Welsh, Jennifer M. "The Morality of 'Drone Warfare.'" In *Drones and the Future of Armed Conflict: Ethical, Legal, and Strategic Implications*, edited by David Cortright, Rachel Fairhurst, and Kristen Wall, 25–45. Chicago: University of Chicago Press, 2015.

West, Bing. "Urban Warfare, Then and Now." *Atlantic*, June 30, 2017. https://www.theatlantic.com/international/archive/2017/06/urban-warfare-hue-mosul/532173/.

Whitehead, David (trans.). *Aineias the Tactician: How to Survive Under Siege*. Oxford: Clarendon Press, 1990.

Willems, Bastiaan. "Defiant Breakwaters or Desperate Blunders? A Revision of the German Late-War Fortress Strategy." *Journal of Slavic Military Studies* 28, no. 2 (2015): 353–378.

Williams, Fionnuala. "Proverbs: The Schools' Collection, Department of Irish Folklore, University College Dublin, as a Source for Research." *Clogher Record* 10, no. 3 (1981): 357–363.

Williams, Jeffery. *The Long Left Flank: The Hard Fought Way to the Reich, 1944–1945*. Toronto: Stoddart Publishing, 1988.

Williams, Phil, and Werner Selle. *Military Contingencies in Megacities and Sub-megacities*. Carlisle, PA: Strategic Studies Institute and US Army War College Press, December 2016.

Williamson, Murray, and Wayne Wei-siang Hsieh. *A Savage War: A Military History of the Civil War*. Princeton, NJ: Princeton University Press, 2016.

Wilner, Alex S. *The Best Defense Is a Terrific Offense: Four Approaches to Countering Modern Terrorism*. Halifax, NS: Atlantic Institute for Market Studies, July 3, 2007.

———. "Targeted Killing in Afghanistan: Measuring Coercion and Deterrence in Counterterrorism and Counterinsurgency." *Studies in Conflict and Terrorism* 33, no. 4 (April 2010): 307–329.

Wittke, Cindy. "The Minsk Agreements—More than 'Scraps of Paper'?" *East European Politics* 35, no. 3 (2019): 264–290.

Wollacott, Emma. "Social Media Platforms Easy to Manipulate, NATO Advisers Find." *Forbes Magazine*, December 6, 2019. https://www.forbes.com/sites/emmawoollacott/2019/12/06/youtube-is-easiest-platform-to-manipulate-nato-advisers-find/#753fbe933255.

XIV Corps (US). "Japanese Defense of Cities as Exemplified by the Battle for Manila." A Report by XIV Corps, Assistant Chief of Staff, Intelligence (G-2), Headquarters Sixth Army, July 1, 1945.

Zinets, Natalia. "Ukraine Struggles to Recruit Soldiers for War in East." Reuters, February 4, 2016. https://uk.reuters.com/article/uk-ukraine-crisis-military/ukraine-struggles-to-recruit-soldiers-for-war-in-east-idUKKCN0VD21Q.

Zverev, Anton. "Artillery Explosions Shake Donetsk Airport in East Ukraine." Reuters, November 17, 2014. https://www.reuters.com/article/us-ukraine-crisis-donetsk/artillery-explosions-shake-donetsk-airport-in-east-ukraine-idUSKCN0J115520141117.

About the Authors

Patrick Finnegan is associate lecturer in strategic studies at the University of Saint Andrews.

Lieutenant General Sir Andrew Graham CBE served in the British Army from 1975 to 2011 as a brigade commander in Northern Ireland, director of army resources and plans, deputy commanding general of multinational corps Iraq, director general of army training and recruiting, and director general of the Defence Academy.

Tyrone L. Groh is associate professor of global security and intelligence at Embry-Riddle Aeronautical University and a decorated US Air Force officer.

John Hardy is assistant dean of graduate studies and assistant professor of homeland security at Rabdan Academy, Abu Dhabi. He is also an adjunct at Charles Sturt University and a fellow of the Research Network for a Secure Australia.

Paul Latawski is senior lecturer in the Department of War Studies at the Royal Military Academy Sandhurst.

Paul Lushenko is a US Army lieutenant colonel and a PhD student in international relations at Cornell University. He was commissioned as an intelligence officer in 2005 following his graduation from West Point as a distinguished honor graduate.

Alex Neads is assistant professor in international security in the School of Government and International Affairs at Durham University, and a member of the Durham Global Security Institute.

Wendy Pullan is professor of architecture and urban studies at the University of Cambridge, where she is director of the Centre for Urban Conflicts Research.

Maj. (ret) John W. Spencer is chair of urban warfare studies at the Modern War Institute and codirector of the Urban Warfare Project. He served in the US Army for twenty-five years.

Steve Tatham retired from the Royal Navy in 2014 with the rank of commander after a career in psychological operations (PsyOps), information operations, and influence operations. He has led UK PsyOps teams in Sierra Leone, Iraq, and Afghanistan and was the UK's director of advanced communications research at the Defence Academy of the United Kingdom from 2007 to 2014.

Louise A. Tumchewics is a senior research fellow at the British Army's Centre for Historical Analysis and Conflict Research. She has assisted the UN and the International Committee of the Red Cross in the development of best practices for military-civilian interface in urban areas.

Matthew S. Wiseman is a Banting Fellow in the Department of History at Saint Jerome's University in the University of Waterloo, Canada.

Index

About the Book

"AVOID CITIES OR DIE WITHIN" HAS BEEN THE PREVAILING ATTITUDE IN the military when it comes to waging war in urban areas. So why do armies continue to fight there? What tactical advantages do they seek? What pitfalls do they face, and how can they achieve success?

The authors of *Small Armies, Big Cities* tackle these strategic questions, drawing on a range of cases to explore how today's professional armies can overcome the challenges of—and even find advantages in—conducting urban operations.

Louise A. Tumchewics is senior research fellow at the British Army's Centre for Historical Analysis and Conflict Research.